LIQUID COOLING OF ELECTRONIC DEVICES BY SINGLE-PHASE CONVECTION

WILEY SERIES IN THERMAL MANAGEMENT OF MICROELECTRONIC AND ELECTRONIC SYSTEMS
Allan D. Kraus and Avram Bar-Cohen, Series Editors

An Introduction to Heat Pipes: Modeling, Testing, and Applications
G.P. Peterson

Design and Analysis of Heat Sinks
Allan D. Kraus
Avram Bar-Cohen

Liquid Cooling of Electronic Devices by Single-Phase Convection
Frank P. Incropera

LIQUID COOLING OF ELECTRONIC DEVICES BY SINGLE-PHASE CONVECTION

FRANK P. INCROPERA
College of Engineering
University of Notre Dame
Notre Dame, Indiana

A Wiley-Interscience Publication
JOHN WILEY & SONS, INC.
New York / Chichester / Weinheim / Brisbane / Singapore / Toronto

This book is printed on acid-free paper. ∞

Copyright © 1999 by John Wiley & Sons, Inc. All rights reserved.

Published simultaneously in Canada.

No part of this publication may be reproduced, stored in a retrieval system or transmitted in any form or by any means, electronic, mechanical, photocopying, recording, scanning or otherwise, except as permitted under Sections 107 or 108 of the 1976 United States Copyright Act, without either the prior written permission of the Publisher, or authorization through payment of the appropriate per-copy fee to the Copyright Clearance Center, 222 Rosewood Drive, Danvers, MA 01923, (978) 750-8400, fax (978) 750-4744. Requests to the Publisher for permission should be addressed to the Permissions Department, John Wiley & Sons, Inc., 605 Third Avenue, New York, NY 10158-0012, (212) 850-6011, fax (212) 850-6008. E-Mail: PERMREQ@WILEY.COM.

This publication is designed to provide accurate and authoritative information in regard to the subject matter covered. It is sold with the understanding that the publisher is not engaged in rendering professional services. If professional advice or other expert assistance is required, the services of a competent professional person should be sought.

Library of Congress Cataloging-in-Publication Data:
Incropera, Frank P.
    Liquid cooling of electronic devices by single-phase convection / by Frank P. Incropera.
      p. cm.
    "A Wiley-Interscience publication."
    Includes bibliographical references and indexes.
    ISBN 0-471-15986-7 (alk. paper)
    1. Electronic apparatus and appliances—Temperature control.
  2. Heat—Convection.   3. Liquids—Thermal properties.
  4. Microelectronic packaging.   I. Title.
TK7870.25.I54   1999
621.381′044—dc21   98-53676

Printed in the United States of America

10 9 8 7 6 5 4 3 2 1

To Andrea,
    My best friend and life's partner.
        With Love.

CONTENTS

1 INTRODUCTION   1

    1.1 Introduction / 1
    1.2 Indirect Liquid Cooling / 2
    1.3 Direct (Immersion) Liquid Cooling / 5
    1.4 Fluid Selection / 8
    1.5 Summary / 13

2 FUNDAMENTALS OF HEAT TRANSFER AND FLUID FLOW   15

    2.1 Introduction / 15
    2.2 Conduction / 16
       2.2.1 Heat and Rate Equations / 16
       2.2.2 Plane Wall / 16
       2.2.3 Thermal Resistances / 18
       2.2.4 Extended Surfaces / 23
       2.2.5 Transient Conduction / 34
    2.3 Equations of Motion / 34
    2.4 Similarity Parameters / 36
    2.5 Natural (Free) Convection / 40
       2.5.1 Heated Vertical and Horizontal Surfaces / 40
       2.5.2 Vertical, Parallel-Plate Channels / 43
       2.5.3 Rectangular Cavities / 45

2.6 Forced Convection: External Flow / 47
    2.6.1 Flat Plate in Parallel Flow / 47
    2.6.2 Circular Cylinders in Cross Flow / 51
2.7 Forced Convection: Impinging Jets / 55
    2.7.1 Introduction / 55
    2.7.2 Unconfined, Free-Surface Jets / 55
    2.7.3 Unconfined, Submerged Jets / 64
    2.7.4 Effects of Confinement and Multiple Jets / 67
2.8 Forced Convection: Internal Flow / 69
    2.8.1 General Considerations / 70
    2.8.2 Energy Balances / 74
    2.8.3 Friction Effects and Pressure Losses / 75
    2.8.4 Heat Transfer / 79
2.9 Mixed Convection / 87
2.10 Summary / 87

3 NATURAL CONVECTION   89

3.1 Introduction / 89
3.2 Geometrical Features of a Rectangular Cavity with Discrete Heat Sources / 90
3.3 Mathematical Model / 91
3.4 Vertical Cavities: Theoretical Results / 92
    3.4.1 Flush-Mounted Heat Sources with Negligible Substrate Conduction / 92
    3.4.2 Flush-Mounted Heat Sources with Substrate Conduction / 102
    3.4.3 Protruding Heat Sources / 106
3.5 Vertical Cavities: Experimental Results / 112
    3.5.1 Flush-Mounted Heat Sources / 112
    3.5.2 Protruding Heat Sources / 117
3.6 Horizontal and Inclined Cavities / 118
3.7 Summary / 122

4 CHANNEL FLOWS   125

4.1 Introduction / 125
4.2 Forced Convection for Discrete, Flush-Mounted Heat Sources / 125
    4.2.1 Theoretical Considerations / 125
    4.2.2 Experimental Results / 131

4.3 Mixed Convection for Discrete, Flush-Mounted Heat Sources / 135
    4.3.1 Physical Features / 135
    4.3.2 Experimental Results / 138
4.4 Protruding Heat Sources / 145
    4.4.1 Forced Convection / 145
    4.4.2 Mixed Convection / 148
4.5 Microchannels / 151
    4.5.1 General Considerations / 151
    4.5.2 Channel Pressure Drop and Heat Transfer / 153
    4.5.3 System Performance / 155
4.6 Summary / 163

5 JET IMPINGEMENT COOLING   165

5.1 Introduction / 165
5.2 Circular, Unconfined, Free-Surface Jets / 165
    5.2.1 Stagnation Zone / 165
    5.2.2 Boundary Layer Development Regions / 170
    5.2.3 Average Heat Transfer Coefficients / 172
5.3 Rectangular, Unconfined, Free-Surface Jets / 177
    5.3.1 Stagnation Zone / 177
    5.3.2 Boundary Layer Development Regions / 181
5.4 Circular, Unconfined, Submerged Jets / 183
    5.4.1 Stagnation Zone / 183
    5.4.2 Boundary Layer Development Regions / 183
    5.4.3 Average Heat Transfer Coefficients / 185
5.5 Rectangular, Unconfined, Submerged Jets / 190
    5.5.1 Stagnation Zone / 190
    5.5.2 Boundary Layer Development Regions / 191
    5.5.3 Average Heat Transfer Coefficients / 191
5.6 Unconfined Multiple Jets / 195
    5.6.1 Free-Surface Jets / 195
    5.6.2 Submerged Jets / 199
5.7 Semiconfined Jets / 200
    5.7.1 Rectangular Jets / 200
    5.7.2 Circular Jets / 211
5.8 Special Considerations / 214
5.9 Summary / 215

6 HEAT TRANSFER ENHANCEMENT 217

- 6.1 Introduction / 217
- 6.2 Natural Convection / 218
  - 6.2.1 Parallel-Plate Fins / 218
  - 6.2.2 Pin Fins / 223
- 6.3 Channel Flow / 226
  - 6.3.1 Discrete Heat Sources in Forced Convection / 226
  - 6.3.2 Discrete Heat Sources in Mixed Convection / 235
  - 6.3.3 Microchannels / 235
- 6.4 Impinging Jets / 237
  - 6.4.1 Circular Jets / 237
  - 6.4.2 Rectangular Jets / 247
- 6.5 Summary / 251

APPENDIX A NOMENCLATURE 253

APPENDIX B THERMOPHYSICAL PROPERTY FUNCTIONS 261

REFERENCES 265

AUTHOR INDEX 279

SUBJECT INDEX 283

PREFACE

In 1986, the U.S. National Science Foundation sponsored a 3-day workshop to assess *Research Needs in Electronic Cooling* (Incropera, 1986). At that time, the need for improved electronic cooling technologies was drawing the attention of a large number of industrial and university researchers, who attended the workshop in nearly equal numbers. The principal objectives of the workshop were to strengthen industry/university research interfaces, to critically assess past and existing research activities, and to identify areas of future research having the greatest potential to advance cooling technologies.

At the workshop, considerable attention was given to *air cooling*, which was the predominant option of the day, and to identifying research initiatives that would enhance its efficacy. However, there was also recognition that very large scale integrated (VLSI) circuits would continue to become denser and faster, increasing power dissipation on a single chip, and that, in multiprocessor computing systems, chips would be clustered in closer proximity, thereby increasing power densities at the system level. These trends were stimulating interest in *liquid cooling*, and a good deal of the workshop was devoted to establishing a roadmap for future research on the subject. Distinctions were made between *direct* and *indirect cooling*, for which contact between the liquid coolant and the electronics is and is not permitted, respectively, as well as between *single-phase liquid cooling* and cooling with phase change (*boiling* and *two-phase flow*).

In the years following the 1986 workshop, a large amount of research has been performed to delineate fundamental principles associated with liquid cooling and to develop a knowledge base that can be used to design related cooling systems. The research has been disseminated in the form of numerous publications that have appeared in the open literature, and the objective of this book is to codify results obtained for direct and indirect liquid cooling by single-phase convection. The book serves two purposes: first, to clarify physical mechanisms associated with single-phase convection heat transfer in the context of electronic cooling, and second, to provide results that may be used for the rational design of liquid cooling systems.

In Chapter 1, the rationale for liquid cooling is established, and different liquids are assessed in terms of *figures of merit* for convection heat transfer. Although water provides superior heat transfer properties, contact with electronic devices is precluded, and it can only be used with indirect cooling schemes. In contrast, contact is permitted for dielectric liquids, such as the *Fluorinerts*, which may, therefore, be used for direct (immersion) cooling of electronic devices. Although dielectric liquids are less effective than water as a coolant, their cooling properties are superior to those of air.

Heat transfer fundamentals related to liquid cooling are reviewed in Chapter 2. Emphasis is placed on principles of *natural* and *forced convection*, although some attention is given to *conduction*. Even if the principal heat transfer agent is a liquid, there will always be pathways for conduction heat transfer from the electronics to the coolant. Moreover, even with liquids, cooling may be enhanced using extended surfaces (fins), which are combined convection/conduction devices. Because radiation does not occur in liquid cooling systems, it is not treated in Chapter 2. In addition, because this book is concerned exclusively with cooling by single-phase convection, there is no discussion of boiling or two-phase flow.

Liquid cooling by natural convection is discussed in Chapter 3. Emphasis is placed on flow and heat transfer in rectangular cavities that are filled with a dielectric liquid and on one of whose walls chiplike devices are attached. Although allowable heat rates are at the low end of the range associated with liquid cooling, a desirable feature of the scheme is that direct liquid cooling is effected by passive means.

Chapters 4 and 5 deal with liquid cooling by forced convection. In Chapter 4, the geometry of principal interest is a rectangular channel through which liquid is pumped and on one or more of whose walls chiplike devices are attached. Low flow limiting conditions corresponding to *mixed convection* are also considered, as are conditions associated with flow through microchannels machined in a chip or heat sink. In Chapter 5, attention is directed to impinging jets, which are capable of dissipating heat rates at the high end of the range associated with liquid cooling.

Whether one is interested in a liquid cooling technology at the low or high end of the efficacy spectrum, opportunities for significantly increasing heat rates, while maintaining acceptably low device temperatures, may be pursued through the use of extended surfaces or fins. As applied to electronic cooling, fins typically take the form of arrays of cylindrical pins or parallel plates, which increase the surface area available for convection. In Chapter 6, the use of fins under conditions corresponding to natural convection (Chapter 3) and forced convection associated with channel flow (Chapter 4) or jet impingement (Chapter 5) are considered.

Although much has been done in the past decade to provide a framework for the rational design of liquid cooling systems, actual implementation of such systems has been inhibited by two major trends. One trend has to do with advancements in circuit design that reduce heat dissipation, as, for example, through the replacement of bipolar technologies with complementary metal oxide semiconductor (CMOS) devices. Such developments extend the applicability of air cooling and diminish incentives to seek alternatives. The second trend pertains to the intensely competitive market for electronic products and the attendant need to minimize production costs. The response to this trend has been to focus on means of enhancing air cooling technologies, while resisting the use of liquid cooling.

Despite the foregoing trends, advancements in electronic technologies will continue in the direction of increased miniaturization of devices and packages, with increased levels of power dissipation. In some such cases, liquid cooling may well provide the only means of maintaining acceptable product temperatures, and this book is intended to assist the engineer in developing suitable thermal management systems. In using its contents, the engineer is strongly urged to develop an understanding of pertinent fundamentals before proceeding to quantitative considerations.

Much of the emphasis and many of the perspectives provided in this book are attributable to personal experiences over a 10-year period from approximately the mid-1980s to the mid-1990s. I am grateful to Professor Satish Ramadhyani and to our former students, Diane Besserman, Reiner Brinkman, Ted Heindel, Frank Kelecy, Jim Kerby, Vic Mahaney, Dave Moffatt, Mark Polentini, Diane Schafer, Phil Sullivan, Kevin Teuscher, and Dave Womac, who have contributed much to the evolution of the field and to my knowledge of the subject. I am also grateful to other graduate students with whom I have worked, namely, Andrea Knox, James Maughan, Dan Osborne, Steve Slayzak, Dave Vader, Dave Wolf, and Dave Zumbrunnen. Although their work was not explicitly linked to electronic cooling, it involved related issues in forced and mixed convection, which have contributed significantly to my understanding of single-phase convection in channel and jet impingement flows. Finally, I am deeply indebted to Dr. Richard Chu of IBM, who nucleated my earliest interests in electronic cooling and provided many useful insights over the duration of my involvement with the subject.

FRANK P. INCROPERA

Notre Dame, Indiana

LIQUID COOLING OF ELECTRONIC DEVICES BY SINGLE-PHASE CONVECTION

CHAPTER 1

INTRODUCTION

1.1 INTRODUCTION

Since the development of the first digital computers using silicon integrated circuits, reliable operation has depended on the ability to dissipate heat while maintaining acceptably low circuit temperatures. Historical trends that have made this problem progressively more difficult involve the development of ever larger scales of circuit integration on a single chip and the arrangement of multiple chips in closer proximity on a module. Such trends have made the role of heat transfer and thermal design increasingly more important, and the development of future computers may well be limited by the inability to maintain effective cooling.

For decades, air has been the preferred fluid for cooling electronic packages ranging from printed circuit boards to chips and multichip modules. Its desired attributes relate to simplicity, low cost, ease of maintenance, and high reliability. Accordingly, an extensive knowledge base has been developed to facilitate the design of related cooling systems. However, relative to liquids, the thermophysical properties of air render it less attractive as a coolant. Its comparatively low thermal conductivity ($k \sim 0.026$ W/m $\cdot$ K) and Prandtl number ($Pr \sim 0.70$) diminish its efficacy for heat removal by convection. Moreover, its low density ($\rho \sim 1.20$ kg/m^3) and specific heat ($c_p \sim 1.0$ kJ/kg $\cdot$ K) diminish its thermal capacitance and, hence, its ability to store thermal energy without incurring an unacceptable temperature rise, ΔT_c. Although heat transfer may be enhanced by attaching a finned *heat sink* to the electronic component, thereby increasing the surface area for convection, the extent to which benefits may be realized from this option is limited by manufacturing constraints. Limits are also associated with the unacceptable noise levels and pressure losses that would accompany large flow rates of air passing through the heat sink.

The functional relationship between the power dissipation Q and temperature T_d of an electronic device depends on the temperature T_c of the coolant to which heat is ultimately rejected and the *thermal resistance* R_{th} to heat transfer between the device and the coolant. With $Q \sim (T_d - T_c)/R_{th}$, it follows that, for a prescribed maximum allowable temperature difference, the maximum amount of heat that can be dissipated decreases with increasing resistance. It is the comparatively large thermal resistances associated with air cooling and the need to maintain low device temperatures in the face of ever-increasing power dissipation that have stimulated interest in other cooling options.

Current trends suggest that semiconductor feature sizes will approach 100 nm, permitting the fabrication of silicon chips with many millions of devices. Moreover, the large wiring densities achievable with thin-film deposition on ceramic surfaces, as well as the high signal-carrying capability of multilayer substrates, will continue to encourage the development of multichip modules. Collectively, these trends point to heat fluxes well in excess of 100 W/cm^2 for a single chip, 25 W/cm^2 for a multichip module, and 10 W/cm^2 for a printed circuit board (Bar-Cohen, 1991). Because chip junction temperatures must typically be maintained below $T_d \sim 85°$C, the corresponding *unit* or *areal thermal resistance* must be well below $R''_{th} = 1$ cm$^2 \cdot °$C/W, with future requirements suggesting the need for values as low as 0.1 cm$^2 \cdot °$C/W.

Because liquids have superior thermal properties (larger thermal conductivities and Prandtl numbers), their ability to transfer heat by convection, and, hence, to provide smaller thermal resistances, substantially exceeds that of gases. A distinction may be made between *indirect* cooling, for which the heat-dissipating electronic components are physically separated from the liquid, and *direct* cooling, for which the components are wetted by the liquid.

1.2 INDIRECT LIQUID COOLING

In the past two decades, indirect liquid cooling systems were devised to exceed the thermal performance of air cooling, while avoiding the difficulties that were perceived to be associated with direct liquid cooling. As applied to multichip modules, indirect liquid cooling involves attachment of a *cold plate* to the module. Heat removed from a chip or chip package is transferred through an intermediate structure before reaching the cold plate, through which the liquid coolant is pumped.

A well-known example of indirect liquid cooling is the thermal conduction module (TCM) used on the IBM 3080X/3090 series of computers (Chu et al., 1982). As shown in Figure 1.1, coolant is pumped through channels in an aluminum cold plate that is attached to a multichip module. Because there is no contact with the chip, water may be used as the coolant, thereby taking advantage of its superior thermal properties. As shown in the inset of Figure 1.1, contact is made between each chip and the spherical head of a spring-loaded aluminum piston. The pistons are seated in helium-filled cavities of an aluminum housing, which is bolted to the cold plate and, hence, readily detached for serviceability. Heat transfer from a chip occurs sequentially by conduction across the helium gap at the chip/piston interface, conduction

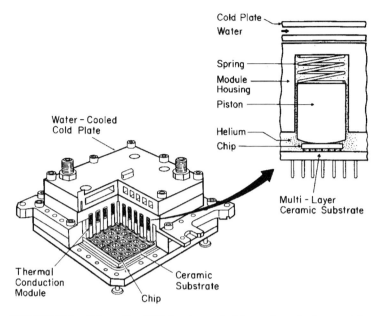

FIGURE 1.1 Schematic of IBM water-cooled thermal conduction module.

along the piston and across the helium gap separating the piston from the housing, conduction through the housing and its interface with the cold plate, and, finally, by conduction in the cold plate and convection to the water. Although the thermal resistance associated with convection heat transfer to the water is small, the total thermal resistance is determined primarily by internal components, of which the helium gap resistances are the most significant.

Internal thermal resistances associated with indirect liquid cooling may be reduced by routing the coolant closer to the chips. One approach (Fig. 1.2) involves mounting chips to both sides of a printed wiring board and using a flexible bellows to link the system with a cold plate (Yamamoto et al., 1988a, b). Each chip is separated from the bellows by a stack consisting of an elastomeric (compliant) material sandwiched between conductive (heat transfer or spreader) plates. Cooling is provided by water jet impingement on the stack, and the chip-to-coolant thermal resistance includes contributions due to conduction and interfacial contact within the stack, as well as convection to the jet.

Interface resistances between a cold plate and a multichip module may be eliminated by making the cold plate an integral part of the module (Kishimoto and Ohsaki, 1986). As shown in Figure 1.3, water is pumped through small (submillimeter) channels in a mutilayer alumina substrate to which an array of very large scale integration (VLSI) chips is mounted. Sasaki and Kishimoto (1986) determined optimum values of the channel width for prescribed values of the channel height and pressure drop, while Kishimoto and Sasaki (1987) found that heat transfer could be enhanced by

4 INTRODUCTION

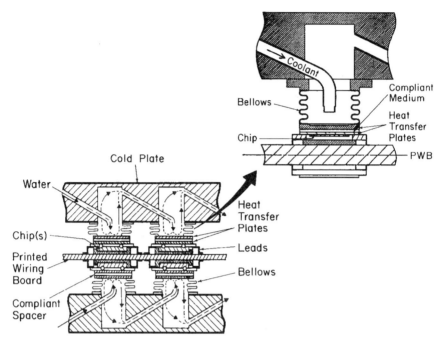

FIGURE 1.2 Schematic of Fujitsu FACOM M-780 water-cooled module.

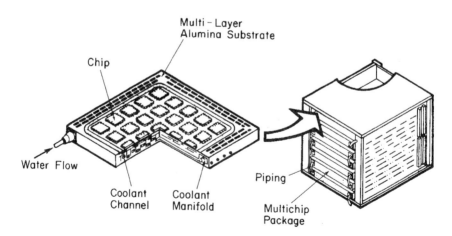

FIGURE 1.3 Schematic of water-cooled, multilayer substrate.

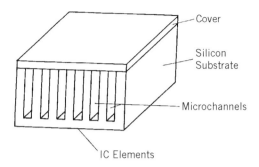

FIGURE 1.4 Silicon chip with microchannel cooling.

fabricating an array of interrupted, diamond-shaped fins within the channels. The modules may be stacked to achieve a large volumetric packing density.

Indirect liquid cooling may also be effected by fabricating a miniature cold plate with microchannels and attaching it to a chip (Phillips, 1990). However, the limit to which indirect liquid cooling may be taken corresponds to fabrication of the chip and cold plate as an integral unit. This concept is illustrated in Figure 1.4. With heat-dissipating integrated-circuit (IC) components on one side of a chip, microchannels may be precision machined or chemically etched on the opposite side (Tuckerman and Pease, 1981, 1982). Closure of the channels is achieved with a cover plate, and liquid is continuously pumped through the channels. Hence, the conduction and interface resistances associated with conventional indirect cooling schemes are virtually eliminated, and areal thermal resistances of $R''_{th} < 0.1$ cm$^2 \cdot ^\circ$C/W have been reported for $(T_d - T_c) \approx 80^\circ$C. Ultimately, however, application of microchannel cooling techniques will depend on the ability to establish flow connections that are readily manufactured and reliable. To minimize the crippling effects of fouling and/or blockage of flow passages, stringent requirements must be placed on coolant purity and, hence, on the coolant distribution unit.

1.3 DIRECT (IMMERSION) LIQUID COOLING

Because direct or immersion liquid cooling maintains physical contact between the coolant and the electronic components, the coolant must have a very large dielectric strength and good chemical compatibility with the components. Typically, such coolants are characterized by low boiling points, and their use could involve cooling by pool or forced convection boiling, as well as by single-phase convection (natural, forced, or mixed). In this book, attention is focused on the many options associated with liquid cooling by single-phase convection.

Immersion cooling has been implemented in the Cray-2 supercomputer (Danielson et al., 1986). As shown in Figure 1.5, vertical flow of a dielectric fluid between stacks of circuit modules precedes horizontal flow through the modules. Memory and logic chips attached to the modules are immersed in the liquid and are cooled by

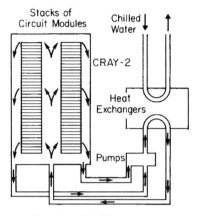

FIGURE 1.5 Cray-2 liquid immersion cooling system.

convection heat transfer directly to the liquid. The coolant is transported elsewhere for heat removal and returned to the modules at a reduced temperature to maintain continuous cooling.

Although there were few applications of direct liquid cooling in the 1980s and 1990s, a significant enabling knowledge base was developed during that period and prospects for future use are good. The technology is usually associated with the need to dissipate large heat fluxes and, hence, with the use of single-phase forced convection or boiling. However, applications involving natural convection are not precluded. One such application could involve a liquid-encapsulated module, as illustrated in Figure 1.6. Heat is transferred from the multichip module to the dielectric liquid and from the liquid to an opposing wall, which is cooled by air or water routed through channels in the wall. Buoyancy forces induced by heating and cooling the liquid in

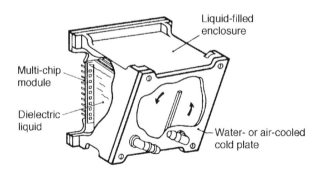

FIGURE 1.6 Immersion cooling by natural convection for a multichip module forming one wall of a liquid-filled enclosure.

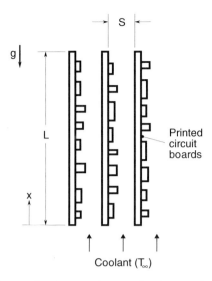

FIGURE 1.7 Immersion cooling by natural convection for a vertical array of PCBs open at the top and bottom.

this manner drive recirculation in the cavity, with liquid ascending and descending along the multichip module and cold plate, respectively. Alternatively, the fluid need not be encapsulated, as for an array of printed-circuit boards (PCBs) configured to provide vertical, parallel-plate channels that are open at the top and bottom (Fig. 1.7). If the array is submerged in a dielectric liquid, heat transfer from the array induces buoyancy forces that sustain an ascending flow through the channels. Fluid exiting the top of the array would be routed to an external heat exchanger, where it would be cooled before returning to the bottom of the array to sustain the upward flow.

If larger heat rates must be dissipated and smaller thermal resistances are needed, the motion of the liquid may be externally driven by a pump (*forced convection*), rather than internally driven by buoyancy forces (*natural convection*). One option that is well suited for compact packaging arrangements involves mounting chips to a substrate that comprises one wall of a rectangular channel through which liquid is pumped (Fig. 1.8). The thermal resistance associated with heat flow to the liquid would decrease with increasing flow rate and with disruption of flow development, as, for example, by insertion of transverse ribs in the channel. Alternatively, forced-convection heat transfer from a chip could be maintained by liquid jet impingement in a manner similar to that illustrated in Figure 1.2 for indirect cooling. However, with direct liquid cooling, the conduction and interface resistances associated with the heat transfer and compliant plates of Figure 1.2 would be eliminated. In fact, extremely low thermal resistances ($R''_{th} \sim 0.25 \text{ cm}^2 \cdot \text{K/W}$) may be achieved using a dielectric liquid with a modest flow rate ($\dot{V} \sim 1$ lpm).

Irrespective of whether liquid immersion cooling is effected by natural or forced convection, heat transfer may be enhanced by using *extended surfaces* with the heat

8 INTRODUCTION

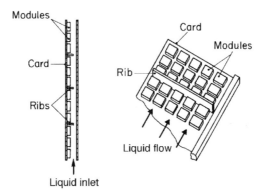

FIGURE 1.8 Immersion cooling by forced-convection flow through a rectangular channel with transverse ribs.

source. Their principal function is to increase the surface area available for heat transfer by convection, and they may take the form of millimeter-sized studs or plates. They must be fabricated from a material of large thermal conductivity, and attachment to the heat source must be characterized by a small interfacial resistance. Extended surfaces may enhance heat transfer by disrupting the flow, as well as by increasing the surface area available for convection.

1.4 FLUID SELECTION

For indirect cooling, chemical and electrical compatibility between the liquid and the electronic components is not an issue, and a liquid with excellent heat transfer properties, such as water, may be used. However, for immersion cooling, stringent chemical and electrical requirements are imposed on the liquid. Compatibility with chip, substrate, and PCB materials such as silicon, silicon nitride, gold, solder, epoxy/glass, and assorted plastics requires that the liquid be chemically inert. Because it must provide electrical isolation between closely spaced conductors, the liquid should also have a dielectric constant close to unity.

If a liquid is to be used for immersion cooling, it must be stable, nontoxic, and nonflammable, as well as inert and of high dielectric strength. Although *chlorofluorocarbons* (CFCs), such as the family of *Freons*, have such attributes, concerns for their deleterious environmental effects preclude their use for electronic cooling. Capable of ascending to the stratosphere, they release chlorine atoms that convert ozone to ordinary oxygen in a chain reaction. However, *fluorocarbons* (FCs), such as the *Fluorinerts*, have no such restrictions. Termed *perfluorinated liquids* because all carbon-bound hydrogen atoms are replaced by fluorine, they contain no other halogen atoms such as chlorine. Because the carbon atoms are joined to the most reactive element in the periodic table (fluorine), FCs are characterized by very strong chemical bonds and, hence, by excellent chemical stability. They are, therefore, unaffected by prolonged exposure to electronic components. Moreover, they have dielectric constants

(~ 1.7) that are close to that of air (~ 1.0) and far superior to that of water (~ 80). The corresponding dielectric strengths are approximately 15,000 V/mm.

The relation of a fluid's properties to its efficacy as a coolant is linked to a quantity termed the *convection heat transfer coefficient*. The transfer of thermal energy by convection from a heated surface to a fluid in motion over the surface is determined by *Newton's law of cooling*, which may be expressed as

$$q = \bar{h} A_s (T_s - T_f) \tag{1.1a}$$

or

$$q'' = h(T_s - T_f) \tag{1.1b}$$

Equation 1.1a relates the *rate of heat transfer q* from an entire surface of area A_s to its *average convection heat transfer coefficient* $\bar{h}$ and the difference between the surface temperature T_s and an appropriate reference temperature T_f for the fluid. In contrast, Eq. 1.1b relates the *heat flux* q'' at a particular point on the surface to the *local convection coefficient h* at that point.

In electronic cooling, the goal is one of maximizing the ratio of the heat rate or flux to the temperature difference. That is, for an allowable temperature difference, it is desirable to maximize the amount of heat that may be dissipated. Alternatively, for a fixed heat dissipation and reference temperature, it is desirable to minimize the surface temperature and thereby maximize the reliability of the electronic device. In either case, it follows that, if cooling is by convection, a large value of $\bar{h}$ or h is desired. A large convection coefficient corresponds to a small *convection resistance*, and the two quantities are related by expressions of the form

$$R_{th} = \frac{1}{\bar{h} A_s} \tag{1.2a}$$

or

$$R''_{th} = \frac{1}{h} \tag{1.2b}$$

according to whether concern is for the entire surface or a specific location on the surface, respectively. Mouromtseff (1942) and Saylor et al. (1988) have developed *figures of merit* for comparing the effectiveness of different fluids. If cooling is by forced convection, a dimensionless form of the convection coefficient, termed the *Nusselt number*, typically depends on a dimensionless flow parameter, termed the *Reynolds number*, and the *Prandtl number*. That is,

$$Nu_{L_o} \sim Re_{L_o}^m Pr^n \tag{1.3}$$

where m and n are positive exponents less than unity, $Nu_{L_o} \equiv h L_o / k$, $Re_{L_o} \equiv U_o L_o / \nu$, and L_o and U_o are the characteristic length and velocity, respectively, associated with the flow. The convection coefficient is, therefore, governed by the following relation:

$$h \sim \frac{F_F U_o^m}{L_o^{1-m}} \quad (1.4)$$

where the *figure of merit*, which combines the influence of all pertinent fluid properties, is

$$F_F = \frac{k Pr^n}{\nu^m} \quad (1.5a)$$

The convection coefficient increases with increasing F_F and, hence, with increasing thermal conductivity k and Prandtl number Pr, as well as with decreasing kinematic viscosity ν. Because $\nu \equiv \mu/\rho$ and $Pr \equiv \nu/\alpha = c_p \mu/k$, F_F may also be expressed as

$$F_F = \frac{k^{1-n} \rho^m c_p^n}{\mu^{m-n}} \quad (1.5b)$$

Hence, the convection coefficient also increases with increasing density ρ and specific heat c_p, as well as with decreasing dynamic viscosity μ.

Values of n are typically in the range $0.33 \leq n \leq 0.40$, whereas $m = 0.80$ for *turbulent flow*. For *laminar, external flow* over a flat surface, $m = 0.5$, whereas for the special case of *laminar, fully developed, internal flow* in a duct, $m = n = 0$ and $F_F = k$.

If cooling is by natural convection, the Nusselt number typically depends on a dimensionless parameter termed the *Rayleigh number*,

$$Nu_{L_o} \sim Ra_{L_o}^m \quad (1.6)$$

where

$$Ra_{L_o} \equiv \frac{g \beta \Delta T L_o^3}{\alpha \nu} \quad (1.7)$$

Hence,

$$h \sim \frac{F_N \Delta T^m}{L_o^{1-3m}} \quad (1.8)$$

where the figure of merit is

$$F_N = k \left(\frac{\beta}{\alpha \nu} \right)^m \quad (1.9a)$$

The convection coefficient increases with increasing F_N, which, in turn, increases with increasing thermal conductivity and expansion coefficient β, as well as with decreasing thermal diffusivity α and kinematic viscosity. Because $\alpha = k/\rho c_p$ and $\nu = \mu/\rho$, F_N may be expressed as

$$F_N = \frac{k^{1-m} \rho^{2m} c_p^m \beta^m}{\mu^m} \tag{1.9b}$$

in which case the convection coefficient also increases with increasing density and specific heat, as well as with decreasing dynamic viscosity. Values of m are typically equal to 0.25 and 0.33 for laminar and turbulent flow, respectively.

Equations 1.4 and 1.8 also reveal a dependence of h on the characteristic length L_o of the flow configuration, which is of the form $h \sim L_o^{m-1}$ and $h \sim L_o^{3m-1}$ for forced and natural convection, respectively. Hence, with $m < 1$ for forced convection, h increases with decreasing L_o. The same behavior applies for laminar natural convection, as $m \approx 0.25$. This trend is advantageous in electronic cooling because of the inherently small sizes of the electronic components. However, for turbulent natural convection, $m \approx 0.33$ and h is approximately independent of L_o.

Equations 1.4 and 1.8 also tell us that h increases with increasing flow velocity U_o in forced convection and with increasing temperature difference ΔT in natural convection. In natural convection, flow velocities increase with the buoyancy force, which is proportional to the temperature difference.

Room temperature property values of selected fluids are presented in Table 1.1. In view of the prominent role of the thermal conductivity in Eqs. 1.5a and 1.9a, the large value for water clearly establishes it as the superior heat transfer fluid. Hence, water is the fluid of choice for indirect liquid cooling.

The thermal conductivities of the three fluorocarbons (FCs) are much smaller than that of water but substantially larger than that of air. Although they are also smaller than the conductivities of R-113 and silicone oil, their superior electrical, chemical, and environmental attributes make them a preferred fluid for immersion cooling. Moreover, if cooling is by natural convection, heat transfer is favored by the large thermal expansion coefficients of the FCs.

The FCs are also characterized by large values of the *volumetric heat capacity*, ρc_p, which is beneficial when cooling is effected in an open-flow system. For a fixed coolant flow rate and heat dissipation, the coolant temperature rise decreases with increasing ρc_p. The comparatively small values of the dynamic viscosity also reduce the resistance to flow, thereby enhancing heat transfer and reducing pump power requirements.

For liquids, properties such as μ and Pr depend strongly on temperature, whereas ρ, c_p, and k are only weakly to moderately dependent. Analytical expressions for the temperature dependence of these properties are provided in Appendix B for water, FC-72, and FC-77. Of the two fluorocarbons, FC-77 has the larger boiling point and, hence, would be preferred in applications for which the fluid temperature exceeds 56°C (the boiling point of FC-72).

TABLE 1.1 Representative Fluid Properties at Room Temperature and Atmospheric Pressure.

Property	FC-43	FC-72	FC-77	R-113	Silicone Oil	Water	Air
Density, ρ (kg/m^3)	1850	1680	1780	1510	950	997	1.17
Specific heat, c_p (J/kg·K)	1050	1050	1050	980	1630	4179	1007
Thermal conductivity, k (W/m·K)	0.066	0.057	0.063	0.070	0.160	0.600	0.026
Viscosity, μ (N·s/m^2)	4.68×10^{-3}	7.20×10^{-4}	1.44×10^{-3}	5.03×10^{-4}	0.029	8.91×10^{-4}	1.85×10^{-5}
Prandtl number, Pr	75.6	13.8	24.0	7.0	295	6.2	0.71
Thermal expansion coefficient, β (1/K)	0.0012	0.0016	0.0014	0.0017	0.0010	0.00026	0.0034
Figure of merit, F_F^a (W·s$^{0.8}$/K·m$^{2.6}$)	8200	16,600	13,400	19,900	4300	75,700	161
Figure of merit, F_N^b (W·s$^{0.5}$/K$^{1.25}$·m^2)	22.7	33.2	30.0	40.1	21.4	127.2	1.5
Boiling point (°C)	174	56	97	48	290	100	—

[a] F_F is evaluated for turbulent flow with $m = 0.8$ and $n = 0.33$.
[b] F_N is evaluated for laminar flow with $m = 0.25$.

TABLE 1.2 Representative Ranges of Convection Coefficient and Thermal Resistance[a]

	h (W/m$^2 \cdot$ K)	R_{th} (K/W)
Natural convection		
Air	2–25	5000–400
Oils	20–200	500–50
Fluorinerts	25–250	400–40
Water	100–1,000	100–10
Forced convection		
Air	20–200	500–50
Oils	200–2,000	50–5
Fluorinerts	250–2,500	40–4
Water	1,000–10,000	10–1

[a] The values of R_{th} pertain to a surface of area $A_s = 1$ cm^2, where R''_{th}(cm$^2 \cdot$ K/W) $= (10^4$ cm^2/m$^2) \div h$ (W/m$^2 \cdot$ K) and $R_{th} = R''_{th}/A_s$.

Approximate ranges of the convection coefficient are provided in Table 1.2 for heat transfer by natural or forced convection with different fluids. The corresponding values of the thermal resistance pertain to a surface of area $A_s = 1$ cm^2. The convection coefficient is larger and, hence, the thermal resistance is smaller for liquids than for air, for water than for oils or the Fluorinerts, and for forced convection than for natural convection. Note that cooling by natural convection with a Fluorinert is comparable to cooling by forced convection with air.

If, for a particular heat dissipation, the desired temperature of an electronic package cannot be maintained by liquid immersion cooling, it may still be possible to operate below this temperature by attaching a finned heat sink to the package. By increasing the surface area from which heat is transferred by convection to the liquid, the heat sink can reduce the total resistance to heat transfer, thereby reducing the temperature of the package. Although the heat sink and its interface with the package add conduction and contact resistances, respectively, the total thermal resistance may still be reduced by exercising care in specifying the fin geometry and the method of attachment.

1.5 SUMMARY

The relentless advancement in the performance of electronic equipment will, for some time, continue to place stringent demands on related cooling technologies. For computing and data processing equipment, problems may involve transferring large heat fluxes from a single chip in a high-performance PC or workstation or the management of heat rates dissipated by multiple processors in a larger, more complex system. In both cases, operation below the maximum allowable temperature of each device is critical to reliability, as is maintaining nearly uniform temperatures for interdependent devices.

For many years, air cooling, with or without extended surfaces, has played a major role in the thermal management of electronic systems. Applications have been facilitated by the availability of an extensive knowledge base in the form of relations for natural- and forced-convection heat transfer from complex geometries, including finned heat sinks. However, in many respects, air cooling technologies may be viewed as mature, with limited opportunities for further advancement. It is, therefore, likely that increased emphasis will be placed on liquid-cooling technologies.

As for air cooling, the development of advanced liquid cooling systems depends on the existence of a substantial knowledge base, much of which has been developed over a period beginning in the mid-1980s and extending through the 1990s. Results have been obtained for natural-, forced-, and mixed-convection heat transfer from isolated chips, as well as from arrays of chips. Complications related to issues such as three-dimensional flows, multiple length scales associated with the flow and electronic devices, and the coupled effects of heat transfer by convection and conduction from the devices to the liquid and substrate, respectively, have been addressed. In the following chapters, this knowledge base is reviewed with two objectives in mind: first, to provide the reader with an appreciation of the fundamentals and, second, to provide heat transfer relations that may be used for the rational design of liquid cooling systems.

In Chapter 2, the fundamentals of heat transfer pertinent to electronic cooling are presented. Chapter 3 deals with electronic cooling by natural convection, and Chapters 4 and 5 are concerned with cooling by forced and mixed convection. In Chapter 4, attention is given to situations for which liquid is routed through a channel and cooling may be by direct or indirect means. In Chapter 5, cooling by impinging liquid jets is considered. Whether cooling is effected by natural, forced, or mixed convection, heat transfer may be enhanced by using fins (extended surfaces), and applications to electronic devices are considered in Chapter 6.

CHAPTER 2

FUNDAMENTALS OF HEAT TRANSFER AND FLUID FLOW

2.1 INTRODUCTION

In developing a liquid immersion cooling system, it is beneficial, if not imperative, to have a good understanding of related flow and heat transfer phenomena. In combination with associated computational tools, this understanding can do much to reduce testing requirements and the time devoted to achieving an optimal system design. Heat transfer fundamentals pertinent to liquid immersion cooling are reviewed in this chapter.

Because immersion cooling depends on the *convection* mode of heat transfer, emphasis is placed on the fundamentals of flow and heat transfer associated with *free, forced*, and *mixed* convection. Because of its compatibility with other packaging requirements and/or its ability to dissipate large heat rates, forced convection is considered for *internal flows* and *impinging jets*. Despite providing lower levels of heat dissipation, free and mixed convection may be used, respectively, in passive cooling schemes and applications requiring low pump power requirements.

Heat-dissipating components will always have interfaces with solid materials, such as substrates and heat sinks, to which and through which heat is transferred by *conduction*. Hence, relative to overall cooling system design, an understanding of both modes of heat transfer (conduction and convection) is important.

In this chapter, emphasis is placed on reviewing the fundamentals of flow and heat transfer that are germane to liquid immersion cooling. However, space limitations do not permit a comprehensive treatment of the subject, and readers wishing more detailed treatments of particular topics are referred to Incropera and DeWitt (1996).

2.2 CONDUCTION

2.2.1 Heat and Rate Equations

The central objective of a conduction analysis is to determine the manner in which temperature is distributed in a stationary medium resulting from conditions imposed at the boundaries of the medium. This distribution may be obtained by solving what is known as the *heat equation*. Once the temperature distribution is obtained, it may be used with *Fourier's law* to determine the heat flux at any point within the medium.

The heat equation may be derived by applying conservation of energy to a differential control volume, across whose boundaries heat is transferred by conduction and within which thermal energy may be generated and changes in energy storage may occur over time. In Cartesian coordinates, the equation is of the form (Incropera and DeWitt, 1996)

$$\frac{\partial}{\partial x}\left(k\frac{\partial T}{\partial x}\right) + \frac{\partial}{\partial y}\left(k\frac{\partial T}{\partial y}\right) + \frac{\partial}{\partial z}\left(k\frac{\partial T}{\partial z}\right) + \dot{q} = \rho c_p \frac{\partial T}{\partial t} \qquad (2.1)$$

where the first three terms on the left-hand side account for conduction in each of the coordinate directions and the fourth term, $\dot{q}$, accounts for the volumetric rate at which thermal energy is generated because of conversion from some other energy form. In an electronic component, the term would account for the conversion from electrical to thermal energy. In a heat sink or a substrate, there would be no generation and $\dot{q} = 0$. The term on the right-hand side of the equation accounts for changes in thermal energy storage resulting from transient effects and reduces to zero under steady-state conditions.

If the solution, $T(x, y, z, t)$, to Eq. 2.1 is known, the heat flux at any location in the medium may be determined by applying *Fourier's law*. The heat flux is a vector quantity, and its component in the x direction is given by

$$q_x'' = -k\frac{\partial T}{\partial x} \qquad (2.2)$$

Similar expressions may be written for the y and z directions, and forms of the heat equation and Fourier's law may also be developed for cylindrical and spherical coordinates (Incropera and DeWitt, 1996).

2.2.2 Plane Wall

The *plane wall* is an idealization that approximates many situations for which electronic cooling is influenced by conduction. Heat transfer by conduction occurs in a single coordinate direction across a material whose area perpendicular to the direction of conduction is invariant in this direction. Such a situation is illustrated in Figure 2.1. With temperatures prescribed at the two surfaces of the wall ($x = 0, L$), the temperature distribution within the wall depends on whether heat is or is not being generated.

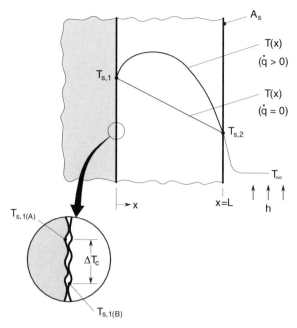

FIGURE 2.1 Temperature distributions for one-dimensional, steady-state conduction in a plane wall with surface contact and convection conditions at $x = 0$ and $x = L$, respectively.

Of course, the wall cannot exist in isolation, and representative interface conditions are shown. At $x = L$, the wall is presumed to be wetted by a fluid to which heat is transferred by convection. At $x = 0$, the wall is presumed to be in contact with another solid to which or from which heat is transferred, depending on the temperature distribution in the wall.

For one-dimensional, steady-state conduction with heat generation, Eq. 2.1 reduces to

$$\frac{\partial}{\partial x}\left(k\frac{\partial T}{\partial x}\right) + \dot{q} = 0 \tag{2.3}$$

Assuming constant thermal conductivity and uniform heat generation, Eq. 2.3 may be integrated twice to yield

$$T(x) = -\frac{\dot{q}x^2}{2k} + C_1 x + C_2 \tag{2.4}$$

Applying the boundary conditions, $T(0) = T_{s,1}$ and $T(L) = T_{s,2}$, it follows that

$$T(x) = \frac{\dot{q}}{2k}\left(\frac{x}{L}\right)\left(1 - \frac{x}{L}\right) - (T_{s,1} - T_{s,2})\frac{x}{L} + T_{s,1} \tag{2.5a}$$

For $\dot{q} > 0$, the temperature distribution is parabolic, and the maximum temperature occurs within the wall. For $\dot{q} = 0$, the distribution is linear,

$$T(x) = T_{s,1} - (T_{s,1} - T_{s,2})\frac{x}{L} \tag{2.5b}$$

and the maximum temperature corresponds to $T_{s,1}$.

The heat flux may be determined by applying Fourier's law, Eq. 2.2, to the foregoing distributions. It follows that, for $\dot{q} > 0$,

$$q_x'' = -\dot{q}L\left(\frac{1}{2} - \frac{x}{L}\right) + \frac{k(T_{s,1} - T_{s,2})}{L} \tag{2.6a}$$

or, for $\dot{q} = 0$,

$$q_x'' = \frac{k(T_{s,1} - T_{s,2})}{L} \tag{2.6b}$$

To obtain the heat rate q_x, the heat flux is multiplied by the area A_s, and, for $\dot{q} = 0$, it follows that

$$q_x = \frac{kA_s(T_{s,1} - T_{s,2})}{L} \tag{2.7}$$

For $\dot{q} > 0$, the heat flux varies with x; for $\dot{q} = 0$, it is a constant, independent of x. Similar results may be obtained for one-dimensional, radial (cylindrical and spherical) systems (Incropera and DeWitt, 1996).

2.2.3 Thermal Resistances

Under steady-state conditions, a first-order thermal analysis of an electronic cooling system may often be performed by identifying related pathways for heat removal, assigning a *thermal resistance* to each path, and analyzing the resultant network of resistances. Thermal resistances are associated with conduction in a material for which $\dot{q} = 0$, as well as interface conditions involving convection to a fluid or conduction to another material. In addition to providing building blocks for a thermal network analysis, the resistances themselves provide useful indicators of the efficacy of a particular cooling option.

Conduction Resistances By analogy to Ohm's law for electrical current, temperature difference and heat rate provide thermal equivalents to voltage and current. Hence, a thermal resistance may be defined as

$$R_{th} = \frac{\Delta T}{q} \tag{2.8}$$

For one-dimensional, steady-state conduction through a plane wall, Eq. 2.7 then yields a *conduction resistance* of the form

$$R_{th(cnd)} = \frac{L}{kA_s} \quad (2.9)$$

Similar expressions may be obtained for one-dimensional conduction through cylindrical and spherical walls (Incropera and DeWitt, 1996).

In some cases, a one-dimensional approximation to heat flow by conduction is inadequate and multidimensional effects must be considered. A case in point involves attachment of a small heat source, such as a chip, to a large *substrate* or *spreader plate* (Fig. 2.2). As indicated by the heat flow lines, thermal energy generated in the source spreads laterally as it is transferred by two- or three-dimensional conduction in the substrate. The problem has been considered by Yovanovich and Antonetti (1988), who obtained the following expression for the *spreading resistance*, $R_{th(sp)} \equiv (T_{s,h} - T_{sub})/q_h$, associated with conduction in the substrate:

$$R_{th(sp)} = \frac{1 - 1.410 A_r + 0.344 A_r^3 + 0.043 A_r^5 + 0.034 A_r^7}{4 k_{sub} (A_{s,h})^{1/2}} \quad (2.10)$$

where $A_r = A_{s,h}/A_{s,sub}$ is the ratio of the heat source area to the substrate area. The expression may be used to a good approximation if the thickness of the substrate is more than five times larger than the square root of the heat source area. In the limit as $A_r \to 0$, Eq. 2.10 reduces to the following expressions for circular and square heaters of diameter D_h and width L_h, respectively:

$$R_{th(sp)} = \frac{1}{2\pi^{1/2} k_{sub} D_h} \quad (2.11a)$$

$$= \frac{1}{4 k_{sub} L_h} \quad (2.11b)$$

Convection Resistance The rate of convection heat transfer q across a solid/fluid interface of area A_s, such as that at $x = L$ in Figure 2.1, is determined from

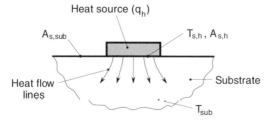

FIGURE 2.2 Heat flow lines for multidimensional conduction in a substrate or spreader plate.

20   FUNDAMENTALS OF HEAT TRANSFER AND FLUID FLOW

Newton's law of cooling,

$$q = \bar{h}A_s(T_s - T_f) \qquad (2.12a)$$

where $\bar{h}$ is the *average convection coefficient* and T_s and $T_f = T_\infty$ are the surface and fluid free-stream temperatures, respectively. Newton's law may also be expressed in terms of the local heat flux q'' at any location on the surface

$$q'' = h\,(T_s - T_f) \qquad (2.12b)$$

where h is the *local convection coefficient*, which varies with location on the surface. From Eq. 2.12a it follows that, for the entire surface, a *convection resistance* may be expressed as

$$R_{\text{th(cnv)}} = \frac{1}{\bar{h}A_s} \qquad (2.13)$$

Contact Resistance   When there is heat transfer across an interface between two materials, as at $x = 0$ in Figure 2.1, there is an attendant temperature change at the interface. The change is associated with a *thermal contact resistance*, $R_{\text{th,c}}$, which is defined as

$$R_{\text{th,c}} = \frac{\Delta T_c}{q} \qquad (2.14)$$

Heat transfer across an interface between two contacting materials depends on many parameters, such as the shape and size of the contact spots, their distribution along the interface, conditions within the voids between spots, and the contact pressure. Models for predicting the contact resistance have been developed (Kraus and Bar-Cohen, 1983; Yovanovich and Antonetti, 1988), but more commonly, accurate results for specific interface conditions are determined experimentally.

Unit Thermal Resistance   As defined by Eq. 2.8, the thermal resistance is associated with a specific area across which heat transfer occurs, and it has units of °C/W or K/W. However, it is often preferable to work with a *unit or areal thermal resistance*, $R''_{\text{th}} \equiv \Delta T/q''$, which has units of $m^2 \cdot {}^\circ$C/W or $m^2 \cdot$ K/W. If the unit thermal resistance due to conduction, convection, or interfacial contact is known, the thermal resistance associated with a prescribed surface area A_s is simply given by $R_{\text{th}} = R''_{\text{th}}/A_s$.

Example 2.1

A chip is attached to a substrate of thickness L_{sub}, thermal conductivity k_{sub}, and equivalent surface area. A unit thermal contact resistance $R''_{\text{th,c}}$ is associated with the chip/substrate interface. The exposed surface of the chip is convectively cooled, while the exposed surface of the substrate is chilled and maintained at a uniform temperature T_c. During operation, heat dissipation within the chip provides a heat flux of q_h'', while the chip is at a uniform temperature T_h.

1. Develop a model that may be used to determine the chip heat flux corresponding to a maximum allowable operating temperature of $T_{h,\max}$.
2. For operating conditions corresponding to $\bar{h} = 500$ W/m$^2 \cdot$ K, $T_f = T_c = 17°$ C, $R_{th,c}'' = 5 \times 10^{-5}$ m$^2 \cdot$ K/W, and an alumina substrate ($k_{sub} = 50$ W/m $\cdot$ K) of thickness $L_{sub} = 10$ mm, what is the chip heat flux corresponding to a maximum allowable temperature of $T_{h,\max} = 85°$C?

Solution

Known: Operating conditions for a chip that is convectively cooled on one surface and attached to a substrate at the opposite surface. Contact resistance associated with the attachment. Substrate thickness, thermal conductivity, and temperature of chilled surface.

Find:

1. Expression relating the chip heat flux to the maximum allowable chip temperature and other system parameters
2. Value of q_h'' corresponding to $T_{h,\max} = 85°$C and prescribed system parameters

Schematic:

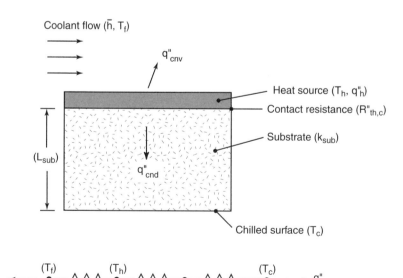

Assumptions:

1. Steady-state conditions
2. One-dimensional conduction in the substrate
3. Constant properties (k_{sub})
4. Isothermal chip

Analysis:

1. The system may be represented by an equivalent thermal circuit that consists of a convection resistance, $R_{th(cnv)}$, the contact resistance, and a substrate conduction resistance, $R_{th(cnd)}$. The resistances are in series, with portions of the chip heat flux routed to the coolant by convection and to the chilled surface by sequential transfer through the contact resistance and conduction in the substrate to the chilled surface. Hence,

$$q_h'' = q_{cnv}'' + q_{cnd}'' = \frac{T_h - T_f}{R_{th(cnv)}''} + \frac{T_h - T_c}{R_{th,c}'' + R_{th(cnd)}''}$$

Substituting from Eqs. 2.9 and 2.13 for a unit area of $A_s = 1 \text{ m}^2$,

$$q_h'' = \frac{T_h - T_f}{(1/\bar{h})} + \frac{T_h - T_c}{R_{th,c}'' + (L_{sub}/k_{sub})}$$

With T_h set equal to a prescribed maximum and with known values of the remaining parameters on the right-hand side of the foregoing equation, the maximum allowable chip heat flux may be determined.

2. For the prescribed quantities,

$$q_h'' = \frac{(85 - 17)°C}{(500 \text{ W/m}^2 \cdot \text{K})^{-1}} + \frac{(85 - 17)°C}{5 \times 10^{-5} \text{ m}^2 \cdot \text{K/W} + (0.010 \text{ m}/50 \text{ W/m} \cdot \text{K})}$$
$$= (34{,}000 + 272{,}000) \text{ W/m}^2 = 306{,}000 \text{ W/m}^2 = 30.6 \text{ W/cm}^2$$

Comments:

1. For the prescribed conditions, the convection resistance,

$$R_{th(cnv)}'' = 0.002 \text{ m}^2 \cdot \text{K/W},$$

greatly exceeds the cumulative resistance, $R_{th,c}'' + R_{th(cnd)}'' = 0.00025 \text{ m}^2 \cdot \text{K/W}$, associated with heat transfer through the substrate. Hence, $q_{cnd}'' \gg q_{cnv}''$. If the convection coefficient were increased to $\bar{h} = 5000 \text{ W/m}^2 \cdot \text{K}$, the corresponding reduction in the thermal resistance, $R_{th(cnv)}'' = 0.0002 \text{ m}^2 \cdot \text{K/W}$, would increase the convection flux, $q_{cnv}'' = 340{,}000 \text{ W/m}^2$, and, hence, the total heat flux, $q_h'' = 612{,}000 \text{ W/m}^2 = 61.2 \text{ W/cm}^2$.

2. A more common arrangement is one for which multiple chips are attached to a substrate, as shown below.

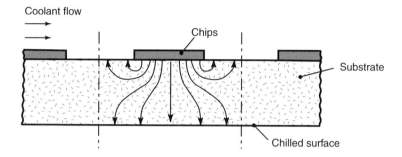

The *spreading* effect associated with the enlarged substrate reduces the conduction resistance and, hence, increases heat transfer by conduction to the chilled surface.

2.2.4 Extended Surfaces

Extended surfaces, or *fins*, are used to enhance convection heat transfer by increasing the surface area available for convection. In effect, they act as appendages to a *base*, or *prime*, surface from which heat is transferred by convection to a coolant. Fin types commonly used in electronic cooling are *rectangular (longitudinal)* or *pin fins* (Fig. 2.3) of uniform rectangular and circular cross sections, respectively.

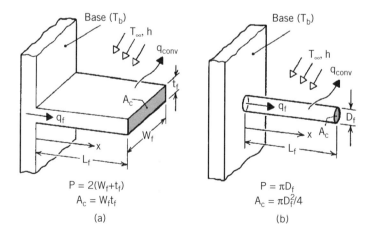

FIGURE 2.3 Straight fins of uniform cross section: (*a*) rectangular or longitudinal fin and (*b*) pin fin.

Fin Temperature Distribution and Heat Rate For one-dimensional conduction through a fin of uniform cross-sectional area A_c and uniform convection coefficient h along the surface of the fin, the differential equation that governs the temperature distribution, $T(x)$, in the fin is of the form (Incropera and DeWitt, 1996)

$$\frac{d^2\theta}{dx^2} - m^2\theta = 0 \qquad (2.15)$$

where $\theta(x) \equiv T(x) - T_\infty$ and $m^2 \equiv hP/kA_c$. The general solution is

$$\theta(x) = C_1 e^{mx} + C_2 e^{-mx} \qquad (2.16)$$

where the constants C_1 and C_2 are determined by applying boundary conditions at the base ($x = 0$) and tip ($x = L_f$) of the fin.

The base condition is determined by the temperature T_b, whereas different conditions may be imposed at the tip. If, for example, heat is transferred by convection from the tip to the fluid, the boundary condition is of the form $-k\,d\theta/dx|_{x=L_f} = h\theta(L_f)$ and the solution yields the following expressions for the temperature distribution and the fin heat transfer rate, where $M = (hPkA_c)^{1/2}\theta_b$.

Tip Convection:

$$\frac{\theta}{\theta_b} = \frac{\cosh m(L_f - x) + (h/mk)\sinh m(L_f - x)}{\cosh mL_f + (h/mk)\sinh mL_f} \qquad (2.17a)$$

$$q_f = M\frac{\sinh mL_f + (h/mk)\cosh mL_f}{\cosh mL_f + (h/mk)\sinh mL_f} \qquad (2.17b)$$

Results for an *adiabatic tip* may be inferred by setting $h = 0$ in the foregoing expressions.

Adiabatic Tip:

$$\frac{\theta}{\theta_b} = \frac{\cosh m(L_f - x)}{\cosh mL_f} \qquad (2.18a)$$

$$q_f = M \tanh mL_f \qquad (2.18b)$$

If, instead, the tip temperature θ_L is known, the temperature and heat rate are given by the following.

Known Tip Temperature:

$$\frac{\theta}{\theta_b} = \frac{(\theta_L/\theta_b)\sinh mx + \sinh m(L_f - x)}{\sinh mL_f} \qquad (2.19a)$$

$$q_f = M\frac{(\cosh mL_f - \theta_L/\theta_b)}{\sinh mL_f} \qquad (2.19b)$$

Performance Parameters The efficacy of an extended surface may be assessed by evaluating its *thermal efficiency*, η_f, and *effectiveness*, ε_f. Its inclusion in a network analysis of a cooling system requires knowledge of the *thermal resistance*, $R_{th,f}$.

The maximum driving potential for convection heat transfer between a fin and the adjoining fluid is the temperature difference between the base and the fluid, $\theta_b = T_b - T_\infty$. Hence, the maximum rate at which a fin could transfer energy is the rate that *would* exist *if* the entire fin were at the base temperature. However, conduction through the fin is associated with a negative temperature gradient along the fin and, hence, with a reduction in $T(x)$ from the base to the tip. It follows that the convection heat flux, $q'' = h(T - T_\infty)$, decreases with increasing x, and the actual rate of heat transfer from the fin, q_f, is less than the maximum possible rate, $q_{f,\max} = hA_{s,f}(T_b - T_\infty)$.

The fin efficiency is defined as the ratio of the actual heat rate to the maximum possible heat rate, in which case,

$$\eta_f = \frac{q_f}{q_{f,\max}} = \frac{q_f}{hA_{s,f}\theta_b} \tag{2.20}$$

Consider the special case corresponding to a longitudinal or pin fin (Fig. 2.3) with an *adiabatic tip* condition. Hence, from Eq. 2.18b,

$$\eta_f = \frac{M \tanh mL_f}{hPL_f\theta_b} = \frac{(hPkA_c)^{1/2}\theta_b \tanh mL_f}{hPL_f\theta_b} = \frac{\tanh mL_f}{mL_f} \tag{2.21}$$

Because $\tanh mL_f \to 1$ as $mL_f \to \infty$, $\eta_f \to 0$ as $mL_f \to \infty$. With $mL_f = (hP/kA_c)^{1/2}L_f$, it follows that η_f decreases with increasing h and L_f, as well as with decreasing k and A_c/P, all of which increase temperature gradients along the fin and the overall base-to-tip temperature difference, $T_b - T(L_f)$. The ratio A_c/P is proportional to the fin thickness or diameter, in which case, temperature gradients increase and η_f decreases with decreasing t_f or D_f. It, of course, makes no sense to extend the fin length L_f to a distance for which $T(L_f) \to T_\infty$, because there would be negligible convection from its surface at this location. In the limit $mL_f \to 0$, $\tanh mL_f \to mL_f$ and $\eta_f \to 1$.

The effectiveness is defined as the ratio of the fin heat transfer rate to the rate at which heat would be transferred from the base without the fin. Hence,

$$\varepsilon_f = \frac{q_f}{hA_{c,b}\theta_b} \tag{2.22}$$

where $A_{c,b}$, the fin cross-sectional area at the base, is equivalent to A_c for a fin of *uniform* cross-sectional area (Fig. 2.3). For such a fin and the adiabatic tip condition, substitution of Eq. 2.18b yields

$$\varepsilon_f = \frac{M \tanh mL_f}{hA_c\theta_b} = \frac{\tanh mL_f}{(hA_c/Pk)^{1/2}} \tag{2.23}$$

Hence, unlike the fin efficiency, which varies over the range $0 \leq \eta_f \leq 1$, the effectiveness is unbounded ($0 \leq \varepsilon_f < \infty$). The fact that ε_f may be less than unity implies the existence of conditions for which the addition of fins would diminish, not enhance, heat transfer from the base. Such conditions, which correspond to large values of h and t_f or D_f, as well as small values of k, are, of course, to be avoided.

From Eq. 2.8, the thermal resistance of a fin is simply

$$R_{\text{th},f} = \frac{\theta_b}{q_f} \tag{2.24}$$

The resistance, efficiency, and effectiveness are related parameters, and, from Eqs. 2.20, 2.22, and 2.24, it follows that

$$R_{\text{th},f} = \frac{1}{\eta_f h A_{s,f}} = \frac{1}{\varepsilon_f h A_{c,b}} \tag{2.25}$$

Hence, the thermal resistance decreases with increasing $hA_{s,f}$ and $hA_{c,b}$, as well as with increasing η_f or ε_f. However, any measure taken to reduce $R_{\text{th},f}$ by increasing $hA_{s,f}$ or $hA_{c,b}$ will be diminished by attendant reductions in η_f or ε_f, respectively. Note that, for an ideal fin ($\eta_f = 1$), Eq. 2.25 reduces to $R_{\text{th},f} = 1/hA_{s,f}$, which is equivalent to the convection resistance of Eq. 2.13 and consistent with the existence of an isothermal fin at T_b. The efficiency η_f provides a measure of the degree to which convection from the fin is diminished and the thermal resistance is increased by nonideal conditions. Note also that the convection resistance of the base surface without the fin is $R_{\text{th},b} = 1/hA_{c,b}$. Hence, if equivalent convection coefficients are assumed for the fin and base surfaces, the following expression for the effectiveness may be inferred from Eq. 2.25:

$$\varepsilon_f = \frac{R_{\text{th},b}}{R_{\text{th},f}} \tag{2.26}$$

If the fin is to enhance heat transfer, its resistance must not exceed that of the exposed base.

Expressions for the performance parameters are also available for fins of nonuniform cross section (Kraus and Bar-Cohen, 1995; Incropera and DeWitt, 1996), and expressions for determining the dimensions or shape of a fin that maximize heat transfer or minimize fin volume are presented by Kraus and Bar-Cohen (1995).

Fin Arrays A fin is rarely used as a single entity, but is, instead, used as the repeating element of an array. Heat sinks used for electronic cooling often consist of arrays of *longitudinal* (Fig. 2.4) or *pin* (Fig. 2.5) fins.

The parallel, longitudinal, or plate fins of Figure 2.4 are attached to a common base, and the geometry of the array is determined by the fin thickness t_f, spacing S_f (or pitch P_f), length L_f, and height H_f. Liquid flow through the channels formed by adjoining fins may be driven by a pump (*forced convection*) or induced by buoyancy forces associated with the temperature difference, $\Delta T = T_b - T_\infty$ (*natural convection*). In both cases, flow is influenced by the distance S_w between the fin tips and

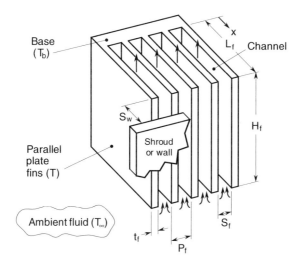

FIGURE 2.4 An array of longitudinal (parallel plate) fins.

an adjoining wall. If $S_w = 0$, all of the fluid is confined to movement through the channels. However, for forced convection and $S_w > 0$, the pressure drop associated with flow through the channels ($x < L_f$) exceeds that associated with flow outside the channels ($L_f < x < L_f + S_w$), and a significant fraction of the fluid can bypass the channels. For natural convection, with the gravitational vector parallel to the fin height, fluid flow through the channels may originate from the region between the fin tips and the wall, as well as from below the fins. For $S_w = 0$, fluid can only be entrained from the region below the fins.

Similar behavior characterizes the fin array of Figure 2.5, for which the pins may be *aligned* or *staggered*. The geometry of the in-line array is determined by the pin length L_f and diameter D_f, as well as by the transverse and longitudinal pitches, $P_{f,T}$ and $P_{f,L}$. For the staggered array, a diagonal pitch $P_{f,D}$ is also prescribed. In both forced and natural convection, flow over the pins is influenced by the tip-to-wall separation S_w.

The nature of fluid flow over the fin arrays of Figures 2.4 and 2.5 determines the manner in which the local convection coefficient h varies with location on the fins, as well as differences between values for the base and fin surfaces. Because such variations are rarely known, they are often assumed to be negligible and a single, average convection coefficient $\bar{h}$ is assigned to the entire fin array, including the base.

Just as the performance of a single fin may be characterized by its efficiency η_f, the performance of an array of fins and the base to which it is attached may be characterized by an *overall surface efficiency* η_o defined as

$$\eta_o = \frac{q_t}{q_{t,\max}} = \frac{q_t}{\bar{h} A_{s,t} \theta_b} \tag{2.27}$$

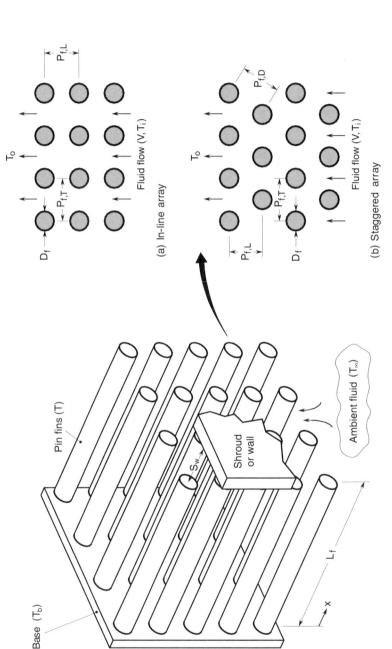

FIGURE 2.5 An array of pin fins: (*a*) in-line array and (*b*) staggered array.

where q_t is the total heat rate from the surface area $A_{s,t}$ associated with both the fins and the *exposed* portion of the base. If there are N fins in the array and the area of the exposed base is designated as $A_{s,b}$, the total surface area is

$$A_{s,t} = NA_{s,f} + A_{s,b} \tag{2.28}$$

The maximum heat rate ($q_{t,\max}$) would be achieved ($\eta_o = 1$) if the entire fin surface were at the base temperature $T_b(\eta_f = 1)$.

If local convection coefficients on the fin and base surfaces are approximated as the average coefficient for the entire array, the total heat rate may be expressed as

$$q_t = N\eta_f \overline{h} A_{s,f} \theta_b + \overline{h} A_{s,b} \theta_b = \overline{h} \left[N\eta_f A_{s,f} + (A_{s,t} - NA_{s,f}) \right] \theta_b \tag{2.29}$$

Rearranging, it follows that

$$q_t = \overline{h} A_{s,t} \left[1 - \frac{NA_{s,f}}{A_{s,t}} (1 - \eta_f) \right] \theta_b \tag{2.30}$$

or, substituting from Eq. 2.27,

$$\eta_o = 1 - \frac{NA_{s,f}}{A_{s,t}} (1 - \eta_f) \tag{2.31}$$

which may be used to calculate the overall surface efficiency from knowledge of the fin efficiency and the geometry of the array.

The thermal *network* representations of fin arrays that are an integral part of the base or are attached to the base are shown in Figure 2.6. If the fins and base are fabricated as a single entity, there is no contact resistance at the base. However, the fins are more commonly manufactured separately and are attached to the base metallurgically, by an adhesive, or by a press fit in slots machined on the base. In such cases, there is a thermal contact resistance at the base.

As indicated by the thermal circuits of Figure 2.6, there are parallel heat flow paths associated with conduction/convection in the fins and convection from the exposed portions of the base. The corresponding thermal resistances are $(N\eta_f \overline{h} A_{s,f})^{-1}$ and $[\overline{h}(A_{s,t} - NA_{s,f})]^{-1}$, respectively, and for fins that are integral with the base, Figure 2.6a, the overall thermal resistance is

$$R_{th,o} = \frac{\theta_b}{q_t} = \frac{1}{\eta_o \overline{h} A_{s,t}} \tag{2.32}$$

where η_o is given by Eq. 2.31. For attached fins, Figure 2.6b, the contact and fin resistances are in series, and the following expression may readily be obtained for the overall thermal resistance:

$$R_{th,o(c)} = \frac{\theta_b}{q_t} = \frac{1}{\eta_{o(c)} \overline{h} A_{s,t}} \tag{2.33}$$

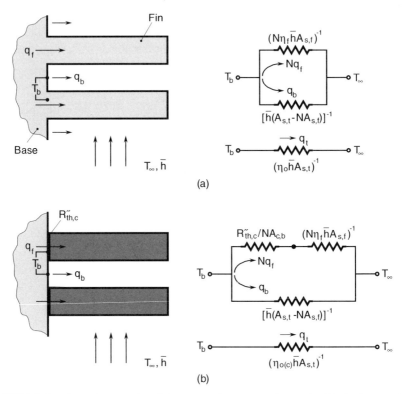

FIGURE 2.6 Thermal networks of fin arrays that are (a) integral with the base and (b) attached to the base.

where the overall surface efficiency is now of the form

$$\eta_{o(c)} = 1 - \frac{NA_{s,f}}{A_{s,t}}\left(1 - \frac{\eta_f}{C_o}\right) \quad (2.34a)$$

with

$$C_o = 1 + \eta_f \bar{h} A_{s,f} \left(\frac{R''_{th,c}}{A_{c,b}}\right) \quad (2.34b)$$

In manufacturing, care must be taken to minimize $R''_{th,c}$. If $R_{th,c} \ll R_{th,f}$, $C_o \to 1$ and $\eta_{o(c)} \to \eta_o$.

In an electronic package, chips may be separated from the heat sink by a spreader and/or substrate, with additional interfaces between pairs of materials. In such cases, the package may be represented by a thermal circuit that includes additional conduction and contact resistances (Kraus and Bar-Cohen, 1995).

Example 2.2

A heat sink for an electronic cooling application consists of an array of square fins, each of width w_f on a side. The heat sink has a square planform of width $W = 50$ mm, and two possible fin arrangements are considered. Case A corresponds to an 8×8 array of fins, each of width $w_f = 3$ mm, while Case B corresponds to a 24×24 array for which $w_f = 1$ mm. The heat sink is fabricated from an aluminum alloy ($k = 175$ W/m·K), and, for the intended application, the temperature at the base of the fins must not exceed $T_b = 75°C$. Convection cooling is maintained by flow of a dielectric liquid across the fins, with $T_f = 25°C$ and an average convection coefficient of $\bar{h} = 1000$ W/m²·K.

1. If the fin length is $L_f = 10$ mm, determine the maximum allowable heat rate, efficiency, effectiveness, and thermal resistance for a single fin in each array. What are the total heat rate and the overall efficiency and thermal resistance of each array?
2. For Case A, how are the foregoing performance parameters affected by variations in $\bar{h}$ and L_f?

Solution

Known: Configuration and thermal conductivity of a finned heat sink. Temperature and convection coefficient of coolant flow.

Find:

1. Fin and array performance parameters
2. Effect of convection coefficient and fin length on performance for Case A

Schematic:

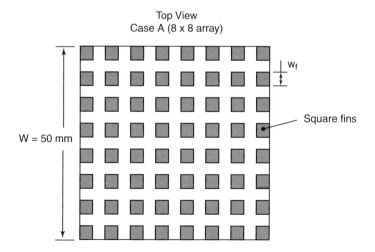

32 FUNDAMENTALS OF HEAT TRANSFER AND FLUID FLOW

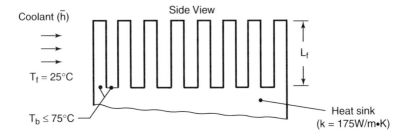

Assumptions:

1. Steady-state conditions
2. One-dimensional conduction along fins
3. Uniform convection coefficient ($h = \bar{h}$)
4. Constant thermal conductivity

Analysis:

1. With convection at the fin tip, the fin heat rate is given by Eq. 2.17b:

$$q_f = M \frac{\sinh mL_f + (\bar{h}/mk)\cosh mL_f}{\cosh mL_f + (\bar{h}/mk)\sinh mL_f}$$

where $m = (\bar{h}P/kA_c)^{1/2} = (4\bar{h}/kw_f)^{1/2}$ and $M = (\bar{h}PkA_c)^{1/2}\theta_b = (4\bar{h}kw_f^3)^{1/2}(T_b - T_f)$. From knowledge of q_f, the fin efficiency, effectiveness, and resistance may be computed from Eqs. 2.20, 2.22, and 2.24, respectively, where

$$\eta_f = \frac{q_f}{\bar{h}A_{s,f}\theta_b} = \frac{q_f}{\bar{h}(4w_f L_f + w_f^2)(T_b - T_f)}$$

$$\varepsilon_f = \frac{q_f}{\bar{h}A_{c,b}\theta_b} = \frac{q_f}{\bar{h}w_f^2(T_b - T_f)}$$

$$R_{th,f} = \frac{\theta_b}{q_f} = \frac{T_b - T_f}{q_f}$$

Substituting the prescribed numerical values, it follows that

Case	m (m^{-1})	M (W)	q_f (W)	η_f	ε_f	$R_{th,f}$ (K/W)
A ($w_f = 3$ mm)	87.3	6.87	5.05	0.78	11.2	9.90
B ($w_f = 1$ mm)	151.2	1.32	1.21	0.59	24.2	41.36

The total heat rate and overall efficiency of the array are given by Eqs. 2.27 and 2.31, respectively,

$$q_t = \eta_o \bar{h} A_{s,t} \theta_b$$

$$\eta_o = 1 - \frac{N A_{s,f}}{A_{s,t}}(1 - \eta_f)$$

where $A_{s,t} = N A_{s,f} + A_{s,b}$ and $A_{s,b} = W^2 - N A_{c,b}$. For Cases A and B, $N = 64$ and $N = 576$, respectively. It follows that

Case	$A_{s,b}$ (m²)	$A_{s,f}$ (m²)	$A_{s,t}$ (m²)	η_o	q_t (W)	$R_{t,o}$ (K/W)
A ($w_f = 3$ mm)	1.92×10^{-3}	1.29×10^{-4}	1.018×10^{-2}	0.824	419.3	0.119
B ($w_f = 1$ mm)	1.92×10^{-3}	4.10×10^{-5}	2.554×10^{-2}	0.621	792.4	0.063

From the foregoing results, it is clear that, although superior performance for a single fin is provided by the larger value of w_f, the inverse is true for the array. That is, larger and smaller values of q_t and $R_{t,o}$, respectively, are obtained for $w_f = 1$ mm. In general, improved performance is associated with using a larger number of thinner fins, rather than a smaller number of thicker fins.

2. The following results were obtained from parametric calculations based on the foregoing model:

Case A: $L_f = 10$ mm

$\bar{h}$ (W/m² · K)	ε_f	η_f	$R_{th,f}$ (K/W)	q_f (W)	$R_{t,o}$ (K/W)	q_t (W)
100	13.9	0.97	79.8	0.6	1.01	50
5000	6.6	0.46	3.4	14.9	0.04	1440

Case A: $\bar{h} = 1000$ W/m² · K

L_f (mm)	ε_f	η_f	$R_{th,f}$ (K/W)	q_f (W)	$R_{t,o}$ (K/W)	q_t (W)
1	2.3	0.99	48.0	1.0	0.31	163
50	15.3	0.23	7.3	6.9	0.09	536

Although ε_f and η_f decrease with increasing $\bar{h}$, the reduction is more than offset by the increase in $\bar{h}$, yielding a substantial reduction in the thermal resistance and an increase in heat rate for both a single fin and the fin array. Although η_f decreases with increasing L_f, the reduction is more than offset by the increase in $A_{s,f}$, again yielding a substantial reduction in thermal resistance and an increase in heat rate for both a single fin and the array. Similar results would be obtained for Case B.

Comments: Because the temperature at the tip of the fin approaches T_f with increasing $\bar{h}$ and/or L_f, diminishing returns are associated with such increases, for which there would be practical limits. Such limits depend on the fin width and thermal conductivity, and would increase with increasing w_f and k.

2.2.5 Transient Conduction

Electronic cooling problems are typically considered for steady-state conditions, but situations may arise for which starting (or shutdown) transients are of interest. Although detailed predictions of temperature as a function of position and time during the transient could be obtained by solving Eq. 2.1, considerable effort would be required, particularly for multiple materials and interface conditions. A simpler procedure, which may be used if temperatures within the heat-generating component are nearly uniform during the transient, is termed the *lumped-capacitance* method.

If the total thermal resistance between a heat-generating electronic component and the fluid used to cool the component is designated as $R_{th,t}$, its transient thermal response, $T(t)$, during startup may be expressed as (Incropera and DeWitt, 1996)

$$\frac{T - T_\infty}{T_i - T_\infty} = \exp(-at) + \frac{b/a}{T_i - T_\infty}[1 - \exp(-at)] \qquad (2.35)$$

where T_i is the initial temperature of the component, $a \equiv (R_{th,t}\rho \forall c_p)^{-1}$, $b \equiv \dot{E}_g/\rho \forall c_p$, and $\dot{E}_g$ is the rate at which electrical energy is converted to thermal energy in the component. By setting $\dot{E}_g = 0$ and, hence, $b = 0$, the foregoing expression may also be used for shutdown.

Conditions for which the lumped-capacitance method may be used to a good approximation are discussed by Incropera and DeWitt (1996).

2.3 EQUATIONS OF MOTION

A detailed description of the velocity field associated with flow of a liquid coolant may be obtained by solving what are known as the *equations of motion*. For *steady, three-dimensional, laminar* flow with *constant properties*, the equations may be expressed in the following form for rectangular coordinates:

$$\frac{\partial u}{\partial x} + \frac{\partial v}{\partial y} + \frac{\partial w}{\partial z} = 0 \qquad (2.36)$$

$$\rho\left(u\frac{\partial u}{\partial x} + v\frac{\partial u}{\partial y} + w\frac{\partial u}{\partial z}\right) = \mu\left(\frac{\partial^2 u}{\partial x^2} + \frac{\partial^2 u}{\partial y^2} + \frac{\partial^2 u}{\partial z^2}\right)$$

$$-\frac{\partial p}{\partial x} + \rho g_x \beta(T - T_{\text{ref}}) \qquad (2.37)$$

$$\rho\left(u\frac{\partial v}{\partial x} + v\frac{\partial v}{\partial y} + w\frac{\partial v}{\partial z}\right) = \mu\left(\frac{\partial^2 v}{\partial x^2} + \frac{\partial^2 v}{\partial y^2} + \frac{\partial^2 v}{\partial z^2}\right)$$
$$-\frac{\partial p}{\partial y} + \rho g_y \beta(T - T_{\text{ref}}) \quad (2.38)$$

$$\rho\left(u\frac{\partial w}{\partial x} + v\frac{\partial w}{\partial y} + w\frac{\partial w}{\partial z}\right) = \mu\left(\frac{\partial^2 w}{\partial x^2} + \frac{\partial^2 w}{\partial y^2} + \frac{\partial^2 w}{\partial z^2}\right)$$
$$-\frac{\partial p}{\partial z} + \rho g_z \beta(T - T_{\text{ref}}) \quad (2.39)$$

The first equation imposes the conservation of mass requirement at each location in the flow domain, whereas Eqs. 2.37 to 2.39 express Newton's second law of motion for each of the coordinate directions. Terms on the left-hand side of each equation account for changes in the component of the momentum flux resulting from bulk fluid motion (*advection*) in each of the three directions, whereas terms in the parentheses on the right-hand side account for changes resulting from molecular motion (*diffusion*). The remaining terms on the right-hand side account for pressure and buoyancy forces.

In the foregoing formulation of the momentum equations, the fluid is assumed to be incompressible, except for the effect of density variations on the buoyancy force. In formulating this force, the *Boussinesq approximation* is invoked, and the density is assumed to vary linearly with temperature, where T_{ref} is an appropriate reference temperature for the system of interest. The quantities g_x, g_y, and g_z represent components of the gravitational acceleration in the three coordinate directions.

Evaluation of the buoyancy force terms in the momentum equations requires knowledge of the temperature field, which may be determined by solving the following form of the energy equation:

$$\rho c_{p,f}\left(u\frac{\partial T}{\partial x} + v\frac{\partial T}{\partial y} + w\frac{\partial T}{\partial z}\right) = k_f\left(\frac{\partial^2 T}{\partial x^2} + \frac{\partial^2 T}{\partial y^2} + \frac{\partial^2 T}{\partial z^2}\right) \quad (2.40)$$

The equation applies for steady, laminar, three-dimensional flow with constant properties and negligible viscous dissipation. Terms on the left-hand side account for transport of energy resulting from bulk fluid motion (*advection*), whereas terms on the right-hand side account for transport resulting from molecular motion (*conduction* or *diffusion*).

In *forced convection*, motion is driven by an *imposed pressure field*. In *natural convection*, it is driven by the body force associated with the *gravitational field*. In *mixed convection*, the imposed pressure field and body force make comparable contributions to fluid motion.

The foregoing equations may be solved using finite-difference or finite-element methods to obtain three-dimensional velocity and temperature fields in the fluid, as well as the three-dimensional temperature field in the solid over which the fluid moves. Numerous commercial *computational fluid dynamic* (CFD) codes are available for this purpose and may be used to treat the complex geometries and flow patterns associated with applications in electronic cooling.

2.4 SIMILARITY PARAMETERS

Similarity parameters are *dimensionless parameters* that facilitate generalization of results obtained for a particular surface geometry and application of the generalization over a range of conditions, which may vary according to the size of the surface, the fluid velocity, and the nature of the fluid. Similarity parameters associated with forced, natural, and mixed convection may be obtained by nondimensionalizing the equations of motion.

For *forced convection*, the pertinent dimensionless parameters are the *Reynolds number*,

$$Re_{L_o} \equiv \frac{U_o L_o}{\nu} \qquad (2.41)$$

where V_o and L_o are the characteristic velocity and length of the flow and geometry, respectively, and the *Prandtl number*,

$$Pr \equiv \frac{c_{p,f}\mu}{k_f} = \frac{\nu}{\alpha_f} \qquad (2.42)$$

The Reynolds number provides a measure of the strength of inertia forces relative to viscous forces in the flow. The Prandtl number is a fluid property that determines the relative effectiveness of momentum transfer by diffusion to heat transfer by diffusion.

For *natural convection*, the *Rayleigh number* provides a measure of the strength of buoyancy forces relative to viscous forces in the flow. According to whether a temperature difference, $\Delta T = T_s - T_{\text{ref}}$, or a surface heat flux, q_s'', is prescribed, the Rayleigh number is defined as

$$Ra_{L_o} \equiv \frac{g\beta \Delta T L_o^3}{\alpha_f \nu} \qquad (2.43)$$

or a *modified Rayleigh number* is defined as

$$Ra_{L_o}^* \equiv \frac{g\beta q_s'' L_o^4}{k_f \alpha_f \nu} \qquad (2.44)$$

Natural convection can also be generalized in terms of a *Grashof number* or a *modified Grashof number*, which, respectively, are defined as

$$Gr_{L_o} \equiv \frac{g\beta \Delta T L_o^3}{\nu^2} \qquad (2.45)$$

$$Gr_{L_o}^* \equiv \frac{g\beta q_s'' L_o^4}{k_f \nu^2} \qquad (2.46)$$

Note that the product of the Grashof and Prandtl numbers yields the Rayleigh number.

In *mixed convection*, both the Reynolds and the Rayleigh (or Grashof) numbers are needed to characterize fully the effect of flow conditions on heat transfer.

The foregoing dimensionless parameters are independent parameters, on which heat transfer depends. The *dependent* dimensionless parameter is termed the *Nusselt number*, which may be defined in terms of the average convection coefficient

$$\overline{Nu_{L_o}} = \frac{\overline{h} L_o}{k_f} \qquad (2.47a)$$

or the local convection coefficient

$$Nu_{L_o} = \frac{h L_o}{k_f} \qquad (2.47b)$$

The convection coefficients, which are defined by Newton's law of cooling, Eqs. 2.12a and 2.12b, are important characteristics of the *thermal boundary layer* that develops as fluid moves over a heated surface.

Thermal boundary layer development is illustrated in Figure 2.7 for flow over a flat plate whose temperature T_s differs from that of the fluid. The condition is one of forced convection, with flow externally imposed in a direction parallel to the plate. At the leading edge of the plate, the fluid is at a uniform temperature T_∞ and velocity u_∞. However, fluid making contact with the plate achieves thermal equilibrium at the plate's temperature, thereby establishing temperature gradients and, hence, heat transfer within the fluid. The thickness δ_{th} of the thermal boundary layer is the distance normal to the surface over which the fluid temperature transitions from the surface (T_s) to the free-stream (T_∞) value. With increasing distance x from the leading edge, temperature gradients penetrate further into the free stream and the thermal boundary layer grows.

The convection heat transfer coefficient may be related to conditions in the thermal boundary layer by applying Fourier's law, Eq. 2.2, to the fluid at $y = 0$. Hence, for the y-coordinate direction,

$$q_s'' = -k_f \frac{\partial T}{\partial y}\bigg|_{y=0} \qquad (2.48)$$

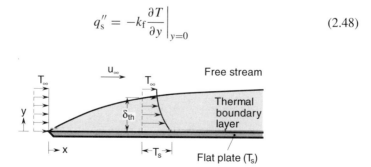

FIGURE 2.7 Thermal boundary layer development for flow over an isothermal flat plate.

This expression may be used to evaluate the heat flux because, at the surface, there is no fluid motion and heat transfer occurs only by conduction. Combining Eqs. 2.12 and 2.48, where $q_s'' = q_s/A_s$, the following expression is obtained for the local convection coefficient:

$$h = \frac{-k_f \, \partial T/\partial y \big|_{y=0}}{T_s - T_\infty} \tag{2.49}$$

An important implication of this result is that, because the overall temperature difference, $T_s - T_\infty$, is fixed and δ_{th} increases with increasing x, temperature gradients, including the value at $y = 0$, must decrease with increasing x. Hence, boundary layer development in the streamwise direction is associated with a corresponding reduction in the local convection coefficient. If the manner in which the local convection coefficient varies over a surface is known, the average coefficient may be determined from the following integration:

$$\overline{h} = \frac{1}{A_s} \int_{A_s} h \, dA_s \tag{2.50}$$

In the interest of maintaining large heat fluxes for a prescribed temperature difference $(T_s - T_\infty)$, or small temperature differences for a prescribed heat flux q_s'', it is desirable to have thin boundary layers and to avoid situations for which there is continuous boundary layer development over large surfaces. When such development occurs, the only event that will interrupt the attendant decay in h is transition from a laminar to a turbulent boundary layer, which is accompanied by a significant increase in h. However, following such a transition and subsequent development of the turbulent boundary layer, the local convection coefficient once again decreases with increasing x. Fortunately, electronic chips and packages are small, precluding extensive boundary layer development, and they are often characterized by protuberances that disrupt such development.

Alternative forms of a dimensionless, dependent heat transfer parameter are the *Stanton number* and the *Colburn j factor*. In terms of local conditions, they are defined as

$$St \equiv \frac{h}{\rho U_o c_{p,f}} = \frac{Nu_{L_o}}{Re_{L_o} Pr} \tag{2.51}$$

and

$$j_H \equiv St \, Pr^{2/3} \tag{2.52}$$

Forms corresponding to average conditions $(\overline{St}, \overline{j}_H)$ may also be defined in terms of the average coefficient $\overline{h}$.

Dimensionless dependent parameters are also associated with the work required to overcome forces that resist movement of fluid over a surface. The forces are due to

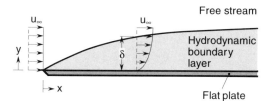

FIGURE 2.8 Hydrodynamic (velocity) boundary layer development for flow over a flat plate.

shear stresses induced by frictional effects at the fluid/surface interface and can also be due to pressure differentials associated with separation of the fluid from a bluff surface.

Friction forces are associated with the *hydrodynamic* or *velocity* boundary layer that develops when a moving fluid interacts with a surface, as for a flat plate in parallel flow (Fig. 2.8). At the leading edge of the plate, the fluid is at a uniform velocity u_∞. However, by virtue of its viscosity, fluid making contact with the plate assumes zero velocity (the *zero-slip* condition), thereby establishing velocity gradients and a shear stress distribution within the fluid. The thickness δ of the hydrodynamic boundary layer is the distance normal to the surface over which the velocity transitions from $u = 0$ at the surface to $u = u_\infty$ in the free stream. With increasing distance x from the leading edge, viscous effects penetrate further into the free stream and the hydrodynamic boundary layer grows.

The frictional resistance to flow imposed by the surface is determined by the surface shear stress, τ_s, which, for a *Newtonian fluid*, is given by

$$\tau_s = \mu \left.\frac{\partial u}{\partial y}\right|_{y=0} \tag{2.53}$$

where μ is the dynamic viscosity of the fluid. The velocity gradient at the surface, and, hence, the surface shear stress, depends on conditions within the boundary layer and decreases with increasing boundary layer thickness.

A dimensionless dependent parameter that is often used to characterize flow resistance is the *friction coefficient* or *Fanning friction factor*

$$C_f = f_F \equiv \frac{\tau_s}{\rho U_0^2/2} \tag{2.54}$$

Alternatively, for internal flows, as through a circular tube or rectangular channel, a *Darcy* or *Moody friction factor* of the following form is often used:

$$f_D \equiv \frac{-(dp/dx)D_h}{\rho U_0^2/2} \tag{2.55}$$

where the pressure gradient in the channel depends on the surface shear stress. A comparable form of the friction factor is commonly used for the pressure drop associated with an array of obstacles, such as the pin fin array of Figure 2.5,

$$f \equiv \frac{\Delta p}{\rho U_o^2/2} \tag{2.56}$$

2.5 NATURAL (FREE) CONVECTION

In forced convection, fluid motion is due to an external agent, as, for example, a fan or a pump. A comparative advantage of natural or free convection is that fluid motion is induced by buoyancy forces and, hence, does not require an external agent. However, because velocities are much smaller than those associated with forced convection, the corresponding convection coefficients and heat rates are also smaller.

In this section, we consider three special cases of heat transfer by natural convection. The first case corresponds to a heated vertical or horizontal surface in an *extensive* and *quiescent* ambient fluid. An extensive medium is one for which other surfaces are sufficiently removed from the heated surface to have a negligible effect on the buoyancy-induced flow. A quiescent fluid is one for which there is no motion at locations well removed from the surface. The second case corresponds to adjoining vertical surfaces, one or both of which are heated. Termed a *parallel-plate channel*, the surfaces are open to the ambient fluid at opposite ends, permitting continuous through-put of the buoyancy-driven flow. The third case corresponds to a completely enclosed *rectangular cavity* within which buoyancy-driven fluid motion is due to differences between wall temperatures.

More detailed treatments of the foregoing natural convection flows are provided by Yang (1987), Gebhart et al. (1988), and Ostrach (1988).

2.5.1 Heated Vertical and Horizontal Surfaces

Laminar velocity boundary layer development on a heated vertical surface is shown in Figure 2.9. The plate is immersed in an extensive, quiescent fluid, and with $T_s > T_\infty$ fluid close to the surface is lighter (less dense) than that in the quiescent region. Buoyancy forces, therefore, induce a boundary layer in which fluid ascends along the surface. The boundary layer thickens with increasing x, and the velocity profile, $u(y)$, at any x goes to zero at the edge of the boundary layer (in principle, as $y \to \infty$), as well as at $y = 0$. This situation differs from that of forced convection (Fig. 2.8), where the velocity increases from zero at the surface to a free-stream value u_∞ at the edge of the boundary layer. However, as in forced convection (Fig. 2.7), the temperature still decays monotonically from T_s to T_∞ across the thermal boundary layer.

Because the boundary layer flow of Figure 2.9 is two dimensional ($w = 0$), the z-direction momentum equation, Eq. 2.39, is not pertinent. In addition, if *velocity boundary layer approximations* of the form $u \gg v$ and $\partial u/\partial y \gg \partial u/\partial x$, $\partial v/\partial x$, $\partial v/\partial y$ are made, the y-momentum equation, Eq. 2.38, simply reduces to $\partial p/\partial y = 0$. Hence, conditions in the velocity boundary layer are described by Eqs. 2.36 and

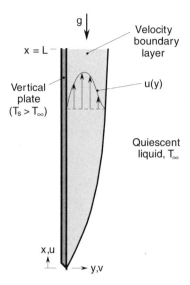

FIGURE 2.9 Velocity boundary layer development for buoyancy-driven flow over a vertical surface ($g_x = -g$, $g_y = g_z = 0$).

2.37, where, for natural convection, p is a dynamic pressure whose variation in the x direction may also be neglected. The equations of motion, therefore, reduce to

$$\frac{\partial u}{\partial x} + \frac{\partial v}{\partial y} = 0 \tag{2.57}$$

$$u\frac{\partial u}{\partial x} + v\frac{\partial u}{\partial y} = v\frac{\partial^2 u}{\partial y^2} + g\beta(T - T_\infty) \tag{2.58}$$

To obtain the temperature T associated with the buoyancy term of the foregoing equation, the energy equation, Eq. 2.40, must also be solved, and if a *thermal boundary layer approximation* of the form $\partial T/\partial y \gg \partial T/\partial x$ is made, the equation reduces to

$$u\frac{\partial T}{\partial x} + v\frac{\partial T}{\partial y} = \alpha_f \frac{\partial^2 T}{\partial y^2} \tag{2.59}$$

The foregoing partial differential equations, Eqs. 2.57 to 2.59, may be transformed to two ordinary differential equations by introducing a *similarity parameter* of the form

$$\eta \equiv \frac{y}{x}\left(\frac{Gr_x}{4}\right)^{1/4} \tag{2.60}$$

The differential equations have been solved numerically, and results for dimensionless forms of the velocity and temperature profiles are represented exclusively in terms of η and Pr (Incropera and DeWitt, 1996).

Correlations for the average Nusselt number associated with natural-convection heat transfer from a vertical plate depend on whether transition from laminar to turbulent flow occurs on the plate. Conditions for which boundary layer disturbances may be amplified, leading to turbulence, depend on the relative magnitude of buoyancy and viscous forces in the fluid, and, hence, on the Rayleigh number. For vertical plates, the *critical* Rayleigh number for the onset of turbulence is $Ra_{x,c} = g\beta(T_s - T_\infty)/\alpha_f \nu \approx 10^9$. Accordingly, for a plate of length L, the following correlations are commonly used to obtain the average Nusselt number (Incropera and DeWitt, 1996).

Heated Vertical Plate:

$$\overline{Nu}_L = 0.59 Ra_L^{1/4} \qquad Ra_L < 10^9 \tag{2.61a}$$

$$\overline{Nu}_L = 0.10 Ra_L^{1/3} \qquad Ra_L > 10^9 \tag{2.61b}$$

Alternatively, the following correlation has been recommended over the entire Rayleigh number range (Churchill and Chu, 1975):

$$\overline{Nu}_L = \left\{ 0.825 + \frac{0.387 Ra_L^{1/6}}{\left[1 + (0.492/Pr)^{9/16}\right]^{8/27}} \right\}^2 \tag{2.62}$$

A heated vertical plate is aligned with the gravity vector, and buoyancy forces induce fluid motion along the plate. For a heated horizontal plate, buoyancy forces are oriented normal to the plate and induce fluid motion toward and/or away from its bottom and top surfaces. As shown in Figure 2.10, the positive buoyancy associated with fluid in proximity to the top surface of the plate yields parcels of warm fluid ascending from the top surface. Conservation of mass dictates that this fluid be re-

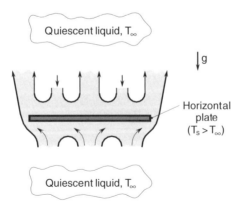

FIGURE 2.10 Buoyancy-driven flows associated with top and bottom surfaces of a heated horizontal plate.

placed by cooler fluid descending from the ambient, and the average Nusselt number is correlated by the following expressions (Incropera and DeWitt, 1996).

Upper Surface of Heated Horizontal Plate:

$$\overline{Nu}_{L_o} = 0.54 Ra_{L_o}^{1/4} \qquad 10^4 < Ra_L < 10^7 \qquad (2.63a)$$

$$\overline{Nu}_{L_o} = 0.15 Ra_{L_o}^{1/3} \qquad 10^7 < Ra_L < 10^{11} \qquad (2.63b)$$

Heat transfer is less effective from the bottom surface, which impedes the motion of fluid ascending from below. The fluid must move horizontally before it can ascend from the edges of the plate, thereby reducing local convection coefficients. The corresponding expression for the average Nusselt number is as follows.

Lower Surface of Heated Horizontal Plate:

$$\overline{Nu}_{L_o} = 0.27 Ra_{L_o}^{1/4} \qquad 10^5 < Ra_L < 10^{10} \qquad (2.64)$$

Equations 2.63 and 2.64 apply for an arbitrary plate geometry, with the characteristic length defined as

$$L_o \equiv \frac{A_s}{P} \qquad (2.65)$$

where A_s and P are the plate surface area and perimeter, respectively.

2.5.2 Vertical, Parallel-Plate Channels

A common free-convection geometry involves adjoining vertical plates that are open to the ambient fluid at opposite ends (Fig. 2.11). If the plates are heated and their temperatures exceed that of the ambient liquid, positive buoyancy forces induce an ascending flow within the channel, which entrains fluid from the ambient at the inlet ($x = 0$) and discharges it to the ambient at the exit ($x = H$). If the plate width W (normal to the page) is much larger than the spacing S, the effect of conditions at the edges of the channel, which may be open or closed, is negligible and two-dimensional flow may be assumed within the channel.

Depending on the ratio of the plate spacing and height, S/H, three different flow regimes are possible. Short channels and/or large spacings provide an *isolated-plate limit* ($S/H \to \infty$), for which conditions at each plate correspond to those of an isolated plate in an infinite liquid (Fig. 2.9). In contrast, long channels and/or small spacings provide a *fully developed limit* ($S/H \to 0$), for which boundary layers developing on adjoining plates merge near the entrance and fully developed conditions may be assumed to exist throughout the channel. In a *developing flow* (Fig. 2.11), neither of the foregoing limits is achieved.

For symmetrically heated, isothermal plates ($T_{s,1} = T_{s,2} = T_s$), the following semiempirical correlation (the *Elenbaas correlation*) has been obtained for the aver-

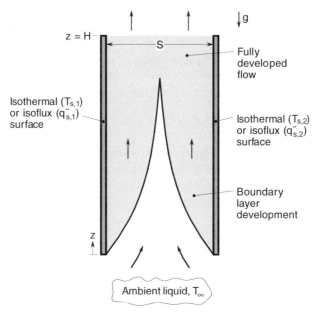

FIGURE 2.11 Developing natural-convection flow between heated vertical plates with opposite ends open to a quiescent, ambient liquid.

age Nusselt number associated with each plate (Elenbaas, 1942):

$$\overline{Nu}_S = \frac{Ra_S/(H/S)}{24}\left\{1 - \exp\left[-\frac{35}{Ra_S/(H/S)}\right]\right\}^{3/4} \qquad (2.66)$$

where

$$\overline{Nu}_S = \frac{\overline{h}S}{k_f} = \left[\frac{q_S/A_s}{(T_s - T_\infty)}\right]\frac{S}{k_f} \qquad (2.67)$$

Hence, from knowledge of the average Nusselt number $\overline{Nu}_S$, the total rate of heat transfer q_s from the plate may be determined. The Nusselt number decreases with increasing H/S, and, in the fully developed limit,

$$\overline{Nu}_{S(\text{fd})} = \frac{Ra_S/(H/S)}{24} \qquad (2.68)$$

The Elenbaas correlation does not reduce to the appropriate form in the isolated-plate limit. To circumvent this deficiency, Bar-Cohen and Rohsenow (1984) combined Eq. 2.68 with Eq. 2.61a for laminar boundary layer flow along an isolated plate to obtain

$$\overline{Nu}_S = \left\{ \frac{576}{[Ra_S/(H/S)]^2} + \frac{2.87}{[Ra_S/(H/S)]^{1/2}} \right\}^{-1/2} \quad (2.69)$$

Bar-Cohen and Rohsenow used this correlation to obtain the following expression for the optimum spacing S_{opt} that maximizes heat transfer from an array of isothermal plates:

$$S_{opt} = \frac{2.71}{(Ra_S/S^3 H)^{1/4}} \quad (2.70)$$

Although heat transfer from each plate decreases with decreasing S, the number of plates that may be placed in a prescribed volume increases, and S_{opt} is the spacing that maximizes the product of $\overline{h}$ and the *total* plate surface area.

For isoflux surfaces, the information of greatest interest is the maximum plate temperature, which occurs at $z = H$. Accordingly, a local Nusselt number is defined as

$$Nu_{S,H} = \frac{q_s''}{(T_{s,H} - T_\infty)} \frac{S}{k_f} \quad (2.71)$$

where the subscript H refers to conditions at $z = H$. For symmetrically heated plates ($q_{s,1}'' = q_{s,2}'' = q_s''$), the corresponding correlation is of the form

$$Nu_{S,H} = \left\{ \frac{48}{[Ra_S^*/(H/S)]} + \frac{2.51}{[Ra_S^*/(H/S)]^{2/5}} \right\}^{-1/2} \quad (2.72)$$

and the optimum spacing is

$$S_{opt} = \frac{2.12}{(Ra_S^*/S^4 H)^{1/5}} \quad (2.73)$$

Expressions for asymmetrical surface conditions are provided elsewhere (Bar-Cohen and Rohsenow, 1984; Incropera and DeWitt, 1996).

2.5.3 Rectangular Cavities

Enclosure of a liquid by a rectangular cavity provides a convenient geometry for passive electronic cooling. Conditions for which opposing vertical surfaces are at uniform (hot and cold) temperatures have been studied extensively (Yang, 1987; Ostrach, 1988; Gebhart et al., 1988) and are illustrated in Figure 2.12. The cavity has major and minor aspect ratios, $A_z = H/S$ and $A_x = W/S$, respectively, and if $A_x \gg 1$, two-dimensional conditions may be assumed for flow within and heat transfer across the cavity.

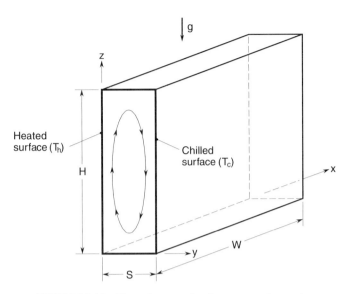

FIGURE 2.12 Natural convection in a rectangular cavity.

For $Ra_S < 10^3$, flow within the cavity is weak to nonexistent and heat transfer from the hot to the cold wall is largely by conduction in the fluid. For $10^3 < Ra_S < 10^5$, ascending and descending thermal boundary layers characterize flow along the hot and cold walls, respectively, and form a single recirculating cell in the cavity. With increasing $Ra_S > 10^6$, secondary and tertiary cells begin to form and transition to turbulence is initiated.

For cavities of moderate aspect ratio ($2 < A_z < 10$), $10^3 < Ra_S < 10^{10}$, and $Pr < 10^5$, Berkovsky and Polevikov (1977) recommend a heat transfer correlation of the form

$$\overline{Nu}_S = 0.22 \left[\frac{Pr}{0.2 + Pr} Ra_S \right]^{0.28} A_z^{-1/4} \tag{2.74}$$

For larger aspect ratios ($1 < A_z < 40$), $10^6 < Ra_S < 10^9$, and $1 < Pr < 20$, MacGregor and Emery (1969) recommend the following correlation:

$$\overline{Nu}_S = 0.046 Ra_S^{1/3} \tag{2.75}$$

which indicates that the average convection coefficient is independent of the cavity spacing, as well as the aspect ratio. This behavior was confirmed by Cowen et al. (1982), who obtained a nearly equivalent correlation for experiments performed with water over the ranges $1.5 < A_z < 61$ and $2 \times 10^5 < Ra_S < 2 \times 10^{11}$. In their correlation, a coefficient of 0.043 was used instead of 0.046.

A horizontal cavity corresponds to rotation of Figure 2.12 by 90°, such that the bottom and top surfaces are heated and cooled, respectively. For Rayleigh numbers less than a critical value of 1708, viscous forces dominate over buoyancy forces, and, with no fluid motion, heat transfer is by conduction from the bottom to the top surface. With increasing $Ra_S > 1708$, there is a progression from steady, two-dimensional roll cells in the cavity to first steady and then transient three-dimensional cells, and, subsequently, turbulent natural convection. The following correlation has been obtained by Globe and Dropkin (1959):

$$\overline{Nu}_s = 0.069 Ra_S^{1/3} Pr^{0.074} \qquad (2.76)$$

and may be used for $0.02 < Pr < 8750$ and $1.5 \times 10^5 < Ra_S < 6.8 \times 10^8$.

In using the foregoing correlations, fluid properties are evaluated at the average surface temperature, $\overline{T} = (T_h + T_c)/2$, and the heat rate is expressed as $q = h \bar{A}_s (T_h - T_c)$.

2.6 FORCED CONVECTION: EXTERNAL FLOW

For forced convection, fluid motion over a surface is driven by an external agent, such as a fan or a pump. If boundary layer development on a surface of interest is not constrained by adjoining surfaces, conditions correspond to *external flow*, and there is a region outside the boundary layer in which velocity and temperature gradients are negligible. This region is termed the *free stream* and related conditions are designated by the subscript ∞. Common examples of external flow include the *flat plate in parallel flow* and *circular cylinders* in *cross flow*, which are discussed in this section, as well as *impinging jets*, which are discussed in the following section.

2.6.1 Flat Plate in Parallel Flow

Thermal boundary layer development and velocity boundary layer development on a flat plate were discussed in Section 2.4 and used to introduce dimensionless parameters for characterizing related surface heat transfer and friction effects. Expressions for these parameters have been developed and their particular forms depend on whether one is interested in local or surface-averaged conditions, whether the flow is laminar or turbulent, and, in the case of heat transfer, whether surface conditions correspond to a uniform temperature, T_s, or heat flux, q_s''.

As shown in Figure 2.13, laminar boundary layer development begins at the leading edge ($x = 0$) of a flat plate in parallel flow. However, defining the Reynolds number in terms of the distance from the leading edge, $Re_x = u_\infty x/\nu$, transition to turbulence will begin at a downstream location x_c if a critical Reynolds number $Re_{x,c}$ is exceeded. In the laminar boundary layer, fluid motion is highly ordered and it is possible to identify streamlines along which the fluid moves. In contrast, fluid motion in the turbulent boundary layer is irregular and characterized by random fluctuations that enhance surface friction and heat transfer by increasing the transfer of momentum and energy within the boundary layer. Transition to a turbulent boundary layer begins with small disturbances in the laminar boundary layer, which may originate from the free stream or may be induced by surface roughness. The ratio of

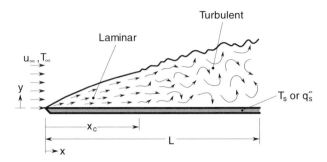

FIGURE 2.13 Transition from a laminar to a turbulent boundary layer for the flat plate in parallel flow.

inertia to viscous forces increases with increasing Reynolds number, and, at the critical Reynolds number, this ratio becomes large enough to amplify the disturbances and sustain the transition to turbulence.

Laminar Flow A laminar boundary layer will exist over the entire surface of a flat plate if the Reynolds number based on the plate length, $Re_L = u_\infty L/\nu$ is less than the critical Reynolds number ($Re_L < Re_{x,c}$). Recognizing that the flow is two dimensional and invoking standard boundary layer approximations ($u \gg v$; $\partial u/\partial y \gg \partial u/\partial x, \partial v/\partial x, \partial v/\partial y; \partial T/\partial y \gg \partial T/\partial x$), appropriate forms of the equations governing mass, momentum, and energy transfer within the boundary layer may be inferred from Eqs. 2.36 to 2.40. The equations are of the form

$$\frac{\partial u}{\partial x} + \frac{\partial v}{\partial y} = 0 \qquad (2.77)$$

$$u\frac{\partial u}{\partial x} + v\frac{\partial u}{\partial y} = \nu\frac{\partial^2 u}{\partial y^2} \qquad (2.78)$$

$$u\frac{\partial T}{\partial x} + v\frac{\partial T}{\partial y} = \alpha_f\frac{\partial^2 T}{\partial y^2} \qquad (2.79)$$

where $\partial p/\partial x = 0$ for a flat plate in parallel flow and buoyancy effects are negligible for forced convection.

The foregoing partial differential equations may be transformed to two ordinary differential equations by introducing a similarity parameter of the form

$$\eta \equiv \frac{y}{x} Re_x^{1/2} \qquad (2.80)$$

The equations have been solved numerically to obtain dimensionless forms of the boundary layer velocity and temperature profiles that depend only on η and the Prandtl number Pr (Incropera and DeWitt, 1996). In turn, the profiles may be used

FORCED CONVECTION: EXTERNAL FLOW

to determine the following expressions for the *local* friction coefficient and Nusselt number:

$$C_{f,x} = \frac{\tau_{s,x}}{\rho u_\infty^2/2} = \frac{0.664}{Re_x^{1/2}} \tag{2.81}$$

$$Nu_x = \frac{h_x x}{k_f} = 0.332 Re_x^{1/2} Pr^{1/3} \quad \text{(Uniform } T_s\text{)} \tag{2.82}$$

where the subscript x refers to conditions at a distance x from the leading edge. Equation 2.82 is restricted to conditions for which the plate is at a uniform temperature, T_s, and $Pr > 0.6$, which is satisfied for all liquids of interest in electronic cooling. If, instead, there is a uniform heat flux, q_s'', at the surface of the plate, the correlation is of the form

$$Nu_x = \frac{h_x x}{k_f} = 0.453 Re_x^{1/2} Pr^{1/3} \quad \text{(Uniform } q_s''\text{)} \tag{2.83}$$

If the local convection coefficient is determined from Eq. 2.82, it may be used with Newton's law of cooling, Eq. 2.12b, to determine the local heat flux from knowledge of T_s; if the convection coefficient is determined from Eq. 2.83, it is used with Eq. 2.12b to determine the local surface temperature from knowledge of q_s''.

Average coefficients associated with the entire plate ($0 \leq x \leq L$) may be obtained from integrations involving the local coefficient. For example, if Eq. 2.82 is substituted into Eq. 2.50, the following expression may be obtained for the average Nusselt number associated with laminar flow over an isothermal flat plate:

$$\overline{Nu}_L = \frac{\overline{h}L}{k_f} = 0.664 Re_L^{1/2} Pr^{1/3} \tag{2.84}$$

The average convection coefficient may then be used with Eq. 2.12a to determine the total heat rate q from the surface. Similarly, the following expression may be obtained for the average friction coefficient:

$$\overline{C}_{f,L} = \frac{\overline{\tau}_s}{\rho u_\infty^2/2} = \frac{1.328}{Re_L^{1/2}} \tag{2.85}$$

where the product $\overline{\tau}_s A_s$ yields the total friction force acting on the plate. For a uniform surface heat flux, the average temperature difference $(\overline{T_s - T_\infty})$ may be expressed as (Incropera and DeWitt, 1996)

$$\overline{(T_s - T_\infty)} = \frac{q_s'' L}{k_f \overline{Nu}_L} \tag{2.86a}$$

where

$$\overline{Nu}_L = 0.680 Re_L^{1/2} Pr^{1/3} \tag{2.86b}$$

Turbulent Flow The local friction coefficient and Nusselt number in a turbulent boundary layer are given by expressions of the form

$$C_{f,x} = 0.0592 Re_x^{-1/5} \tag{2.87}$$

and

$$Nu_x = 0.0296 Re_x^{4/5} Pr^{1/3} \quad \text{(Uniform } T_s\text{)} \tag{2.88}$$

or

$$Nu_x = 0.0308 Re_x^{4/5} Pr^{1/3} \quad \text{(Uniform } q_s''\text{)} \tag{2.89}$$

Although Eqs. 2.88 and 2.89 apply for uniform surface temperature and heat flux, respectively, they differ by only 4%, and Eq. 2.88 may be used to an excellent approximation for either surface condition.

If the Reynolds number based on the plate length exceeds the critical Reynolds number ($Re_L > Re_{x,c}$), transition from laminar to turbulent boundary layer behavior occurs on the plate, as shown in Figure 2.13. From Eq. 2.50, the average convection coefficient for an isothermal plate may be expressed as

$$\overline{h}_L = \frac{1}{L}\left(\int_0^{x_c} h_{x,\text{lam}}\, dx + \int_{x_c}^L h_{x,\text{turb}}\, dx\right) \tag{2.90}$$

where the local coefficients are determined from Eqs. 2.82 and 2.88. Making the appropriate substitutions and evaluating the integrals, the following expression is obtained for the average Nusselt number:

$$\overline{Nu}_L = \left[0.037 Re_L^{4/5} - \left(0.037 Re_{x,c}^{4/5} - 0.664 Re_{x,c}^{1/2}\right)\right] Pr^{1/3} \tag{2.91}$$

Depending on the amount of turbulence in the free stream and surface roughness, the critical Reynolds number may be as small as 10^5 or as large as 5×10^6. For a representative value of $Re_{x,c} = 5 \times 10^5$, Eq. 2.91 reduces to

$$\overline{Nu}_L = \left(0.037 Re_L^{4/5} - 871\right) Pr^{1/3} \tag{2.92}$$

If turbulent boundary layer development is artificially induced at the leading edge, as, for example, by a protuberance extending from the plate (a *boundary layer trip*), $Re_{x,c} = 0$. In such situations, as well as those for which $Re_L \gg Re_{x,c}$, Eq. 2.91 reduces to

$$\overline{Nu}_L = 0.037 Re_L^{4/5} Pr^{1/3} \tag{2.93}$$

Similar results may be obtained for the average friction coefficient, and, for $Re_{x,c} = 5 \times 10^5$,

$$\overline{C}_{f,L} = 0.074 Re_L^{-1/5} - 1742 Re_L^{-1} \qquad (2.94)$$

If $Re_{x,c} = 0$ or $Re_L \gg Re_{x,c}$, this expression reduces to

$$\overline{C}_{f,L} = 0.074 Re_L^{-1/5} \qquad (2.95)$$

A special form of heat transfer from a flat plate is one for which the heated portion of the plate is preceded by an *unheated starting length*, on which velocity boundary layer development occurs before thermal boundary layer growth begins on the heated portion. Treatment of this special case is provided by Incropera and DeWitt (1996).

In all of the foregoing correlations, fluid properties are evaluated at a mean boundary layer temperature, T_f, termed the *film temperature*:

$$T_f = \frac{T_s + T_\infty}{2} \qquad (2.96)$$

2.6.2 Circular Cylinders in Cross Flow

Heat transfer from a bundle of circular rods to a fluid moving in cross flow over the bundle is pertinent to pin fin arrays, which may be used to enhance heat transfer from electronic components (Fig. 2.5). Average convection coefficients for such arrays have been determined for $L_f/D_f \gg 1$, thereby rendering the specific nature of conditions at the fin base and tip negligible.

Flow over a single circular cylinder is complicated by boundary layer *separation*, which may or may not be preceded by boundary layer transition. Separation yields a *wake* (separated flow) region immediately behind the cylinder, and in an array heat transfer is strongly influenced by the interaction of wakes generated by upstream cylinders with downstream cylinders (Incropera and DeWitt, 1996), which, in turn, depends on whether rows in the array are *staggered* or *aligned* (Fig. 2.5).

Flow around pins in the first row of an array corresponds to a single cylinder in cross flow, and, for a uniform surface temperature T_s, the average Nusselt number may be obtained from the following correlation (Zhukauskas, 1972):

$$\overline{Nu}_{D_f} = C Re_{D_f}^m Pr^n \left(\frac{Pr}{Pr_s}\right)^{1/4} \qquad (2.97)$$

All properties are evaluated at T_∞, except Pr_s, which is evaluated at the temperature T_s of the pins. The Prandtl number exponent is $n = 0.37$ or $n = 0.36$ for $Pr \leq 10$ or $Pr > 10$, respectively, whereas the values of C and m depend on the Reynolds number and are provided by Incropera and DeWitt (1996). The average Nusselt number for an array with multiple rows of isothermal pins may be obtained from either of the following correlations due to Grimison (1937) or Zhukauskas (1972), respectively:

$$\overline{Nu}_{D_f} = 1.13 C_1 C_2 Re_{D,\max}^m Pr^{1/3} \tag{2.98}$$

and

$$\overline{Nu}_{D_f} = C C_2 Re_{D,\max}^m Pr^{0.36} \left(\frac{Pr}{Pr_s}\right)^{1/4} \tag{2.99}$$

where

$$Re_{D,\max} \equiv \frac{V_{\max} D_f}{\nu} \tag{2.100}$$

and $V_{\max}$ is the maximum fluid velocity within the array. For an in-line array, the maximum velocity occurs in a transverse plane passing through the pin centers and is given by

$$V_{\max} = \frac{P_{f,T}}{P_{f,T} - D_f} V \tag{2.101}$$

where V is the fluid velocity upstream of the array. For a staggered array, the maximum velocity occurs in a transverse plane passing through the pin centers and is given by Eq. 2.101, if the inequality $P_{f,D} > (P_{f,T} + D_f)/2$ is satisfied. If the inequality is not satisfied, the maximum velocity occurs in a diagonal plane between pin centers and is given by

$$V_{\max} = \frac{P_{f,T}}{2 \left(P_{f,D} - D_f\right)} V \tag{2.102}$$

In Eq. 2.98, all fluid properties are evaluated at an arithmetic mean value of the film temperature, $\overline{T}_f \equiv (T_s + \overline{T})/2$, where $\overline{T} = (T_i + T_o)/2$ is the arithmetic mean of the fluid inlet and outlet temperatures. In Eq. 2.99, all properties except Pr_s are evaluated at $\overline{T}$. The coefficients C and C_1, as well as the exponent m, in Eqs. 2.98 and 2.99 depend on the geometry of the array and/or the Reynolds number, whereas the coefficient C_2 depends on the number of rows N_L in the array, increasing from $C_2 \approx 0.65$ for $N_L = 1$ to $C_2 = 1$ for $N_L > 20$. Values of the coefficients and exponents are provided by Incropera and DeWitt (1996).

For an array of isothermal pins, the total heat rate per unit length of the pins may be expressed as

$$q' = N \left(\overline{h} \pi D_f \Delta T_{\text{lm}}\right) \tag{2.103}$$

where N, the total number of pins, is the product of the number of longitudinal and transverse rows in the array, $N = N_L N_T$. The *log-mean temperature difference* is

$$\Delta T_{lm} = \frac{(T_s - T_i) - (T_s - T_o)}{\ln\left(\dfrac{T_s - T_i}{T_s - T_o}\right)} \quad (2.104)$$

where the temperature of fluid leaving the array, T_o, may be obtained from the expression

$$\frac{T_s - T_o}{T_s - T_i} \approx \exp\left[-\frac{\pi D_f N \overline{h}}{\rho V N_T P_{f,T} c_{p,f}}\right] \quad (2.105)$$

The pressure drop associated with flow across the array may be obtained from a modified form of Eq. 2.56, which is expressed as (Zhukauskas, 1972)

$$\Delta p = N_L \chi \left(\frac{\rho V_{max}^2}{2}\right) f \quad (2.106)$$

The correction factor χ and the friction factor f depend on the Reynolds number and the array geometry, and results are provided by Incropera and DeWitt (1996). If the pressure drop is known, the corresponding pump power requirement may be determined from the following expression:

$$\dot{W} = \frac{\dot{\forall} \Delta p}{\eta_p} \quad (2.107)$$

where η_p is the pump efficiency.

Example 2.3

Heated ($T_h = 85°C$) and chilled ($T_c = 15°C$) isothermal surfaces of length and width $L = W = 100$ mm are separated by a distance of $S = 20$ mm in a rectangular cavity (Fig. 2.12), which is filled with a dielectric liquid (FC-77). Evaluate the heat rate for vertical and horizontal orientations. If, instead, the heated surface is cooled by parallel flow over the surface (Fig. 2.13), what is the heat rate corresponding to a liquid velocity and temperature of $u_\infty = 3$ m/s and $T_\infty = 15°C$, respectively?

Solution

Known: Temperature and dimensions of a heated surface. Temperature of cold plate or fluid used to cool the surface by free convection in a cavity or by forced convection resulting from a parallel flow of prescribed velocity.

Find: Heat rates associated with vertical and horizontal cavity orientations and with forced flow.

54 FUNDAMENTALS OF HEAT TRANSFER AND FLUID FLOW

Assumptions:

1. Steady-state conditions
2. Isothermal surfaces
3. Transition Reynolds number of $Re_{x,c} = 5 \times 10^5$ for forced flow

Properties: FC-77 ($\bar{T} = T_f = 50°C$). From Appendix B: $\rho = 1716$ kg/m^3, $c_{p,f} = 1087$ J/kg·K, $k_f = 0.061$ W/m·K, $\alpha_f = k_f/\rho c_{p,f} = 3.27 \times 10^{-8}$ m^2/s, $\mu = 9.25 \times 10^{-4}$ kg/s·m, $\nu = \mu/\rho = 5.39 \times 10^{-7}$ m^2/s, $Pr = \nu/\alpha_f = 16.5$, and $\beta = 0.00143$ K^{-1}.

Analysis: For free convection in the rectangular cavity, the Rayleigh number is

$$Ra_S = \frac{g\beta(T_h - T_c)S^3}{\alpha_f \nu} = \frac{9.8 \text{ m/s}^2 \times 0.00143 \text{ K}^{-1}(70 \text{ K})(0.02 \text{ m})^3}{3.27 \times 10^{-8} \text{ m}^2/\text{s} \times 5.39 \times 10^{-7} \text{ m}^2/\text{s}}$$

$$= 4.45 \times 10^8$$

From Eq. 2.75, the average Nusselt number for the vertical cavity is

$$\overline{Nu}_S = 0.046 Ra_S^{1/3} = 0.046\left(4.45 \times 10^8\right)^{1/3} = 35.1$$

from which it follows that

$$\bar{h} = \left(\frac{\overline{Nu}_S k_f}{S}\right) = \frac{35.1(0.061 \text{ W/m·K})}{0.02 \text{ m}} = 107 \text{ W/m}^2 \cdot \text{K}$$

and

$$q = \bar{h}A_s(T_h - T_c) = 107 \text{ W/m}^2 \cdot \text{K}(0.10 \text{ m})^2 (70 \text{ K}) = 74.9 \text{ W}$$

For the horizontal cavity, Eq. 2.76 yields

$$\overline{Nu}_S = 0.069 Ra_S^{1/3} Pr^{0.074} = 0.069(4.45 \times 10^8)^{1/3}(16.5)^{0.074} = 64.4$$

Hence,

$$\bar{h} = \frac{\overline{Nu}_S k_f}{S} = \frac{64.4(0.061 \text{ W/m·K})}{0.02 \text{ m}} = 198 \text{ W/m}^2 \cdot \text{K}$$

$$q = \bar{h}A_s(T_h - T_c) = 198 \text{ W/m}^2 \cdot \text{K}(0.10 \text{ m})^2 (70 \text{ K}) = 138.4 \text{ W}$$

For the flat plate in parallel flow,

$$Re_L = \frac{u_\infty L}{\nu} = \frac{3 \text{ m/s}(0.10 \text{ m})}{5.39 \times 10^{-7} \text{ m}^2/\text{s}} = 557{,}000$$

From Eq. 2.92, it follows that

$$\overline{Nu}_L = (0.037 Re_L^{4/5} - 871) Pr^{1/3}$$
$$= \left[0.037(5.57 \times 10^5)^{4/5} - 871\right](16.5)^{1/3} = 1504$$
$$\bar{h} = \frac{\overline{Nu}_L k_f}{L} = \frac{1504(0.061 \text{ W/m} \cdot \text{K})}{0.10 \text{ m}} = 917 \text{ W/m}^2 \cdot \text{K}$$
$$q = \bar{h} A_s (T_s - T_\infty) = 917 \text{ W/m}^2 \cdot \text{K}(0.10 \text{ m})^2 (70 \text{ K}) = 642 \text{ W}$$

Comments:

1. The differences in convection coefficients for the three cases ($\bar{h}_{fp} > \bar{h}_{hc} > \bar{h}_{vc}$) are representative of the relative efficacy of forced convection and natural convection in horizontal and vertical cavities, respectively. Note that, if the horizontal cavity were rotated 180° (hot surface over cold surface), the fluid would be stably stratified and heat transfer would be by conduction. Replacing L by S in Eq. 2.7, the heat rate for this case would be $q = k_f A_s (T_h - T_c)/S = 0.061 \text{ W/m} \cdot \text{K}(0.1 \text{ m})^2 (70 \text{ K})/0.02 \text{ m} = 2.1 \text{ W}$.
2. Note that properties such as α_f, μ, and ν depend strongly on the temperature at which they are evaluated. For values of $25°C < \bar{T} = T_f < 75°C$, the corresponding ranges of the pertinent dimensionless parameters are $2.76 \times 10^8 < Ra_S < 6.35 \times 10^8$, $3.71 \times 10^5 < Re_L < 7.44 \times 10^5$, and $23.9 > Pr > 12.7$.

2.7 FORCED CONVECTION: IMPINGING JETS

2.7.1 Introduction

An impinging liquid jet may be classified according to whether it is a *free surface* or a *submerged* jet and whether its cross section is *circular* or *rectangular (planar)*. A free-surface jet is discharged into an ambient gas, whereas a submerged jet is discharged into a liquid of the same type. In both cases, conditions may be influenced by whether the jets are *unconfined* or *semiconfined* and, in the case of multiple jets, by *cross-flow* effects associated with interactions between neighboring jets. Because of the thin hydrodynamic and thermal boundary layers that form at the impingement surface, convection heat transfer coefficients are large and jet impingement is well suited for large heat flux applications. In this section, we consider the hydrodynamic features associated with pre- and postimpingement jet behavior. In Chapter 5, related heat transfer effects are considered.

2.7.2 Unconfined, Free-Surface Jets

Preimpingement Hydrodynamics Normal impingement of an unconfined, circular, free-surface jet on a discrete heat source is shown schematically in Figure

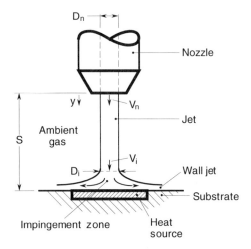

FIGURE 2.14 Schematic of an unconfined, circular, free-surface jet.

2.14. The jet emerges from a nozzle of diameter D_n with a mean velocity V_n, which is determined from knowledge of the volumetric flow rate, $V_n = \dot{\forall}/A_n$, where the cross-sectional area of the nozzle is $A_n = \pi D_n^2/4$. If the jet is oriented vertically downward, gravitational acceleration causes its velocity and diameter to increase and decrease, respectively. Neglecting surface tension and viscous effects, conservation of mass and Bernoulli's equation may be used to obtain the following expressions for the impingement velocity, V_i, and diameter, D_i:

$$\frac{V_i}{V_n} = \left(1 + \frac{2gS}{V_n^2}\right)^{1/2} = \left(1 + \frac{2}{Fr_S^2}\right)^{1/2} \tag{2.108}$$

$$\frac{D_i}{D_n} = \left(\frac{V_n}{V_i}\right)^{1/2} = \left(1 + \frac{2}{Fr_S^2}\right)^{-1/4} \tag{2.109}$$

where the Froude number, $Fr_S \equiv V_n/(gS)^{1/2}$, provides a measure of the ratio of inertia to gravity forces. For a planar jet (also termed a rectangular or slot jet) emerging from a nozzle of width w_n and cross-sectional area A_n, the impingement velocity is still given by Eq. 2.108, whereas the impingement width is

$$\frac{w_i}{w_n} = \frac{V_n}{V_i} = \left(1 + \frac{2}{Fr_S^2}\right)^{-1/2} \tag{2.110}$$

Changes in jet velocity and diameter (or width) are only significant for small discharge velocities V_n and/or large nozzle-to-plate spacings, S. For electronic cooling applications, however, the values of V_n and S are likely to be large and small, respectively, rendering $V_i \approx V_n$ and $D_i(w_i) \approx D_n(w_n)$.

The nozzle-to-plate spacing S can influence jet behavior in other ways. Although the shear stress of the liquid/gas interface is small, jet turbulence and interfacial shear can combine to induce interfacial waves or roughness, which, in turn, can lead to *atomization* of the jet (Grassi and Magrini, 1991) and/or *splattering* of liquid droplets from the impingement surface (Bhunia and Lienhard, 1993). Both processes become more influential with increasing S and have a deleterious effect on surface heat transfer.

If the jet emerges from the nozzle with a nonuniform velocity profile, the elimination of wall friction imposed by the nozzle and the effects of turbulence and viscosity within the jet combine to render the profile more uniform with increasing distance from the nozzle exit. The rate at which the velocity profile relaxes depends on whether the jet is laminar or turbulent. Although laminar circular jets may require up to 20 nozzle diameters to achieve a nearly uniform profile, turbulent jets reach such a condition within approximately 5 diameters (Lienhard, 1995). For a fully developed turbulent jet emerging from a tubular nozzle, Stevens and Webb (1992) found that the free-surface velocity increased to 90% of the average velocity, V_n, within 4 nozzle diameters.

For fully turbulent rectangular jets discharged from a parallel-plate channel, Wolf et al. (1995a, b) also observed relaxation of the velocity profile, with a highly uniform profile achieved within $y/D_h = 10$, where the hydraulic diameter is $D_h = 2w_n$. For a converging channel, Wolf et al. (1995b) report a nearly uniform profile at the nozzle exit.

In any transition from a nonuniform to a uniform velocity profile, there is a corresponding reduction in the jet centerline (midplane) velocity. As we shall find, such changes affect free-stream velocity gradients at the impingement surface, which, in turn, influence local heat transfer.

Heat transfer is also influenced by turbulence within the jet, and the general trend is one of decreasing turbulence intensity with increasing y/D_h. From measurements performed for rectangular turbulent jets produced by both parallel-plate and converging channels, with and without turbulence-enhancing wire meshes or grids inserted at the nozzle exit, the turbulence intensity was found to decay from values as large as 12% at the nozzle exit to levels of less than 2% at $y/D_h = 15$ (Wolf et al., 1995a). An exception to this behavior was observed for large Reynolds numbers ($Re_{D_h} \approx 100,000$), where the intensity increased slightly for $0 < y/D_h < 5$, before it began to decay. The turbulence intensity was also found to become more homogeneous (more uniformly distributed over the jet cross section) with increasing distance from the nozzle exit. Similar trends were recorded by Mansour and Chigier (1994) for a cylindrical jet and can be explained in terms of the relative effects of turbulence production, diffusion, and advection (Wolf et al., 1995a).

As implied in the preceding discussion, the nozzle from which the jet emerges can influence important features such as the velocity profile and turbulence intensity. Nozzle types and effects are reviewed by Lienhard (1995). Many designs are possible, although three basic configurations are widely used (Fig. 2.15).

The *sharp-edged orifice* provides a sharp contraction for flow from a comparatively large upstream plenum and can be used to produce laminar jets of nearly

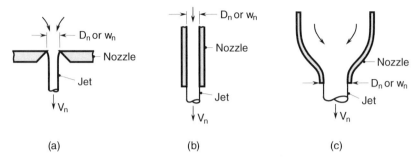

FIGURE 2.15 Basic nozzle configurations: (*a*) sharp-edged orifice, (*b*) circular tube or parallel-plate channel, and (*c*) converging nozzle.

uniform velocity profile. An additional contraction is experienced as the jet exits the nozzle, rendering the jet diameter (or width) less than D_n (or w_n). Nonuniform velocity profiles result from boundary layer development along the walls of a nozzle and, for a sufficiently long *circular tube* or *parallel-plate channel*, fluid will exit the nozzle with a fully developed laminar or turbulent velocity profile. A length-to-diameter ratio of $L/D_h \approx 100$ is sufficient to provide a fully developed (parabolic) profile for laminar flow ($Re_{D_h} < 2300$), whereas $L/D_h \approx 20$ is sufficient to achieve fully developed turbulent flow. *Converging nozzles* suppress boundary layer growth along the nozzle wall, as well as turbulence generation within the nozzle. These effects increase with increasing contraction ratio and streamlining of the interior surface. Conversely, turbulence may be enhanced by roughening the interior surface of the nozzle and/or by attaching a screen or grid to the nozzle exit. Nozzle discharge conditions vary with Reynolds number, and generalized relations between nozzle design and turbulence are lacking.

Stagnation Zone As the jet approaches an impingement surface, it is concurrently decelerated and accelerated in directions normal and parallel to the surface, respectively. These changes occur in a stagnation zone (Fig. 2.16), which is also characterized by a strong favorable pressure gradient parallel to the surface. If the jet is turbulent, the pressure gradient acts to laminarize flow in the stagnation zone. However, the pressure gradient decays to zero, and, in the downstream wall jet region, transition to turbulence may occur.

The extent of the stagnation zone is determined by the diameter (or width) and velocity profile of the impinging jet. For jets of uniform velocity profile, deceleration in the axial direction begins at $y/D_i(w_i) \approx 0.5$, and acceleration in the radial direction extends as far as $r/D_i \approx 0.7$ and $x/w_i \approx 1.1$ for circular (Fig. 2.16*a*) and rectangular (Fig. 2.16*b*) jets, respectively. Within the stagnation zone, there is a thin boundary layer of nearly uniform thickness δ and an external, inviscid flow characterized by a free stream velocity u_∞ at the edge of the boundary layer. Heat transfer in the stagnation zone is strongly influenced by the manner in which u_∞ varies with r or x.

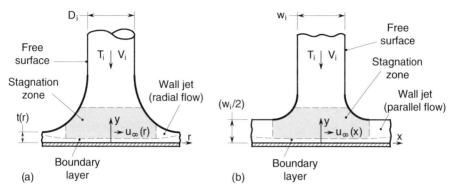

FIGURE 2.16 Stagnation zone of impinging free-surface (*a*) circular and (*b*) rectangular jets.

Hydrodynamic conditions at the stagnation point ($r = 0$) of a circular jet or the stagnation line ($x = 0$) of a rectangular jet are commonly characterized by the constants $C_r \equiv du_\infty/dr|_{r=0}$ or $C_x \equiv du_\infty/dx|_{x=0}$, respectively, which provide the magnitude of flow acceleration in the r or x direction. Heat transfer in the stagnation zone is known to increase with increasing C_r or C_x. In dimensionless form, the constants may be expressed as

$$G_r = \frac{D_i}{V_i} C_r = \frac{d(u_\infty/V_i)}{d(r/D_i)}\bigg|_{(r/D_i)=0} \qquad (2.111\text{a})$$

$$G_x = \frac{w_i}{V_i} C_x = \frac{d(u_\infty/V_i)}{d(x/w_i)}\bigg|_{(x/w_i)=0} \qquad (2.111\text{b})$$

For selected conditions, the velocity gradients may be determined theoretically. For a circular jet of uniform velocity profile and negligible surface tension effects, Liu et al. (1993) predicted a value of $G_r = 0.916$, as well as the following linear variation of the streamwise velocity at the edge of the boundary layer:

$$\frac{u_\infty(r)}{V_i} = \left(\frac{r}{D_i}\right) G_r \qquad (2.112)$$

The expression applies to an excellent approximation for $r/D_i < 0.5$ and may be extended to $r/D_i \approx 0.7$ as a reasonable approximation. In contrast, for a parabolic velocity profile and $S/D_i < 0.5$, Scholtz and Trass (1970) predicted a value of $G_r = 4.646$, if V_i is defined as the average jet velocity, or $G_r = 2.323$, if V_i is defined as the maximum (centerline) velocity. They also predicted a linear variation of u_∞ with r for $r/D_i < 0.4$.

Keeping in mind that heat transfer in the stagnation zone increases with increasing G_r, the parabolic profile is clearly superior to the uniform profile, providing a free-stream velocity gradient that is approximately five times larger. However, remember that the extent of the stagnation zone is correspondingly smaller.

Using laser–Doppler velocimetry and a transparent impingement plate, Stevens et al. (1992) and Stevens and Webb (1993) obtained measurements of $u_\infty(r)$ in the stagnation zone for jets generated by a sharp-edged orifice, a contoured nozzle, and a tubular nozzle that provided a fully developed turbulent flow. In each case, u_∞ increased linearly with r for $r/D_i < 0.5$. Although G_r was independent of Re_{D_n} and S/D_n for $8000 \leq Re_{D_n} \leq 62{,}000$ and $S/D_n < 4$, it did depend on the type of nozzle, achieving its largest and smallest values of approximately 2.2 and 1.2 for the sharp-edged orifice and contoured nozzle, respectively.

Similar results have been obtained for rectangular jets (Fig. 2.16b). For a laminar rectangular jet of uniform velocity profile, the following linear variation of free-stream velocity with distance from the stagnation line was predicted by Milne-Thomson (1955):

$$\frac{u_\infty(x)}{V_i} = \left(\frac{x}{w_i}\right) G_x \qquad \frac{x}{w_i} < 1 \tag{2.113}$$

where

$$G_x = \frac{\pi}{4} = 0.785 \tag{2.114}$$

However, as for the circular jet, a parabolic velocity profile yields a significantly larger value of G_x, with Sparrow and Lee (1975) having predicted values in the range $3.03 \leq G_x \leq 3.48$ for $1.5 \geq S/w_i \geq 0.25$. Beyond the stagnation zone, du_∞/dx approaches zero, as u_∞ approaches V_i.

Wolf et al. (1990) determined the free stream velocity distribution $u_\infty(x)$ from static pressure measurements made along the impingement surface for turbulent jets issuing from a parallel-plate channel. For both plane and circular jets, the pressure and velocity are related through Bernoulli's equation:

$$p = p_\infty + \frac{\rho}{2}\left\{V_i^2 - u_\infty^2\right\} \tag{2.115}$$

where p_∞ is the ambient pressure and V_i refers to the centerline or midplane velocity for a nonuniform velocity profile. The measured value of $G_x = 0.968$ exceeds the theoretical value of 0.785 for a uniform velocity profile. Additional measurements (Wolf et al., 1995b) indicate that, although G_x is independent of Reynolds number, it does depend on the nozzle-to-plate spacing, decreasing from $G_x \approx 1.13$ for $S/w_n = 2$ to $G_x \approx 0.81$ for $S/w_n = 3.0$. This trend is attributed to development of a more uniform velocity profile with increasing S/w_n and is consistent with other results. That is, the streamwise velocity gradient decreases with decreasing nonuniformity in the velocity profile of the impinging jet. However, although Sparrow and Lee (1975) report a value of G_x for a parabolic (fully developed laminar) profile that exceeds that of a uniform profile by approximately a factor of 4, one should not expect such a large difference for a fully developed turbulent profile, which is much flatter.

With respect to electronic cooling, a key feature of the foregoing discussion is that heat transfer in the stagnation zone depends strongly on the streamwise velocity gradient, which, in turn, depends on the velocity profile of the impinging jet.

Theoretical expressions have been obtained for the hydrodynamic boundary layer thickness, which is uniform in the stagnation zone, and the following results have been reported (Lienhard, 1995) for circular jets:

$$\frac{\delta}{D_i} = \frac{1.95}{(G_r Re_{D_i})^{1/2}} \quad (2.116a)$$

and planar jets:

$$\frac{\delta}{w_i} = \frac{2.40}{(G_x Re_{w_i})^{1/2}} \quad (2.116b)$$

Choosing representative values of $G_x = 0.785$, $w_i = 5$ mm, and $Re_{w_i} = 10^4$ for a planar jet of uniform velocity profile, Eq. 2.116b yields $\delta = 0.135$ mm $= 135\mu$m, confirming that the boundary layer is, indeed, thin.

Relative to boundary layer effects in the stagnation zone, it is pertinent to note the special influence of jet turbulence. In particular, heat transfer may be enhanced by the existence of a source of turbulence in the free-stream fluid outside the boundary layer, *under conditions for which there is a favorable (negative) streamwise pressure gradient* (Kestin, 1966). Such conditions characterize the stagnation zone for an impinging jet, and Sutera et al. (1963) proposed that the mechanism for heat transfer enhancement is *amplification of vorticity* associated with free-stream turbulence.

Wall Jet Region For a laminar, circular jet, regions associated with flow along the impingement surface are shown in Figure 2.17. Within the stagnation zone, hydrodynamic and thermal boundary layers are of uniform thicknesses. Beyond the stagnation region ($r > r_s$), the boundary layers begin to grow, as radial velocity gradients in the inviscid region decrease and the free-surface velocity V_s approaches the impingement velocity V_i. In this region, there are competing effects on the film thickness $t(r)$, with radial spreading acting to thin the film, while friction acts to thicken it because of fluid deceleration in the developing boundary layers. For Prandtl numbers characteristic of liquids ($Pr > 1$), the hydrodynamic boundary layer reaches the free surface, $\delta(r) = t(r)$, before the thermal boundary layer, $\delta_{th}(r) = t(r)$, at which point ($r = r_v < r_{th}$), viscous effects extend throughout the liquid and the surface velocity V_s begins to decrease with increasing radius. In this region, velocity profiles at different radial locations are *similar* to each other. The similarity region ends at $r = r_c$, where transition to turbulence begins, and a fully turbulent flow is achieved at $r = r_t$.

With the extent of the *stagnation region* approximated as

$$0.4 < \frac{r_s}{D_i} < 0.7 \quad (2.117a)$$

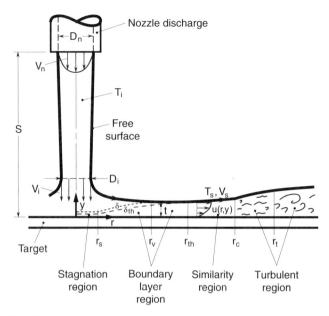

FIGURE 2.17 Flow regimes associated with impingement of a laminar, circular, free-surface jet.

the extents of the *boundary layer*, *similarity*, and *turbulent* regions correspond to $r_s \leq r < r_v$, $r_v < r < r_c$, and $r > r_c$, respectively, where (Lienhard, 1995)

$$\frac{r_v}{D_i} \approx 0.177 Re_{D_i}^{1/3} \qquad (2.117b)$$

$$\frac{r_c}{D_i} \approx 1200 Re_{D_i}^{-0.422} \qquad (2.117c)$$

$$\frac{r_t}{D_i} \approx 28{,}600 Re_{D_i}^{-0.68} \qquad (2.117d)$$

The extent of the stagnation zone, r_s/D_i, depends on the velocity profile of the impinging jet, ranging from approximately 0.4 for a parabolic profile to approximately 0.7 for a uniform profile. At $r/D_i \approx 2.23$, V_s reaches V_i, while boundary layer development continues to r_v/D_i.

Transition to turbulence has been observed for Reynolds numbers as low as $Re_{D_i} \approx 13{,}000$ (Liu et al., 1991). For $Re_{D_i} \approx 13{,}000$, the different zones are bounded by $r_v/D_i \approx 4.17$, $r_c/D_i \approx 22.0$, and $r_t/D_i \approx 45.6$. If a surface is large enough to experience transition, the stagnation zone clearly comprises a very small portion of the total surface.

In that portion of the boundary layer region for which $2.23 < (r/D_i) < (r_v/D_i)$, the boundary layer thickness grows approximately as (Lienhard, 1995)

$$\frac{\delta(r)}{D_i} = 2.68 \left(\frac{r/D_i}{Re_{D_i}}\right)^{1/2} \tag{2.118}$$

For $r > r_v$, the film thickness varies as

$$\frac{t(r)}{D_i} = \frac{0.1713}{(r/D_i)} + 5.147\frac{(r/D_i)^2}{Re_{D_i}} \tag{2.119}$$

whereas the free-surface velocity decreases as

$$V_s(r) = \frac{V_i D_i^2}{5rt(r)} \tag{2.120}$$

In the fully turbulent region, the film thickness may be approximated as

$$\frac{t(r)}{D_i} = \frac{0.0209}{Re_{D_i}^{1/4}}\left(\frac{r}{D_i}\right)^{5/4} + C\left(\frac{D_i}{r}\right) \tag{2.121a}$$

where

$$C = 0.171 + 5.147\frac{(r_c/D_i)}{Re_{D_i}} - 0.0209\frac{(r_c/D_i)^{1/4}}{Re_{D_i}^{1/4}} \tag{2.121b}$$

Conditions associated with a planar jet are less complicated. Following bifurcation at the stagnation line, the jet yields oppositely directed flows for which the film thickness is fixed at $t = w_i/2$ and the free-surface velocity is $V_s = V_i$. Boundary layer growth occurs outside the stagnation zone, with δ reaching $w_i/2$ before or after transition to turbulence, depending on the Reynolds number, velocity profile, and turbulence intensity of the impinging jet. Beyond the stagnation region, $(x/w_i) \geq 1$, the velocity at the edge of the boundary layer may be approximated as (Vader et al., 1991)

$$\frac{u_\infty(x)}{V_i} = \tanh\left(\frac{x}{w_i}\right) \qquad \frac{x}{w_i} > 1 \tag{2.122}$$

where $u_\infty(x) \sim V_i$ for $x/w_i > 3$.

Splattering A phenomenon that can influence the hydrodynamics of an impinging, free-surface, circular jet in the radial flow region is *splattering*. The term refers to droplet ejection from the wall jet, and the phenomenon depends on the nozzle-to-surface separation, S/D_n, and the Weber number, $We_{D_n} \equiv \rho V_n^2 D_n/\sigma$, which is a measure of the ratio of inertia to surface tension forces in the impinging jet. It is linked to disturbances in the free surface of a turbulent jet, which may be amplified as the jet forms a liquid film, causing the film to become unstable and droplets to be discharged. Once discharged, the droplets are lost from the film, adversely affecting heat transfer from the surface.

Defining the onset of splattering as a condition for which 5% of the liquid is lost, Lienhard (1995) correlated onset in terms of a critical nozzle-to-surface spacing S_o and the Weber number. For a particular value of We_{D_n}, splattering occurs if S/D_n exceeds the value determined from the expression

$$\frac{S_o}{D_n} = \frac{130}{1 + 5 \times 10^{-7} We_{D_n}^2} \qquad (2.123)$$

Lienhard (1995) also reported the following correlation for the fraction of the impinging jet flow rate that is splattered:

$$\xi = -0.258 + 7.85 \times 10^{-5} \omega - 2.51 \times 10^{-9} \omega^2 \qquad (2.124a)$$

where

$$\omega = We_{D_n} \exp\left[\frac{0.971}{We_{D_n}^{1/2}}\left(\frac{S}{D_n}\right)\right] \qquad (2.124b)$$

is a scaling parameter for the amplitude of jet free-surface disturbances that reach the impingement surface. Splattering occurs within a narrow radial band about the radius $r/D_n \approx 4.5$ and can result in the loss of as much as 70% of the liquid film.

Hydraulic Jump Another phenomenon that can degrade heat transfer is termed the *hydraulic jump*. The phenomenon may occur in the wall jet region for both circular and planar free-surface jets and is characterized by an abrupt increase in the film thickness and a corresponding reduction in the average film velocity. Hence, it is accompanied by a significant reduction in the heat transfer coefficient.

Unfortunately, there remains a good deal of uncertainty in predicting the location of the hydraulic jump, which depends on numerous parameters, including appropriately defined Reynolds, Froude, and Weber numbers (Liu and Lienhard, 1993). In electronic cooling, the phenomenon is to be avoided, prospects for which are enhanced by increasing the jet velocity and by removing any downstream obstruction to flow from the target surface. That is, the film should be allowed to flow freely from the surface.

2.7.3 Unconfined, Submerged Jets

As shown in Figure 2.18, a submerged, axisymmetric, impinging jet can be subdivided into three regions: a *free jet*, an *impingement* or *stagnation zone*, and a *wall jet*. Flow in the free-jet region corresponds to that of a jet issuing into an infinite medium of the same fluid in the absence of an impingement surface. In the stagnation region,

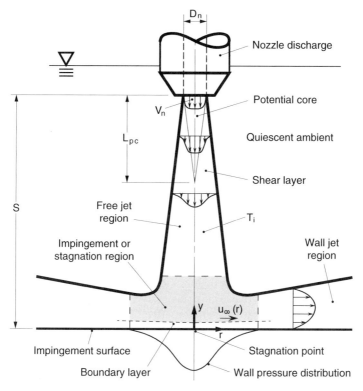

FIGURE 2.18 Flow regimes associated with impingement of a circular, submerged jet.

development of the free jet is altered by the target, which turns the flow from a direction perpendicular to the surface to a direction parallel to the surface. Following impingement, flow spreads radially over the surface in the wall jet region.

The free jet is characterized by lateral spreading resulting from the entrainment of ambient fluid in a shear layer that develops downstream of the nozzle exit, but is unaffected by the target surface. With increasing distance from the nozzle, the shear layer widens, while the *potential core*, in which the fluid velocity is unaffected by the shear layer, shrinks and eventually disappears. Beyond the tip of the potential core, the centerline velocity of the jet decreases, while the velocity profile becomes bell shaped. The length of the potential core, L_{pc}, varies with Re_{D_n}, decreasing from approximately 20 to 5 nozzle diameters as Re_{D_n} increases from 1000 to 4000 and increasing to approximately 6 nozzle diameters for $Re_{D_n} > 4000$ (Hrycak et al., 1970). Relative to free-surface jets, submerged jet impingement heat transfer is more sensitive to the nozzle-to-surface separation distance S, especially if the target surface is beyond the potential core.

The impingement or stagnation zone is often characterized by laminar flow, even if the jet itself is turbulent (Hrycak et al., 1970). It encompasses an axial distance of

1.6 to 2.2 nozzle diameters from the impingement surface and a radial distance of 1.6 to 3.0 nozzle diameters from the stagnation point (Gauntner et al., 1969). Near the surface, the flow is sharply decelerated in the axial direction and begins to accelerate radially. The rapid reduction in axial velocity causes the pressure to increase above that of the ambient fluid, while acceleration parallel to the wall yields a reduction in the pressure difference between the jet and the ambient fluid. Hence, there is a large favorable pressure gradient in the impingement zone that, for turbulent jets, laminarizes the flow.

As fluid leaves the impingement region, there is a marked increase in turbulence, which augments mixing and transition from an accelerating impingement region flow to a decelerating wall jet. Deceleration occurs because of momentum exchange between the jet and the ambient fluid, resulting in the entrainment of zero momentum fluid from the ambient, as well as to radial spreading of the flow. As described by Glauert (1956), the wall jet can be divided into two regions: an inner layer where flow and turbulence are affected by the wall, and an outer region where flow conditions are determined by shear interactions with the quiescent fluid. The maximum velocity in the wall jet occurs at the boundary between these two subregions. Similar conditions characterize planar (rectangular) jets, although deceleration of the lateral flow in the wall jet region is due exclusively to the entrainment of ambient fluid.

Although entrainment and deceleration effects associated with the free-jet and wall jet regions of a submerged jet differ considerably from processes associated with the preimpingement and wall jet regions of a free-surface jet, the stagnation zones are virtually identical if impingement occurs within the potential core of the submerged jet ($S < L_{pc}$). Within the stagnation zone, the boundary layer thickness is constant, and the velocity u_∞ at the edge of the boundary layer increases linearly with r and x for circular and planar jets, respectively (Martin, 1977). That is,

$$\frac{u_\infty(r)}{V_n} = \left(\frac{r}{D_n}\right) \cdot G_r \quad (2.125a)$$

$$\frac{u_\infty(x)}{V_n} = \left(\frac{x}{w_n}\right) \cdot G_x \quad (2.125b)$$

where V_n is the mean velocity at the nozzle exit and

$$G_r = 1.04 - 0.034 \frac{S}{D_n} \quad (2.126a)$$

$$G_x = 1.02 - 0.024 \frac{S}{w_n} \quad (2.126b)$$

The corresponding boundary layer thicknesses are

$$\frac{\delta}{D_n} = \frac{1.95}{(G_r Re_{D_n})^{1/2}} \quad (2.127a)$$

$$\frac{\delta}{w_n} = \frac{2.38}{(G_x Re_{w_n})^{1/2}} \quad (2.127b)$$

Except for the prescribed velocity and jet diameter or width, these results are equivalent to those of Eqs. 2.112, 2.113, 2.116a, and 2.116b for the free-surface jet.

Because of the entrainment of ambient fluid, the accelerating impingement region flow must transition to a decelerating flow in the wall jet region and eventually decay to zero. Although the favorable pressure gradient and acceleration maintain laminar flow in the impingement region, transition to turbulence accompanies the onset of deceleration in the wall jet region.

2.7.4 Effects of Confinement and Multiple Jets

Because modern electronic packages are designed to fit in compact enclosures, a packaging unit will likely have a small volume that confines the flow. Consider a semiconfined rectangular jet for which a confining wall is parallel to the impingement surface in the jet exit plane (Fig. 2.19). Although the three regions associated with an unconfined, submerged jet may still exist (free jet, stagnation zone, and wall jet), the wall restricts entrainment and ultimately confines the liquid to parallel flow between two surfaces. If the length of the confining surface is large relative to the nozzle width and the spacing between surfaces, fluid entrainment will cause recirculation cells to form on both sides of the impinging jet (Fig. 2.19a). As the nozzle-to-plate separation distance increases, fluid is entrained from outside the confining wall

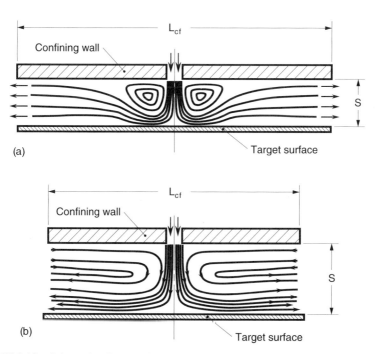

FIGURE 2.19 Schematic of streamlines for semiconfined rectangular jets: (a) ($L_{cf}/S \gg 1$) and (b) ($L_{cf}/S)0(1)$.

(Fig. 2.19b), and the wall has less influence on fluid flow and heat transfer under the jet. If the nozzle-to-plate spacing is increased further, the confining wall will have no effect and the system behaves as an unconfined jet. As for the unconfined jet, there is a bell-shaped pressure distribution along the target surface. However, for $L_{cf} \gg S$ (Fig. 2.19a), constriction of the flow by the recirculation zones reduces the pressure below the ambient value on both sides of the stagnation line, causing strong acceleration along the surface. The pressure recovers in downstream regions of the flow, creating an adverse gradient that can promote transition to turbulence or boundary layer separation.

An alternative, semiconfined configuration is one for which an impinging circular or rectangular jet is turned 180° and discharged through an annular passage (Fig. 2.20). The scheme is designed to maintain high stagnation zone heat transfer coefficients over much of the target and is similar to that deployed in the Fujitsu indirect liquid cooling module (Fig. 1.2).

With electronic packages often consisting of multiple discrete heat sources, thermal management schemes may involve the use of arrays of impinging jets. Even for a single heat source, multiple jets may be favored by virtue of creating multiple stagnation zones on the target surface, thereby increasing the overall average heat transfer coefficient. However, conditions may be adversely affected by interactions between adjoining jets and/or by ineffective drainage of liquid from the surface.

Consider an array of semiconfined planar jets, as shown in Figure 2.21. If the spent fluid is forced to leave the system in the lateral (x) direction (Fig. 2.21a), the hydrodynamic conditions of one jet will clearly be influenced by the cross flow imposed by neighboring jets. For small nozzle-to-target separations, the effect can be significant, with the cross flow confined to lower portions of the channel, preventing jet penetration to the target and thereby degrading convection heat transfer. Conditions are entirely different if the outflow is routed through slots in the confining wall (Fig. 2.21b). In this case, secondary stagnation zones, which are characterized by

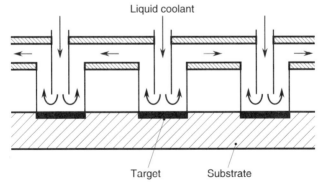

FIGURE 2.20 Semiconfined circular or rectangular jets with counterflow collection of the spent fluid.

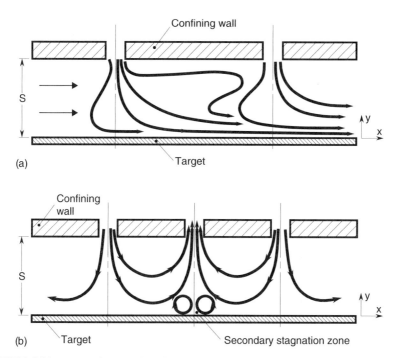

FIGURE 2.21 Interacting, semiconfined planar jets with (*a*) single-outflow location in cross-flow direction and (*b*) multiple outflows at intermediate locations on the confining wall.

boundary layer separation and fluid recirculation, are created by the interaction of wall jets associated with adjacent nozzles.

Another configuration is one for which spent liquid cannot flow upward between the nozzle slots (in the y direction) or be discharged by cross flow in the x direction, but must instead be routed in a direction perpendicular to the plane of the schematic in Figure 2.21 (the z direction). Again, heat transfer at the target surface is affected by outflow conditions.

Plan (top) views of regular arrays of round and slot nozzles are shown in Figure 2.22. Symmetry dictates equivalent conditions for the unit cells delineated by dashed lines. For clusters of square in-line (Fig. 2.22*a*) and equilaterally staggered (Fig. 2.22*b*) circular jets, the unit cells are squares and hexagons, respectively. In each case, P_n represents the nozzle *pitch* and A_r is the ratio of the nozzle cross-sectional area to the surface area of a cell.

2.8 FORCED CONVECTION: INTERNAL FLOW

An internal flow is one for which fluid motion is completely confined by a surface, in which case boundary layers are unable to develop without eventually being constrained. The most common example is flow through a circular tube, although flows

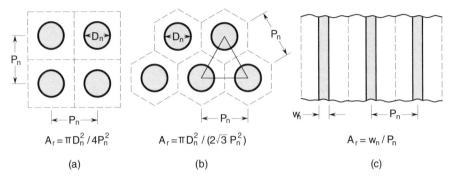

FIGURE 2.22 Plan view of nozzle geometries for (*a*) square, in-line array of circular jets, (*b*) staggered, equilateral array of round jets, and (*c*) array of slot jets.

through rectangular channels may be more pertinent to electronic cooling applications. Cooling channels may, for example, be machined in a heat sink or cold plate to which electronic devices are attached. In such cases, as there is no contact between the devices and the coolant, any liquid, including water, may be used. Alternatively, electronic components may be mounted to one or more interior walls of the channel, in which case there is direct contact and the liquid must be a dielectric.

2.8.1 General Considerations

Internal flows are distinguished by whether they are fully developed or developing, as well as by whether they are laminar or turbulent. These conditions are linked to the Reynolds number, which is defined as

$$Re_{D_h} = \frac{\rho w_m D_h}{\mu} = \frac{w_m D_h}{\nu} \tag{2.128a}$$

or

$$Re_{D_h} = \frac{\dot{m} D_h}{\mu A_c} \tag{2.128b}$$

where $D_h = 4A_c/P$ is the channel hydraulic diameter, w_m is the mean longitudinal velocity of the coolant passing through the channel, and $\dot{m} = \rho w_m A_c$ is the coolant flow rate.

Below a critical Reynolds number $Re_{D_{h,c}}$, which marks the onset of turbulence, the flow is laminar. The value of $Re_{D_{h,c}}$ depends slightly on the duct cross section, but to a good approximation, the accepted value for flow in a circular tube may be used irrespective of duct geometry. This value is

$$Re_{D_{h,c}} \approx 2300 \tag{2.129}$$

FORCED CONVECTION: INTERNAL FLOW

The Reynolds number required to achieve fully turbulent conditions exceeds $Re_{D_h,c}$ and depends on conditions at the channel inlet. If there is a sudden contraction in the cross-sectional area, transition to turbulence is enhanced and fully turbulent conditions are achieved for values of Re_{D_h} only slightly larger than $Re_{D_h,c}$. In contrast, if the inlet corresponds to a smoothly contoured nozzle, Reynolds numbers as large as $Re_{D_h} \approx 10{,}000$ may be needed to achieve a fully turbulent flow. However, irrespective of the inlet condition, fully turbulent flow is often assumed for any $Re_{D_h} > Re_{D_h,c}$.

As illustrated in Figure 2.23 for laminar flow, in a circular tube *hydrodynamic* and *thermal conditions* may be differentiated in terms of *entrance* and *fully developed* regions. If fluid enters the tube with a uniform velocity (Fig. 2.23a), a hydrodynamic boundary layer begins to develop when contact is made with the wall and grows at the expense of a shrinking inviscid region. Eventually, boundary layer merger occurs at the centerline and a parabolic velocity profile is achieved in the fully developed region. The extent of the hydrodynamic entrance region is commonly associated with

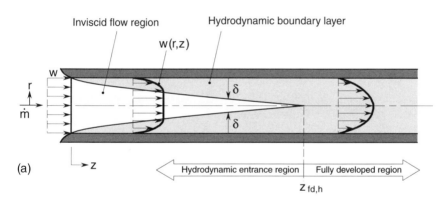

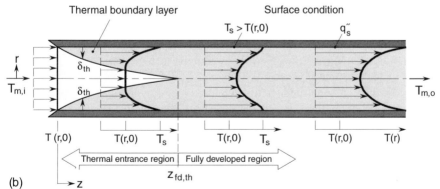

FIGURE 2.23 Entrance and fully developed regions for laminar flow in a circular tube: (*a*) hydrodynamic conditions and (*b*) thermal conditions.

the distance $z_{\text{fd,h}}$ at which the centerline velocity achieves 99% of its fully developed value (Shah and London, 1978).

Defining a dimensionless hydrodynamic entrance length as $z_{\text{fd,h}}^+ = z_{\text{fd,h}}/(D_h Re_{D_h})$, approximate values of 0.055 and 0.011 are associated with laminar flow through a circular tube and *parallel plates*, respectively (Shah and London, 1978). The parallel-plate geometry is a limiting case of the rectangular duct for which the hydraulic diameter is twice the spacing between plates. In principle, the geometry corresponds to an aspect ratio, $A = W/H$, of infinity, where W and H are the channel width and height, respectively. In practice, it is a good approximation for ducts of aspect ratio $0.1 > A > 10$. For aspect ratios in the range $0.2 < A < 5$, a value of $z_{\text{fd,h}}^+ \approx 0.08$ may be assumed.

As shown in Figure 2.23b, entrance and fully developed regions are also associated with thermal conditions. If fluid enters the circular tube at a uniform temperature $T(r, 0)$ that is less than the surface temperature, a thermal boundary layer begins to develop, and a fully developed thermal condition is reached if either a uniform temperature T_s or a heat flux q_s'' is imposed at the surface. The general shape of the fully developed temperature profile differs according to the imposed surface condition.

For internal flows, Newton's *law of cooling* is expressed as

$$q_s'' = h(T_s - T_m) \tag{2.130}$$

where T_m is the mean or mixed-mean temperature of the fluid over the duct cross section. With heat transfer to the fluid, this temperature increases in the z direction. The *local convection coefficient*, which corresponds to a particular location on the surface, is, in principle, infinite at the tube inlet, where the thickness of the thermal boundary layer is zero, and decreases with boundary layer development in the z direction, approaching a constant value, h_{fd}, as fully developed thermal conditions are achieved (Fig. 2.24a). The extent of the thermal entrance region is commonly associated with the distance $z_{\text{fd,th}}$ at which h is equal to $1.05 h_{\text{fd}}$ (Shah and London, 1978), and a dimensionless form is defined as $z_{\text{fd,th}}^* = z_{\text{fd,th}}/(D_h\, Re_{D_h}\, Pr)$.

The extent of the thermal entrance region depends on whether there is *simultaneous* hydrodynamic and thermal boundary layer development from the tube inlet or hydrodynamic development precedes thermal development. The second case corresponds to a situation for which the tube has an *unheated starting length*, allowing fully developed hydrodynamic conditions to be reached before heating is initiated and a thermal boundary layer begins to develop. Such a condition is also achieved without an unheated starting length in the limit of very large Prandtl numbers ($Pr \to \infty$). In this case, which is termed a *thermally developing flow*, fully developed hydrodynamic conditions are reached well before fully developed thermal conditions, and thermal boundary layer growth essentially occurs in the presence of a fully developed velocity profile.

The thermal entrance length is slightly larger for a simultaneously developing flow than for a thermally developing flow and for a uniform surface heat flux than for a uniform surface temperature. For the circular tube, values of $z_{\text{fd,th}}^*$ are 0.033 and 0.043 for thermally developing flows with uniform surface temperature and heat

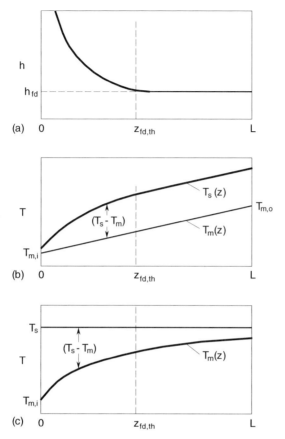

FIGURE 2.24 Characteristic axial distribution along a tube of length L: (a) local convection coefficient, (b) surface and fluid mean temperatures for a uniform surface heat flux, and (c) surface and fluid mean temperatures for an isothermal surface.

flux, respectively. For simultaneously developing flows, $z^*_{fd,th}$ depends on the Prandtl number, decreasing with increasing Pr. However, for liquids and conditions of interest in electronic cooling ($Pr > 3$), the thermal entrance length for simultaneously developing flow is approximately equal to that of a thermally developing flow.

For a thermally developing flow between parallel plates, with equivalent thermal conditions on both plates, $z^*_{fd,th}$ is approximately equal to 0.008 and 0.011 for a uniform surface temperature and heat flux, respectively (Shah and London, 1978). For rectangular ducts with aspect ratios in the range $0.25 \leq A \leq 4$, $z^*_{fd,th} \approx 0.05$ for a uniform surface temperature. For a uniform surface heat flux, $z^*_{fd,th} = 0.066$ for $A = 1$ and decreases to 0.042 as the aspect ratio decreases or increases to values of 0.25 and 4, respectively.

Boundary layers develop much more rapidly for turbulent flow than for laminar flow, and, to a reasonable approximation, fully developed hydrodynamic and thermal conditions are reached within

$$\left(\frac{z}{D_h}\right)_{fd,h} \approx \left(\frac{z}{D_h}\right)_{fd,th} \approx 10 \qquad (2.131)$$

2.8.2 Energy Balances

Because flow in a channel is completely bounded, energy balances may be applied to obtain a relationship between the total heat rate and the fluid inlet and outlet temperatures, as well as the variation of the fluid temperature in the longitudinal direction (Incropera and DeWitt, 1996). From an energy balance applied to a control volume that encompasses the entire flow passage ($0 < z < L$), it may be shown that, for an incompressible fluid, the total heat rate is related to fluid inlet and outlet conditions by the following expression:

$$q = \dot{m}\left[c_v\left(T_{m,o} - T_{m,i}\right) + \frac{(p_o - p_i)}{\rho}\right] \qquad (2.132)$$

which accounts for flow work (p/ρ) effects, as well as changes in thermal energy. Generally, such effects may be neglected, unless the pressure drop is extremely large, as could be the case for microchannel cooling schemes (Section 4.5).

If heating is characterized by a uniform heat flux q_s'' around the perimeter and along the length of the channel, the total heat rate is also given by

$$q = q_s''(P \cdot L) \qquad (2.133)$$

Moreover, if flow work effects are negligible and conservation of energy is applied to a differential control volume (Incropera and DeWitt, 1996), the following expression is obtained for the variation of the mean fluid temperature, $T_m(z)$, in the longitudinal direction:

$$T_m(z) = T_{m,i} + \frac{q_s'' P}{\dot{m} c_p} z \qquad (2.134)$$

As shown in Figure 2.24b, the mean temperature increases linearly over the entire channel length. From Eq. 2.130 and the fact that the local convection coefficient decreases with increasing z in the entrance region (Fig. 2.24a), it follows that the temperature difference, $(T_s - T_m)$, is extremely small at $z = 0$ and increases with increasing z until the fully developed region is reached. For $z > z_{fd,th}$, $(T_s - T_m) = q_s''/h_{fd}$ is constant.

If heating is characterized by a uniform temperature T_s around the periphery and along the length of the channel, it may be shown that the temperature difference decreases exponentially with increasing z (Fig. 2.24c) and is given by the following expression (Incropera and DeWitt, 1996):

$$T_s - T_m(z) = (T_s - T_{m,i}) \exp\left(-\frac{Pz\overline{h}}{\dot{m}c_p}\right) \quad (2.135)$$

where $\overline{h}$ is the *average* convection coefficient over the region extending from the inlet to the z location of interest. This expression may be applied at $z = L$ to obtain the outlet fluid temperature. For this case, the appropriate expression for the total heat rate is

$$q = \overline{h} A_s \Delta T_{lm} \quad (2.136)$$

where $A_s = P \cdot L$ and ΔT_{lm} is the log-mean temperature difference

$$\Delta T_{lm} = \frac{(T_s - T_{m,o}) - (T_s - T_{m,i})}{\ln\left[(T_s - T_{m,o})/(T_s - T_{m,i})\right]} \quad (2.137)$$

2.8.3 Friction Effects and Pressure Losses

Pump power requirements, which may be determined from Eq. 2.107, are typically of little concern in liquid flows because of their large density. However, they can be significant if the pressure loss is large, which may be the case for flow through microchannels.

In general, flow through a channel enters from an upstream plenum and is discharged through a downstream plenum (Fig. 2.25). In dimensionless form, the overall (plenum-to-plenum) pressure drop, $\Delta p = p_u - p_d$, may be expressed as (Kays and London, 1984)

$$\frac{\Delta p}{\rho w_m^2/2} = \left[(1 - \sigma_c^2) + K_c\right] + 4 f_{F,app}\left(\frac{L}{D_h}\right) - \left[(1 - \sigma_e^2) - K_e\right] \quad (2.138)$$

where $\rho w_m^2/2$ is the *dynamic pressure* associated with flow in the channel. The quantities $\sigma_c = A_c/A_{p,u}$ and $\sigma_e = A_c/A_{p,d}$ are area ratios associated with contraction

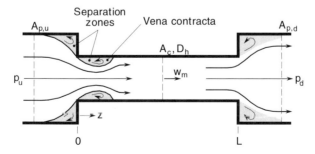

FIGURE 2.25 Flow through a channel with upstream and downstream plenums.

and expansion of the flow at the channel entrance and exit, respectively, while K_c and K_e are *loss coefficients* associated with the corresponding flow irreversibilities. The second term on the right-hand side of Eq. 2.138 accounts for losses resulting from friction forces in the channel, where $f_{F,app}$ is the *apparent Fanning friction factor* associated with a channel of length L.

The quantities $(1 - \sigma_c^2)$ and $-(1 - \sigma_e^2)$ account for the pressure *drop* and *rise*, respectively, resulting from fluid acceleration and deceleration at the entrance and exit. The effects are reversible and cancel each other if $\sigma_c = \sigma_e$. However, the loss coefficients K_c and K_e both account for irreversibilities that contribute to the overall pressure drop. For the sharp contraction shown in Figure 2.25, flow separation occurs both upstream and downstream of the corner, and a *vena contracta* is formed downstream of the contraction. The cumulative effect of the separation phenomena is to yield loss coefficients as large as $K_c \approx 0.5$ for turbulent flow in the channel and $K_c \approx 1.0$ for laminar flow (Kays and London, 1984). However, if the contraction occurs at a *well-rounded* or *bell-mouth* inlet, separation effects are greatly reduced and the loss coefficient may be as low as $K_c \approx 0.04$ (Fox and McDonald, 1985). Loss coefficients associated with an abrupt expansion at the exit may be as large as $K_e \approx 1$ for both laminar and turbulent flows.

The effects of σ and Re_{D_h} on the loss coefficients associated with abrupt contraction and expansion are provided by Kays and London (1984) for circular tubes and parallel-plate channels. Both coefficients decrease with increasing σ, while K_c and K_e decrease and increase, respectively, with increasing Re_{D_h}. Additional losses may be associated with bends, valves, and/or fittings associated with flow passages upstream and downstream of the channel (Fox and McDonald, 1985).

Pressure losses associated with flow through the channel may be determined from knowledge of the Fanning friction factor, Eq. 2.54, which takes the form

$$f_F = \frac{\overline{\tau}_s}{\rho w_m^2/2} \qquad (2.139)$$

where $\overline{\tau}_s$ is the average shear stress around the perimeter of the channel. For a circular tube or parallel-plate channel, $\overline{\tau}_s$ is uniformly distributed around the surface. For a rectangular channel, it varies and approaches zero in the corners.

For *fully developed flow* in a channel, a force balance may be used to relate the shear stress to the pressure gradient, $dp/dz = -4\overline{\tau}_s/D_h$. Integrating over a fully developed section of length L and substituting from Eq. 2.139, it follows that the pressure drop over the section is $\Delta p_{fd} = 4\overline{\tau}_s(L/D_h) = 4f_{F,fd}(\rho w_m^2/2)(L/D_h)$. Hence,

$$\frac{\Delta p_{fd}}{\rho w_m^2/2} = 4f_{F,fd}\frac{L}{D_h} \qquad (2.140)$$

The same result is obtained from Eq. 2.138 if entrance and exit effects are negligible and fully developed conditions exist over the length L.

For laminar fully developed flow, the friction factor is known to vary inversely with the Reynolds number, $f_{F,fd} = C/Re_{D_h}$, where the constant depends on the duct cross section. Representative results are provided in Table 2.1.

TABLE 2.1   Friction Factors and Nusselt Numbers for Laminar, Fully Developed Flow in Rectangular Channels

A^a	$f_F \cdot Re_{D_h}^{b,c}$	$Nu_{D_h}^{b,c}$	
		Uniform $\bar{q}_s''$	Uniform T_s
0.0	24.00	8.24	7.54
0.1	21.17	6.70	5.86
0.2	19.07	5.70	4.80
0.3	17.51	4.97	4.11
0.4	16.37	4.46	3.67
0.5	15.55	4.11	3.38
0.6	14.98	3.88	3.20
0.7	14.61	3.74	3.08
0.8	14.38	3.66	3.01
0.9	14.26	3.61	2.98
1.0	14.23	3.60	2.97

[a] The aspect ratio A is defined as W/H or H/W, whichever is smaller.
[b] The friction factor and Nusselt numbers are related to an average shear stress and an average convection coefficient over the periphery of the channel.
[c] For a circular tube, $f_F \cdot Re_{D_h} = 16$, $Nu_{D_h} = 4.36$ for uniform $\bar{q}_s''$ and $Nu_{D_h} = 3.66$ for uniform T_s.
Adapted from Shah and London (1978) with results for Nu_{D_h} attributed to F. W. Schmidt.

For fully developed turbulent flow through channels with smooth surfaces, the friction factor may be approximated from correlations of the form

$$f_{F,fd} = 0.079 Re_{D_h}^{-1/4} \qquad Re_{D_h} < 2 \times 10^4 \qquad (2.141a)$$

$$f_{F,fd} = 0.046 Re_{D_h}^{-1/5} \qquad Re_{D_h} > 2 \times 10^4 \qquad (2.141b)$$

Although surface roughness has a negligible effect on the friction factor for laminar flow, the friction factor increases with increasing roughness for turbulent flow. Graphical results for $f_{D,fd}$ are available in the form of a *Moody diagram* (Fox and McDonald, 1985; Incropera and DeWitt, 1996), where $f_{D,fd} = 4 f_{F,fd}$.

Although the foregoing results may be used as a first approximation for rectangular channels, improved accuracy may be obtained using a correlation developed by Jones (1976)

$$(f_{F,fd})^{-1/2} = 1.737 \ln \left[Re_{D_e} (f_{F,fd})^{1/2} \right] - 0.4 \qquad (2.142)$$

where

$$Re_{D_e} = Re_{D_h} \frac{D_e}{D_h} \qquad (2.143)$$

and

$$\frac{D_e}{D_h} = \left[\frac{2}{3} + \frac{11}{24}A(2-A)\right] \quad (2.144)$$

The aspect ratio A of the channel is W/H or H/W, whichever is smaller, and D_e is termed a *laminar equivalent diameter*. The result applies for smooth channel walls.

If fully developed flow is not reached or is preceded by an entrance region, Eq. 2.140 is replaced by the expression

$$\frac{\Delta p}{\rho w_m^2/2} = 4 f_{F,app} \frac{L}{D_h} \quad (2.145)$$

where $f_{F,app}$ is an *apparent mean friction factor*, which accounts for flow development over the length of the channel (Kays and London, 1984). Results presented by Shah and London (1978) for laminar flow in a rectangular duct have been interpolated by Phillips (1988) and are summarized in Table 2.2. For developing laminar flow in a circular tube and between infinite parallel plates, Shah and London (1978) recommend the following correlation:

$$f_{F,app} \cdot Re_{D_h} = \frac{3.44}{(z^+)^{1/2}} + \frac{C_1 + C_2/(4z^+) - 3.44/(z^+)^{1/2}}{1 + C_3 (z^+)^{-2}} \quad (2.146)$$

TABLE 2.2 Apparent Mean Friction Factors for Developing Laminar Flow in Rectangular Channels of Different Aspect Ratio[a]

	$f_{F,app} \cdot Re_{D_h}$			
z^+	$A = 1.0$	$A = 0.5$	$A = 0.2$	$A \leq 0.1$
0.001	111.0	111.0	111.0	112.0
0.003	66.0	66.0	66.1	67.5
0.005	51.8	51.8	52.5	53.0
0.007	44.6	44.6	45.3	46.2
0.009	39.9	40.0	40.6	42.1
0.010	38.0	38.2	38.9	40.4
0.015	32.1	32.5	33.3	35.6
0.020	28.6	29.1	30.2	32.4
0.030	24.6	25.3	26.7	29.7
0.040	22.4	23.2	24.9	28.2
0.050	21.0	21.8	23.7	27.4
0.060	20.0	20.8	22.9	26.8
0.070	19.3	20.1	22.4	26.4
0.080	18.7	19.6	22.0	26.1
0.090	18.2	19.1	21.7	25.8
0.100	17.8	18.8	21.4	25.6
0.200	15.8	17.0	20.1	24.7
> 1.000	14.2	15.5	19.1	24.0

[a]The aspect ratio A is defined as W/H or H/W, whichever is smaller.
Adapted from Phillips (1988).

where $C_1 = 16$, $C_2 = 1.25$, and $C_3 = 0.00021$ for a circular tube and $C_1 = 24$, $C_2 = 0.674$, and $C_3 = 0.000029$ for a parallel-plate channel. Note that, in all of the foregoing cases, $f_{F,app} \cdot Re_{D_h}$ approaches $f_{F,fd} \cdot Re_{D_h}$ with increasing z^+.

Although fully developed turbulent flow is reached within $(z/D_h) \approx 10$, a value of $(z/D_h) \approx 100$ must be reached before $f_{F,app} \approx f_{F,fd}$. For circular tubes or rectangular channels with smooth walls, Phillips (1988) recommends the following correlation:

$$f_{F,app} = A\, Re_{D_e}^B \tag{2.147a}$$

where

$$A = 0.0929 + \frac{1.0161}{z/D_h} \tag{2.147b}$$

$$B = -0.2680 - \frac{0.3193}{z/D_h} \tag{2.147c}$$

For a circular tube, $D_e = D_h$ is the tube diameter; for a rectangular channel, D_e is given by Eq. 2.144.

2.8.4 Heat Transfer

To calculate the heat flux at a particular location on the surface of a channel using Eq. 2.130, the local convection coefficient h must be known; to calculate the total heat rate for a channel of uniform surface temperature from Eq. 2.136, the average convection coefficient $\overline{h}$ is needed. The coefficients depend on the specific nature of the surface thermal condition, on whether the flow is laminar or turbulent, and on whether there is need to consider entrance region effects.

Entrance region effects need not be considered if concern is only for conditions in the fully developed region, for which Nusselt numbers associated with laminar flow in a rectangular channel are summarized in Table 2.1. The results for *uniform* $\bar{q}_s''$ apply for a *longitudinally uniform heat flux* and a uniform temperature around the perimeter of the channel. Correspondingly, the local convection coefficient and the local heat flux vary around the perimeter. The term *longitudinally uniform heat flux*, therefore, refers to uniformity of the *average* heat flux around the perimeter, $\bar{q}_s''$, in the longitudinal direction. It follows that the heat rate per unit length of the channel, $q' = \bar{q}_s'' \cdot P$, is also uniform (constant) in the longitudinal direction. The effect of the aspect ratio on the Nusselt numbers has been correlated by Shah and London (1978), and the results are of the form

$$Nu_{D_h} = 8.235\left(1 - \frac{2.0421}{A} + \frac{3.0853}{A^2} - \frac{2.4765}{A^3} + \frac{1.0578}{A^4} - \frac{0.1861}{A^5}\right) \tag{2.148}$$

for a *uniform heat flux*, $\bar{q}_s''$. Results for a *uniform temperature* pertain to situations for which the surface temperature T_s is uniform both peripherally and longitudinally and are correlated by

$$Nu_{D_h} = 7.541\left(1 - \frac{2.610}{A} + \frac{4.970}{A^2} - \frac{5.119}{A^3} + \frac{2.702}{A^4} - \frac{0.548}{A^5}\right) \quad (2.149)$$

The foregoing results presume heat transfer from all four walls of the channel. However, if one or more of the walls is well insulated (adiabatic), there is an attendant effect on the temperature profile and, hence, on the peripheral distribution of the local convection coefficient. Results for such cases are provided by Shah and London (1978), and Table 2.3 contrasts Nusselt numbers for channels heated on all four sides ($Nu_{D_h(4)}$) and three sides ($Nu_{D_h(3)}$).

As illustrated in Figure 2.24a, the convection coefficient decays from an extremely large (in principle, infinite) value at the channel entrance to the fully developed value for $z > z_{\text{fd,th}}$. For a thermally developing laminar flow in a circular tube, Shah and London (1978) provide local Nusselt number correlations of the following form:

TABLE 2.3 Nusselt Numbers for Laminar, Fully Developed Flow in Rectangular Channels with a Longitudinally Uniform Heat Flux $\bar{q}_s''$ Associated with Heat Transfer from (a) All Four Walls and (b) Three Walls (Figure 2.26)

$A = W/H$	$Nu_{D_h(4)}$	$Nu_{D_h(3)}$
0.0	8.24	8.24
0.1	6.70	6.94
0.2	5.70	6.07
0.3	4.97	5.39
0.4	4.46	4.89
0.5	4.11	4.51
0.6	3.88	—
0.7	3.74	3.99
0.8	3.66	—
0.9	3.61	—
1.0	3.60	3.56
1.43	3.74	3.20
2.0	4.11	3.15
2.5	4.46	3.17
3.33	4.97	3.31
5.0	5.70	3.64
10.0	6.70	4.25
∞	8.24	5.39

Adapted from Phillips (1988) and Shah and London (1978), with results attributed to F.W. Schmidt.

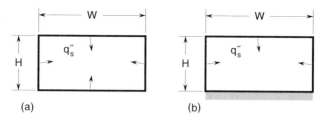

FIGURE 2.26 Cross sections of rectangular channels with heating on (*a*) four sides and (*b*) three sides.

$$Nu_D = 1.302 \left(z^*\right)^{-1/3} - 1.0 \qquad z^* \leq 0.00005 \qquad (2.150a)$$

$$Nu_D = 1.302 \left(z^*\right)^{-1/3} - 0.5 \qquad 0.00005 < z^* \leq 0.0015 \qquad (2.150b)$$

$$Nu_D = 4.364 + 8.68 \left(1000z^*\right)^{-0.506} \exp\left(-41z^*\right) \qquad z^* > 0.0015 \qquad (2.150c)$$

for a *uniform surface heat flux*, and

$$Nu_D = 1.077 \left(z^*\right)^{-1/3} - 0.7 \qquad z^* \leq 0.01 \qquad (2.151a)$$

$$Nu_D = 3.657 + 6.874 \left(1000z^*\right)^{-0.488} \exp\left(-57.2z^*\right) \qquad z^* > 0.01 \qquad (2.151b)$$

for a *uniform surface temperature*. Although Shah and London (1978) also present comparable correlations for simultaneous thermal and hydrodynamic flow development, the foregoing results may be used to a good approximation for liquid coolants ($Pr > 3$), irrespective of the entry region condition.

For *laminar flow* in a *circular tube* of length L with *uniform surface temperature*, the average Nusselt number may be approximated as (Incropera and DeWitt, 1996)

$$\overline{Nu}_D = 3.66 + \frac{0.0668/L^*}{1 + 0.04/(L^*)^{2/3}} \qquad (2.152)$$

for *thermally developing flow* and

$$\overline{Nu}_D = \frac{1.86}{(L^*)^{1/3}} \left(\frac{\mu}{\mu_s}\right)^{0.14} \qquad (2.153)$$

for *simultaneous* (hydrodynamic and thermal) *flow development*, where $L^* = L/(D\,Re_D\,Pr)$. All properties except μ_s are evaluated at the arithmetic mean of the channel inlet and outlet temperatures, $\bar{T}_m = (T_{m,i} + T_{m,o})/2$.

For rectangular channels, Phillips (1988) has interpolated results provided by Shah and London (1978) for a uniform heat flux to obtain the compilation of Table 2.4. The first five columns correspond to conditions for which all four walls are heated, in which case $Nu_{D_h} = Nu_{D_h(4)}$ and the results apply for A or A^{-1}. The

TABLE 2.4 Nusselt Numbers for Thermally Developing, Laminar Flow in Rectangular Channels with a Longitudinally Uniform Heat Flux

	$Nu_{D_h(4)}$					$Nu_{D_h(3)}$
z^*	$A = 1.0$	$A = 2.0$ or 0.5	$A = 3.0$ or 0.333	$A = 4.0$ or 0.25	$0.1 \geq A \geq 10$	$A \geq 10$
0.0001	25.2	23.7	27.0	26.7	31.4	31.6
0.0025	8.9	9.2	9.9	10.4	11.9	11.2
0.005	7.10	7.46	8.02	8.44	10.0	9.0
0.00556	6.86	7.23	7.76	8.18	9.8	8.8
0.00625	6.60	6.96	7.50	7.92	9.5	8.5
0.00714	6.32	6.68	7.22	7.63	9.3	8.2
0.00833	6.02	6.37	6.92	7.32	9.1	7.9
0.01	5.69	6.05	6.57	7.00	8.8	7.5
0.0125	5.33	5.70	6.21	6.63	8.6	7.2
0.0167	4.91	5.28	5.82	6.26	8.5	6.7
0.025	4.45	4.84	5.39	5.87	8.4	6.2
0.033	4.18	4.61	5.17	5.77	8.3	5.9
0.05	3.91	4.38	5.00	5.62	8.25	5.55
0.1	3.71	4.22	4.85	5.45	8.24	5.40
1.0	3.60	4.11	4.77	5.35	8.23	5.38

Adapted from Phillips (1988).

sixth column corresponds to large aspect ratio channels ($A \geq 10$) for which three walls are heated and one of the sides of width W is adiabatic (Fig. 2.26b), in which case $Nu_{D_h} = Nu_{D_h(3)}$. For channels of moderate aspect ratio ($0.25 < A < 4.0$), Phillips (1988) suggests that entry region Nusselt numbers for channels with three heated walls be approximated as

$$Nu_{D_h(3)} \approx Nu_{D_h(4)} \frac{Nu_{D_h(3),fd}}{Nu_{D_h(4),fd}} \quad (2.154)$$

where the Nusselt numbers for fully developed flow ($z^* \to 1$) are obtained from Table 2.4.

For fully developed turbulent flow in a smooth circular tube, the following expression, termed the *Dittus–Boelter equation*, is often used to determine the local Nusselt number for $Re_{D_h} > 10{,}000$ (Incropera and DeWitt, 1996):

$$Nu_{D_h,fd} = 0.023 Re_{D_h}^{4/5} Pr^n \quad (2.155)$$

where $n = 0.4$ for heat transfer to the fluid ($T_s > T_m$). Alternatively, for flows characterized by moderate-to-large temperature differences, ($T_s > T_m$), and, hence, significant property variations, the following correlation due to *Sieder and Tate* is recommended:

$$Nu_{D_h,fd} = 0.027 Re_{D_h}^{4/5} Pr^{1/3} \frac{\mu}{\mu_s} \quad (2.156)$$

where all properties except μ_s are evaluated at the local mean temperature of the fluid. For smaller Reynolds numbers, the correlation due to *Gnielinski* is recommended

$$Nu_{D_h,\text{fd}} = \frac{(f_{F,\text{fd}}/2)(Re_{D_h} - 1000) Pr}{1 + 12.7 (f_{F,\text{fd}}/2)^{1/2} (Pr^{2/3} - 1)} \quad (2.157)$$

and the suggested relation for the friction factor is

$$f_{F,\text{fd}} = (1.58 \ln Re_{D_h} - 3.28)^{-2} \quad (2.158)$$

For turbulent flow, the Nusselt number increases with wall roughness and may be calculated by using Eq. 2.157 with friction factors obtained from the Moody diagram (Fox and McDonald, 1985; Incropera and DeWitt, 1996). Although the foregoing expressions were developed for circular tubes, they may be used to obtain reasonable approximations for channels of noncircular cross section if the Nusselt and Reynolds numbers are defined in terms of the hydraulic diameter.

Because fully developed turbulent flow is achieved within $(z/D_h) \approx 10$, the entrance region is often small relative to the length of the channel and it is reasonable to assume that $\overline{Nu}_{D_h} \approx \overline{Nu}_{D_h,\text{fd}}$. For short channels, $\overline{Nu}_{D_h}$ would be underpredicted by this expression, yielding an underprediction, and, therefore, a conservative estimate, of the total heat rate. In addition, for turbulent flow, conditions are less affected by the existence of one or more adiabatic walls, and the foregoing correlations would provide reasonable estimates of the corresponding convection coefficient. However, to a good approximation, entrance region effects may be treated by modifying the Gnielinski correlation, Eq. 2.157, as follows:

$$\overline{Nu}_{D_h} = \frac{(f_{F,\text{fd}}/2)(Re_{D_h} - 1000) Pr}{1 + 12.7 (f_{F,\text{fd}}/2)^{1/2} (Pr^{2/3} - 1)} \left[1 + \left(\frac{D_h}{L}\right)^{2/3}\right] \quad (2.159)$$

Example 2.4

Heat generated by an array of electronic chips is dissipated by a copper cold plate, within which rectangular cooling channels of height $H_c = 6$ mm and width $W_c = 3$ mm are machined. Ten channels extend over the length $L_c = L_{cp} = 100$ mm of the cold plate, to which a 15 × 10 array of chips, $w_{ch} = 5$ mm on a side, is attached. Water at an inlet temperature of $T_{m,i} = 15°C$ is pumped through each of the channels at a mean velocity of $w_m = 4$ m/s, with channel inlet and outlet conditions characterized by a sharp contraction and expansion, respectively. Because of its large thermal conductivity, the cold plate may be assumed to be at a uniform temperature $T_s > T_{m,i}$ during operation.

1. What is the water outlet temperature, $T_{m,o}$, and the total rate of heat transfer under operating conditions for which $T_s = 70°C$? What is the corresponding heat flux for each chip? If the maximum allowable chip temperature is $T_{ch} = 85°C$, what is the maximum allowable thermal resistance R''_{th} be-

84 FUNDAMENTALS OF HEAT TRANSFER AND FLUID FLOW

tween each chip and the cold plate? As a first approximation, evaluate the properties of water at $T_{m,i} = 15°C$.

2. What is the pressure drop Δp across the cold plate and the corresponding pump power requirement P_p?

Solution

Known: Dimensions and operating conditions of a cold plate. Number, dimensions, and maximum allowable temperature of chip array mounted to the cold plate.

Find:

1. Outlet temperature of coolant and total heat rate. Heat flux associated with each chip and maximum allowable thermal resistance
2. Pressure drop and pump power requirement

Schematic:

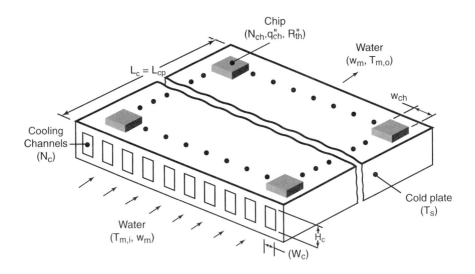

Assumptions:

1. Steady-state conditions
2. Uniform cold plate temperature
3. Equivalent heat dissipation by each chip
4. Equivalent area ratios for contraction and expansion of the flow at the channel entrance and exit
5. Smooth channel surfaces

Properties: Water ($T_{m,i} = 15°C$). From Appendix B: $\rho = 999$ kg/m^3, $\mu = 0.00114$ kg/s · m, $c_p = 4188$ J/kg · K, $k = 0.588$ W/m · K, and $Pr = 8.12$.

Analysis:

1. The outlet temperature may be determined from Eq. 2.135:

$$T_{m,o} = T_s - (T_s - T_{m,i}) \exp\left(-\frac{PL_c \bar{h}}{\dot{m}_1 c_p}\right)$$

where $\dot{m}_1 = \rho w_m A_c = 999$ kg/m^3 × 4 m/s × 0.003 m × 0.006 m = 0.072 kg/s is the flow rate through a single channel. With $D_h = 4A_c/P = 4$ mm,

$$Re_{D_h} = \frac{\rho w_m D_h}{\mu} = \frac{999 \text{ kg/m}^3 \times 4 \text{ m/s} \times 0.004 \text{ m}}{0.00114 \text{ kg/s} \cdot \text{m}} = 14{,}000$$

and the flow in each channel is turbulent. With $L_c/D_h = 25$, fully developed flow is assumed to exist throughout the channel, in which case, $\overline{Nu}_{D_h} \approx Nu_{D_h,fd}$. From Eq. 2.155, it follows that

$$\overline{Nu}_{D_h} \approx 0.023 Re_{D_h}^{4/5} Pr^{0.4} = 0.023(14{,}000)^{4/5}(8.12)^{0.4} = 110$$

$$\bar{h} \approx \overline{Nu}_{D_h} \frac{k}{D_h} = 110 \frac{0.588 \text{ W/m} \cdot \text{K}}{0.004 \text{ m}} = 16{,}200 \text{ W/m}^2 \cdot \text{K}$$

$$T_{m,o} = 70°C - (55°C) \exp\left(-\frac{0.018 \text{ m} \times 0.10 \text{ m} \times 16{,}200 \text{ W/m}^2 \cdot \text{K}}{0.072 \text{ kg/s} \times 4188 \text{ J/kg} \cdot \text{K}}\right)$$

$$= 70°C - 49.2°C = 20.8°C$$

From Eq. 2.137, the log-mean temperature difference is

$$\Delta T_{lm} = \frac{(70 - 20.8)°C - (70 - 15)°C}{\ln\left[(70 - 20.8)°C/(70 - 15)°C\right]} = 52.4°C$$

and, from Eq. 2.136, the rate of heat transfer to coolant flow through a single channel is

$$q_1 = \bar{h} A_s \Delta T_{lm} = 16{,}200 \text{ W/m}^2 \cdot \text{K}(0.018 \text{ m} \times 0.10 \text{ m})52.4°C = 1530 \text{ W}$$

The rate of heat transfer for the entire cold plate is then

$$q = N_c q_1 = 10 \times 1530 \text{ W} = 15{,}300 \text{ W}$$

and the heat flux associated with each chip is

$$q''_{ch} = \frac{q_{ch}}{A_{s,ch}} = \frac{q/N_{ch}}{(w_{ch})^2} = \frac{15{,}300 \text{ W}}{150(0.005 \text{ m})^2} = 4.08 \times 10^6 \text{ W/m}^2$$

The maximum allowable thermal resistance is then

$$R''_{th} = \frac{T_{ch} - T_s}{q''_{ch}} = \frac{15°C}{4.08 \times 10^6 \text{ W/m}^2}$$

$$= 3.67 \times 10^{-6} \text{ m}^2 \cdot \text{K/W}$$

$$= 0.0367 \text{ cm}^2 \cdot °\text{C/W}$$

2. With equivalent area ratios for contraction and expansion at the channel entrance and exit ($\sigma_c = \sigma_e$), Eq. 2.138 reduces to

$$\frac{\Delta p}{\rho w_m^2/2} = K_c + 4 f_{F,\text{app}} \left(\frac{L_c}{D_h}\right) + K_e$$

where $K_c \approx 0.5$, $f_{F,\text{app}} \approx f_{F,\text{fd}}$, and $K_e \approx 1$. Estimating the friction factor from Eq. 2.141a,

$$f_{F,\text{fd}} = 0.079 Re_{D_h}^{-1/4} = 0.079(14{,}000)^{-1/4} = 0.0073$$

it follows that

$$\Delta p \approx 999 \text{ kg/m}^3 (4 \text{ m/s})^2/2(0.5 + 4 \times 0.0073 \times 25 + 1.0)$$

$$= 17{,}800 \text{ N/m}^2 = 0.178 \text{ bar}$$

Neglecting pump inefficiencies, Eq. 2.107, the pump power requirement is then

$$P_p = \dot{W} = \dot{V}\Delta p = \frac{N_c \dot{m}_1 \Delta p}{\rho} = \frac{10(0.072 \text{ kg/s})17{,}800 \text{ N/m}^2}{999 \text{ kg/m}^3} = 12.8 \text{ W}$$

Comments:

1. Because the mean temperature of the water increases by only 5.8°C, little error was introduced by evaluating the properties at $T_{m,i} = 15°C$.
2. Approximating $\bar{h}$ and $f_{F,\text{app}}$ by the convection coefficient and friction factor for fully developed flow causes q and Δp to be underestimated. However, for $L_c/D_h = 25$, associated errors are small.
3. The large value of $\bar{h}$ is attributable to the small hydraulic diameter of the channels.
4. The required value of R''_{th} is small but achievable if care is taken to minimize the chip-to-cold-plate thermal resistance.

2.9 MIXED CONVECTION

The foregoing sections considered natural convection, for which fluid motion is driven exclusively by buoyancy forces, and forced convection, for which fluid flow is driven externally and buoyancy forces are negligible. However, conditions may exist for which natural and forced convection effects are comparable, and neither process may be neglected. A first estimate of the relative significance of the two processes may be made by computing the parameter $Gr_{L_0}/Re_{L_0}^2$, which provides a measure of the ratio of buoyancy to inertia forces. If $(Gr_{L_0}/Re_{L_0}^2) \ll 1$, buoyancy forces are negligible and conditions correspond to forced convection. If $(Gr_{L_0}/Re_{L_0}^2) \gg 1$, buoyancy forces are dominant, and conditions correspond to natural convection. For $(Gr_{L_0}/Re_{L_0}^2) \sim 1$, the combined effects of natural and forced convection (mixed convection) must be considered.

The effect of buoyancy on mixed-convection heat transfer is influenced by the direction of the buoyancy force relative to that of the imposed (forced) flow. The buoyancy-driven and forced flows may move in the same (*assisting* flow) or opposite (*opposing* flow) directions. Upward and downward forced motions over a heated vertical plate are examples of assisting and opposing flows, respectively. Alternatively, the buoyancy-driven and forced motions may be perpendicular (*transverse* flow), examples of which include imposed horizontal motion over a heated horizontal plate or cylinder.

In assisting and transverse flows, buoyancy acts to enhance heat transfer associated with forced convection; in opposing flows, it acts to reduce heat transfer by forced convection. Typically, mixed-convection heat transfer may be correlated by an expression of the form

$$Nu_{L_0}^n = Nu_{L_0,N}^n \pm Nu_{L_0,F}^n \tag{2.160}$$

where $Nu_{L_0,N}$ and $Nu_{L_0,F}$ are determined from existing correlations for natural and forced convection, respectively. The plus sign on the right-hand side of the equation applies for assisting and transverse flows, whereas the minus sign applies for opposing flow. The value of the exponent that provides the best correlation of data is typically $n \sim 3$.

Comprehensive treatments of mixed-convection heat transfer are provided by Chen and Armaly (1987) and Aung (1987) for external and internal flows, respectively.

2.10 SUMMARY

In this chapter, the fundamental principles and relations pertinent to liquid immersion cooling have been considered. Although this cooling option relies on convection heat transfer, electronic components will always have interfaces with solid materials to which and through which heat is transferred by conduction. A first estimate of the effect of one or more conduction pathways on the temperature of an electronic component may be made by formulating an equivalent thermal circuit and evaluating the related conduction resistances. Conduction is also pertinent when extended surfaces (fins) are used to enhance convection heat transfer to the coolant.

Heat transfer to a liquid coolant may occur by natural, forced, or mixed convection, according to the influence of buoyancy forces. If such forces are dominant (natural convection), there are no fluid pumping requirements, and heat rates are comparatively low. If buoyancy forces are negligible and fluid is pumped through the system (forced convection), large heat rates may be achieved, but at the expense of pump power requirements. Internal flows and impinging jets provide attractive cooling options. If a slow flow driven by external means is coupled to a buoyancy-driven flow of comparable strength (mixed convection), buoyancy may act to enhance heat transfer associated with the forced flow.

CHAPTER 3

NATURAL CONVECTION

3.1 INTRODUCTION

With the relentless transition of computing hardware from large, mainframe systems to networked, yet independent, workstations and desktop/laptop computers, each with its own powerful central processing unit (CPU), the limits of conventional cooling technologies are being approached. Although it is customary to use forced-air cooling for heat removal from the CPU, increasing levels of power dissipation necessitate larger air flow rates and, hence, more powerful fans, thereby exacerbating problems associated with system noise and vibration control. An alternative cooling option that circumvents such problems involves natural or free convection. Although related heat transfer coefficients are small compared to forced convection and boiling, higher levels of power dissipation may be achieved by immersing the heat-dissipating components in a dielectric liquid (Incropera, 1988; Bar-Cohen, 1991).

A passive cooling system based on natural convection eliminates the fan or pump assemblies required for forced convection, thereby providing a vibration- and noise-free means of thermal control. However, because heat transfer coefficients are lower than those associated with forced convection, *conjugate effects* are likely to be more significant. For example, if computer chips are mounted in a substrate, with one surface exposed to a dielectric liquid, heat transfer by conduction through the substrate may be comparable to that associated with free convection to the liquid. The term *conjugate heat transfer* refers to the coupling (mutual dependence) of conduction in the substrate and convection in the liquid. The *thermal spreading* associated with substrate conduction provides an additional (indirect) heat transfer path from the chip to the liquid, as well as from the chip to the surroundings. Conjugate effects are sometimes ignored in systems for which cooling is by free convection, and overly simplistic conditions are prescribed at solid/fluid interfaces.

A convenient scheme for using natural convection to cool an array of computer chips involves using the substrate as one wall of a rectangular cavity and filling the cavity with a dielectric liquid. The chips constitute discrete heat sources that sustain buoyancy-induced fluid motion in the cavity and from which three-dimensional heat transfer occurs by convection and conduction in the fluid and substrate, respectively. The contents of this chapter focus on such a situation, with the nominal configuration corresponding to a 3 × 3 array of chips mounted to one of the vertical walls of a rectangular cavity.

3.2 GEOMETRICAL FEATURES OF A RECTANGULAR CAVITY WITH DISCRETE HEAT SOURCES

An electronic package consisting of a 3 × 3 array of discrete sources mounted to one wall of a vertical rectangular cavity is shown in Figure 3.1. The dimensions of W, S, and H in the x, y, and z directions correspond to the fluid enclosure, which is characterized by the *major* and *minor* aspect ratios of $A_z \equiv H/S$ and $A_x \equiv W/S$, respectively. The heat sources are assumed to extend a distance B_{sub} into the substrate and to protrude a distance B_h into the fluid. The three heater rows are numbered sequentially from top to bottom and the columns from left to right.

Although the x and z heater dimensions are typically equivalent, allowance has been made for differences. Three-dimensional flow and heat transfer processes are

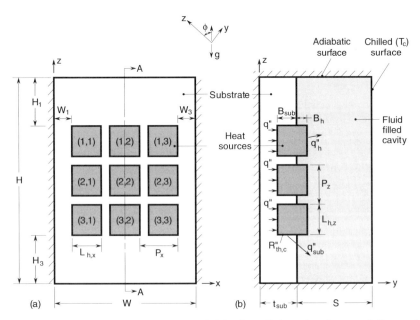

FIGURE 3.1 Plan view (*a*) and section of side view (*b*) for a 3 × 3 array of discrete heat sources mounted to the vertical wall of a rectangular cavity.

associated with the geometry of Figure 3.1, but two-dimensional conditions are approached in the limit as the heater aspect ratio, $A_h \equiv L_{h,x}/L_{h,z}$, approaches infinity. Such an idealization is well approximated by *strip heaters* that extend across the width of a cavity for which $W \gg L_{h,z}$. In this case, discrete behavior is associated with the vertical, but not the horizontal, direction. That is, heaters are no longer arranged in columns, and the total number of heaters corresponds to the number of rows.

A uniform heat flux, q'', is assumed to exist at the back of each heat source, and heat that is transferred by conduction through the source experiences transfer by convection to the fluid (q_h''). However, heat may also be transferred by conduction from the source through the substrate and subsequently by convection from the substrate to the fluid (q_{sub}''). If heat transfer from the back and sides of the substrate to the surroundings is negligible, all of the heat dissipated at the back of a source is transferred by convection to the fluid from either the source or the substrate. The contribution made by the substrate decreases with increasing contact resistance, $R_{th,c}''$, at the source/substrate interface and decreasing substrate thermal conductivity. If the contribution cannot be neglected, convection from the source and conduction in the substrate are coupled (*conjugate*) heat transfer processes.

Under steady-state conditions, heat that is transferred from the sources and substrate to the fluid must, in turn, be transferred from the fluid to the opposing vertical and/or top and bottom surfaces of the enclosure, which are cooled (*chilled*) by external means and thereby maintained at an acceptably low temperature, T_c. Horizontal temperature gradients in the fluid ($\partial T/\partial y < 0$) induce buoyancy forces, which, in turn, generate a clockwise circulation pattern for flow in the enclosure. Because the heater length, $L_{h,z}$, has the strongest influence on the buoyancy force, it is the most appropriate characteristic length for the system. With respect to the recirculating flow, the bottom and top edges of the third and first rows, respectively, constitute *leading* and *trailing edges* of the array and are located at distances of H_3 and H_1 from the bottom and top of the enclosure.

In the vertical orientation of Figure 3.1, the gravity vector is antiparallel to the z axis. An alternative operating condition is one for which the cavity is rotated counterclockwise by 90° and the heat sources are at the bottom of the enclosure. In this horizontal orientation, the gravity vector is antiparallel to the y axis.

3.3 MATHEMATICAL MODEL

Because velocities associated with buoyancy-driven flows are small, the flows are typically laminar, and, except for invoking the Boussinesq approximation, conditions are described by Eqs. 2.36 to 2.40. However, for the enclosure of Figure 3.1, gravity acts exclusively in the vertical direction, in which case $g_x = g_y = 0$ and $g_z = g$. The equations, therefore, reduce to the following forms:

$$\frac{\partial u}{\partial x} + \frac{\partial v}{\partial y} + \frac{\partial w}{\partial z} = 0 \qquad (3.1)$$

$$\rho\left[u\frac{\partial u}{\partial x} + v\frac{\partial u}{\partial y} + w\frac{\partial u}{\partial z}\right] = \mu\left[\frac{\partial^2 u}{\partial x^2} + \frac{\partial^2 u}{\partial y^2} + \frac{\partial^2 u}{\partial z^2}\right] - \frac{\partial p}{\partial x} \quad (3.2)$$

$$\rho\left[u\frac{\partial v}{\partial x} + v\frac{\partial v}{\partial y} + w\frac{\partial v}{\partial z}\right] = \mu\left[\frac{\partial^2 v}{\partial x^2} + \frac{\partial^2 v}{\partial y^2} + \frac{\partial^2 v}{\partial z^2}\right] - \frac{\partial p}{\partial y} \quad (3.3)$$

$$\rho\left[u\frac{\partial w}{\partial x} + v\frac{\partial w}{\partial y} + w\frac{\partial w}{\partial z}\right] = \mu\left[\frac{\partial^2 w}{\partial x^2} + \frac{\partial^2 w}{\partial y^2} + \frac{\partial^2 w}{\partial z^2}\right]$$

$$-\frac{\partial p}{\partial z} + \rho g \beta (T - T_c) \quad (3.4)$$

$$\rho c_{p,f}\left[u\frac{\partial T}{\partial x} + v\frac{\partial T}{\partial y} + w\frac{\partial T}{\partial z}\right] = k_f\left[\frac{\partial^2 T}{\partial x^2} + \frac{\partial^2 T}{\partial y^2} + \frac{\partial^2 T}{\partial z^2}\right] \quad (3.5)$$

If conjugate effects are significant, the solution of the foregoing equations must be coupled to that of the heat equation for conduction in the source and substrate, which is of the form

$$0 = \frac{\partial^2 T}{\partial x^2} + \frac{\partial^2 T}{\partial y^2} + \frac{\partial^2 T}{\partial z^2} \quad (3.6)$$

Procedures for solving Eqs. 3.1 to 3.6 subject to appropriate boundary conditions are described by Heindel (1994). The solutions may be generalized by nondimensionalizing the equations and boundary conditions and representing results in terms of dimensionless parameters such as the modified Rayleigh number, $Ra^*_{L_{h,z}} = g\beta q'' L_{h,z}^4 / k_f \alpha_f \nu$, the Prandtl number, $Pr = \nu/\alpha_f$, the conductivity ratios, $R_h = k_h/k_f$ and $R_{sub} = k_{sub}/k_f$, and assorted length ratios such as A_z and A_h.

3.4 VERTICAL CAVITIES: THEORETICAL RESULTS

3.4.1 Flush-Mounted Heat Sources with Negligible Substrate Conduction

Referring to Figure 3.1, conditions corresponding to discrete heaters that are mounted flush to the substrate and for which substrate conduction is negligible may be approximated by setting $B_h = B_{sub} = t_{sub} = 0$. Hence, all of the heat that is dissipated by the source is transferred directly to the fluid by convection from its surface ($q'' = q''_h$).

Kuhn and Oosthuizen (1986) simulated three-dimensional flow and heat transfer for one or two discrete heat sources mounted to one wall of a vertical cavity with major and minor aspect ratios of $A_z = 3$ and $A_x = 6$, respectively, and for $0 \leq Ra_S \leq 10^5$. The opposing wall was chilled, while the top and bottom surfaces were assumed to be adiabatic. For $H_1/L_{h,z}$ and $H_3/L_{h,z} > 2$, the effect of heater location was shown to be small. However, for $H_1/L_{h,z} < 2$, buoyancy-driven fluid motion was reduced and heat transfer was dominated by conduction across the fluid. Three-dimensional flow near the vertical edges of the heater increased local convection coefficients, causing average Nusselt numbers for the heaters to exceed predictions based on a two-dimensional model.

The most comprehensive calculations for flush-mounted sources with negligible substrate conduction have been performed by Heindel et al. (1995a). Their model was based on the geometry of Figure 3.2, where the plan view acknowledges the plane of symmetry at $x = W/2$ and is restricted to one half of the actual physical region. The computational domain and, hence, the numerical requirements are reduced by a factor of 2. Isoflux conditions were prescribed at each of the square heaters, while the opposing chilled wall was isothermal and the remaining surfaces were assumed to be adiabatic. Equations 3.1 to 3.5 were solved numerically to obtain three-dimensional velocity and temperature distributions for fixed geometric parameters corresponding to $H/S = 7.5$, $(P_x - L_{h,x})/L_{h,z} = 0.25$, $P_z/L_{h,z} = 1.25$, $S/L_{h,z} = 1.0$, $H_1/L_{h,z} = H_3/L_{h,z} = 2.0$, and $W_1/L_{h,z} = 0.5$, where the heater height, $L_{h,z}$, was chosen to be the dominant length scale because of its controlling influence on the buoyancy force and, hence, on flow and thermal conditions. The computed temperature field was used to determine local and average Nusselt num-

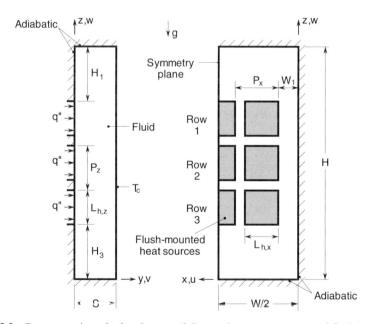

FIGURE 3.2 Representation of a 3 × 3 array of discrete heat sources mounted flush to one wall of a rectangular enclosure with negligible substrate conduction: side view and plan view of one half section bounded by a a symmetry plane. Adapted from Heindel et al. (1995a).

bers for the heaters, where

$$Nu_{L_{h,z}}(x,z) = \frac{q'' L_{h,z}}{[T_h(x,z) - T_c]k_f} \qquad (3.7)$$

$$\overline{Nu}_{L_{h,z}} = \frac{q'' L_{h,z}}{\frac{1}{A_{s,h}} \int_{A_{s,h}} [T_h(x,z) - T_c]k_f \, dA_{s,h}} \qquad (3.8)$$

and $T_h(x, z)$ is the local temperature on a heater surface, which is an output of the computed temperature field. Calculations were performed for $Pr = 5$, heater aspect ratios in the range $0.2 \leq A_h \leq 4.0$, and Rayleigh numbers in the range $10^5 \leq Ra^*_{L_{h,z}} \leq 10^8$.

Fluid motion within the cavity is characterized by a clockwise recirculation for which descending and ascending boundary layer flows are associated with the cooled and heated surfaces, respectively. For $Ra^*_{L_{h,z}} = 10^5$, the boundary layers are thick and extend from the walls to the center ($y/S = 0.5$) of the cavity. However, for fixed values of y and z within the boundary layer on a heated surface, the vertical velocity component (w) is not uniformly distributed in the x direction. Maxima occur in the central region of each heater column, while minima exist in the unheated regions between columns and the velocity decays to zero as the sidewall is approached. In contrast, the spanwise (x) distribution of the vertical velocity component within the descending boundary layer along the chilled wall is highly uniform, except in a thin region near the sidewalls, where w goes to zero.

The foregoing behavior is illustrated in the three-dimensional plot of Figure 3.3a, which provides the distribution of a dimensionless vertical velocity component over the half section of an x–y plane of the enclosure located at the trailing edge of the top row ($z/L_{h,z} = 5.5$) of heaters for which $A_h = 1$. The dark regions on the bottom plane designate locations of the heaters, while the solid and dashed lines designate contours of positive and negative values of the velocity, respectively. Increasing the aspect ratio (Fig. 3.3b) has little effect on the maximum value of the vertical velocity component, but increases the uniformity of flow across the heater face, while intensifying the difference between maximum and minimum velocities across the heated wall. The uniformity of flow across a heater and the difference between maximum and minimum velocities also increase with increasing Rayleigh number. In addition, with increasing $Ra^*_{L_{h,z}}$ there is an increase in velocities and a decrease in boundary layer thicknesses associated with flow along the heated and chilled walls.

Because temperature gradients in the y direction are concentrated in the boundary layers, temperatures are approximately independent of y outside the boundary layers. However, because of heating, fluid temperatures increase with increasing z, creating the *thermally stratified* core shown in Figure 3.4. The region below row 3 is isothermal, while the central region above the leading edge of this row ($z/L_{h,z} > 2.0$) is highly stratified. Ascending and descending boundary layers at the heated and chilled walls, respectively, are distinguished by large temperature gradients in the y direction. The large gradients and thickening thermal boundary layer near the heater

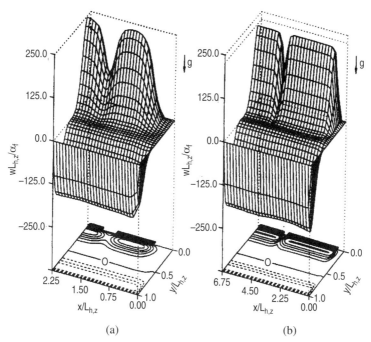

FIGURE 3.3 Distribution of the dimensionless vertical velocity component for an x–y plane at the trailing edge of row 1 for $Ra^*_{L_{h,z}} = 10^6$ and $Pr = 5$: (a) $A_h = 1$ and (b) $A_h = 4$. Adapted from Heindel et al. (1995a).

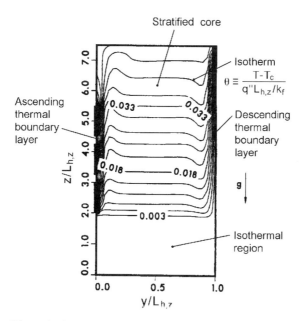

FIGURE 3.4 Dimensionless temperature contours predicted by the three-dimensional model for a y–z plane located at $x/L_{h,z} = 1.0$ ($Ra^*_{L_{h,z}} = 10^8$, $Pr = 5$, $A_z = 7.5$, $A_h = 1$). Adapted from Heindel (1994).

surfaces are revealed by convergence of the isotherms to form the darkened regions. The boundary layer thicknesses increase and the temperature gradients decrease with decreasing $Ra^*_{L_{h,z}}$, eventually causing merger of the boundary layers and departure of the stratified core conditions. For a fixed value of $Ra^*_{L_{h,z}}$, boundary layer merger and elimination of a stratified core may be achieved by increasing the major aspect ratio, as, for example, by decreasing the distance S between the vertical walls.

If flow in the cavity were strictly two dimensional, the x component of the velocity (u) would be zero. Such is not the case, however, with motion in the spanwise (x) direction induced as fluid passes over a heated surface and is thereby accelerated in the vertical direction. This acceleration entrains fluid from adjoining, unheated regions, and the effect becomes more significant with increasing $Ra^*_{L_{h,z}}$ and decreasing A_h.

The effect of entrainment on the local Nusselt number is shown in Figure 3.5, where predictions based on the three-dimensional model are compared with those from a two-dimensional approximation. The three-dimensional results correspond to variations of $Nu_{L_{h,z}}$ with z along the edge and the center of a column of heaters. The larger Nusselt numbers at the edge are attributed to lateral advection (entrainment) of cooler fluid from the adjoining adiabatic region, as well as to diffusion of energy from fluid flowing over the heater to cooler fluid in the adiabatic region. Absence of these effects at the center of the column yields much smaller Nusselt numbers, which are in good agreement with predictions based on the two-dimensional model ($A_h \to \infty$).

Both models predict a reduction in $Nu_{L_{h,z}}$ from the leading to the trailing edge of each heat source, as well as a slight enhancement at the trailing edge for $Ra^*_{L_{h,z}} = 10^5$. This effect is attributed to diffusion of energy from warm fluid at the trailing edge to cool fluid in the adjoining adiabatic region. In transitioning from the trailing edge of an upstream heater to the leading edge of the adjoining downstream heater, there is partial dissipation of the thermal boundary layer in the adiabatic region between heaters and, hence, an increase in the local Nusselt number at the leading edge of the downstream heater. With renewal of boundary layer development on the downstream heater, the local Nusselt number resumes its decay with increasing z. The largest average Nusselt number is associated with row 3, where thermal boundary layer thicknesses and fluid temperatures outside the boundary layer are smallest, and $\overline{Nu}_{L_{h,z}}$ decreases monotonically with decreasing row number. Owing principally to edge entrainment effects, three-dimensional predictions of $\overline{Nu}_{L_{h,z}}$ are from 10% to 15% larger than those of a two-dimensional model. The Nusselt numbers increase with increasing $Ra^*_{L_{h,z}}$ (Fig. 3.5b), while the significance of streamwise diffusion at the trailing (and lateral) edges decreases.

Surface thermal conditions are also revealed by three-dimensional plots of the temperature (Fig. 3.6), as well as the local Nusselt number (Fig. 3.7). In Figure 3.6, the dimensionless surface temperature, $\theta^* \equiv (T - T_c)/q''L_{h,z}k_f$, increases abruptly at the leading edge of each heat source and reaches a maximum on the heater midline near its trailing edge. The reduction in θ^* beyond the trailing edge of each heater is due to the cessation of heating from adiabatic portions of the surface and partial dissipation of the adjoining thermal boundary layer. However, beyond the leading edge of row 3, the temperature of adiabatic portions of the surface increases mono-

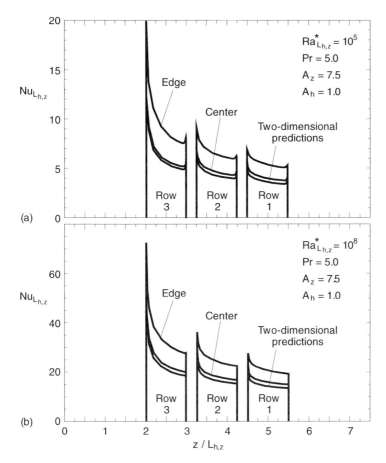

FIGURE 3.5 Vertical distribution of the local Nusselt number at the edge and center of a column of heaters predicted by three-dimensional model and x-independent distribution obtained from two-dimensional model: (a) $Ra^*_{L_{h,z}} = 10^5$ and (b) $Ra^*_{L_{h,z}} = 10^8$. Adapted from Heindel et al. (1995a).

tonically with z because of a corresponding increase in the fluid temperature. With increasing fluid temperature and renewal of thermal boundary layer development on downstream heaters, the maximum heater temperature increases with decreasing row number (from bottom to top).

The increase in θ^* with increasing z on a heater surface (Fig. 3.6) is consistent with a reduction in the local Nusselt number resulting from boundary layer development (Fig. 3.7). Local Nusselt numbers are largest at the leading edge of the first heat source and decrease monotonically to the trailing edge. Similar conditions characterize downstream heaters, which experience a slight increase in $Nu_{L_{h,z}}$ at their leading edge because of partial dissipation of the thermal boundary layer in the preceding adiabatic region. As indicated previously, the increase in $Nu_{L_{h,z}}$ at the lateral edges

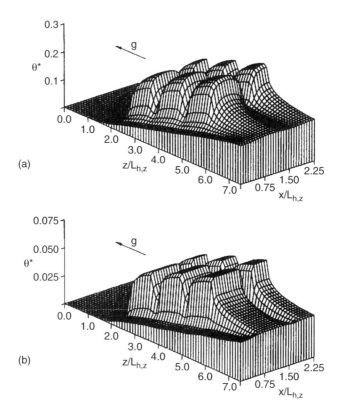

FIGURE 3.6 Distribution of dimensionless surface temperature, $\theta^* = (T - T_c)/q''L_{h,z}k_f$, on the vertical wall of a rectangular cavity with a 3 × 3 array of flush-mounted heaters for $Pr = 5$, $A_h = 1$, and $A_z = 7.5$: (a) $Ra^*_{L_{h,z}} = 10^5$ and (b) $Ra^*_{L_{h,z}} = 10^8$. Adapted from Heindel et al. (1995b).

of a heater is due to the entrainment of cooler fluid from adjoining adiabatic regions, as well as to the diffusion of energy from warmer boundary layer fluid above the heater to cooler fluid in the adiabatic regions. This increase in $Nu_{L_{h,z}}$ corresponds to a decrease in θ^* at the edges. The spanwise distributions of θ^* and $Nu_{L_{h,z}}$ across a heater become more uniform with increasing $Ra^*_{L_{h,z}}$ (Figs. 3.6b and 3.7b), while values of θ^* and $Nu_{L_{h,z}}$ decrease and increase, respectively. The increase in $Nu_{L_{h,z}}$ is due to intensification of the buoyancy-driven flow. Although the actual surface temperature also increases with $Ra^*_{L_{h,z}}$, the increase in $(T - T_c)$ is not as large as the increase in q'' and θ^* decreases.

The effect of the heater aspect ratio on the Nusselt number distribution has also been determined (Heindel et al., 1995a). As shown in Figure 3.8, the spanwise distribution of $Nu_{L_{h,z}}$ becomes more uniform with increasing A_h, thereby diminishing the influence of edge effects on total heat transfer from the source. The maximum

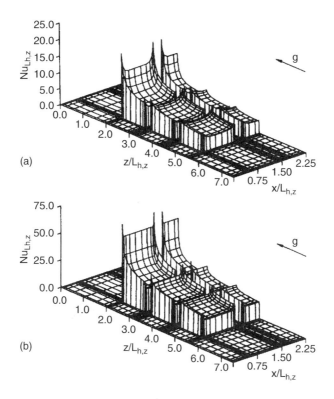

FIGURE 3.7 Distribution of local Nusselt number for a 3×3 array of heat sources mounted flush to the vertical wall of a rectangular cavity for $Pr = 5$, $A_h = 1$, and $A_z = 7.5$: (a) $Ra^*_{L_{h,z}} = 10^5$ and (b) $Ra^*_{L_{h,z}} = 10^8$. Adapted from Heindel et al. (1995b).

value of the Nusselt number, which corresponds to the *spikes* in the distribution at the corners of the leading edge for row 3, also decreases with increasing A_h. This effect is attributed to an increase in the amount of warm fluid over the heater surface relative to the fixed amount of cooler fluid in the adjoining adiabatic regions. The effect of aspect ratio was considered by varying $L_{h,x}$, while fixing $L_{h,z}$ and all other dimensions of the enclosure. Hence, the width of the adiabatic region was fixed at $0.25L_{h,z}$. An increase in this width would reduce the amount by which the maximum value of $Nu_{L_{h,z}}$ decreases with increasing A_h.

The influence of edge effects on the row-averaged Nusselt number is shown in Figure 3.9, where predictions of $\overline{Nu}_{L_{h,z}}$ from the three-dimensional model are plotted as a function of A_h and compared with results from an approximate two-dimensional model. Average Nusselt numbers decrease with decreasing row number because of the growth of the thermal boundary layer in the streamwise (upward) direction. For $A_h = 0.2$, the edge effects are significant, with the three-dimensional prediction exceeding the two-dimensional approximation by 54%, 49%, and 36% for rows 1, 2, and 3, respectively. However, as expected, differences decrease with

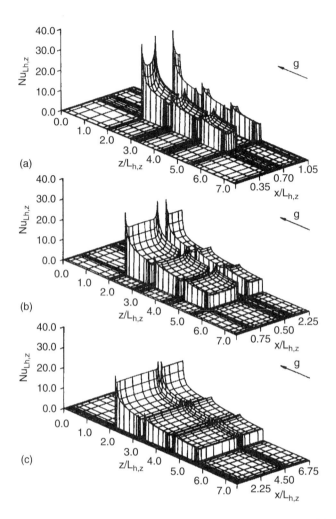

FIGURE 3.8 Distribution of local Nusselt number for a 3×3 array of heat sources mounted flush to the vertical wall of a rectangular cavity for $Pr = 5$, $Ra^*_{L_{h,z}} = 10^6$, $(P_x - L_{h,x})/L_{h,z} = 0.25$, and $A_z = 7.5$: (a) $A_h = 0.2$, (b) $A_h = 1.0$, and (c) $A_h = 4.0$. Adapted from Heindel et al. (1995a).

increasing A_h, and, for $A_h > 3$, the three- and two-dimensional predictions agree to within 10%, rendering the two-dimensional approximation acceptable for design calculations.

Heindel (1994) also considered the effect of the Prandtl number and the major cavity aspect ratio on the three-dimensional predictions. For a fixed geometry and Rayleigh number, local and average results were found to be independent of the Prandtl number over the range $5 < Pr < 500$. The effect of $A_z = H/S$ was considered by varying the spacing S between the vertical walls, while fixing all other

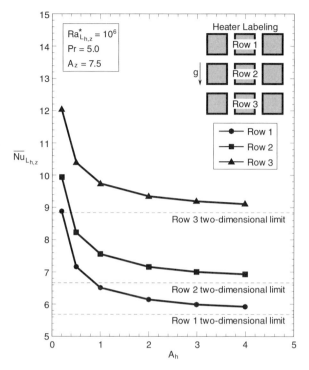

FIGURE 3.9 Effect of heater aspect ratio on row-averaged Nusselt numbers for a 3 × 3 array of heat sources mounted flush to the vertical wall of a rectangular cavity for $Pr = 5$, $Ra^*_{L_{h,z}} = 10^6$, $(P_x - L_{h,x})/L_{h,z} = 0.25$, and $A_z = 7.5$. Adapted from Heindel et al. (1995a).

characteristics of the heater/enclosure geometry. For $A_z < 15$, row-averaged Nusselt numbers were found to be independent of A_z. In contrast, for $Ra^*_{L_{h,z}} = 10^6$ and $A_z > 15$, boundary layers on the adjoining walls begin to interact, reducing circulation within the cavity and, hence, the row-averaged Nusselt numbers. However, A_z may be increased such that convection is totally suppressed and heat transfer to the chilled wall is exclusively by conduction. At this point, the Nusselt numbers would increase with further increases in A_z.

Two-dimensional numerical simulations have also been performed for flow and heat transfer in a rectangular cavity with alternating isoflux and adiabatic regions located on one vertical wall and the opposing wall isothermally chilled. The isoflux regions represented strip heaters that, in principle, extended to infinity in the spanwise direction ($L_{h,x} = W \gg L_{h,z}$). For three alternating adiabatic and flush-mounted isoflux sections of equivalent height, Keyhani et al. (1988b) and Prasad et al. (1990) found that the flow and heat transfer were approximately independent of the Prandtl number over the range $1 < Pr < 166$. An important implication of this result is that heat transfer correlations developed from experiments performed for a fluid whose Prandtl number is in this range could be applied to all fluids with Prandtl numbers in the same range. They also found that temperatures on the heated wall were strongly influenced by thermal stratification within the core region (see Fig. 3.4). For the same configuration, Shen et al. (1989) found that differences in temperatures, and, therefore, Nusselt numbers, between the heated surfaces were reduced by increasing the vertical extent of the adiabatic regions. Such an increase has the effect of enhancing dissipation of the thermal boundary layer in the adiabatic region between heat sources, and the renewal of boundary layer development at the leading edge of downstream heaters more closely approaches initial growth at the leading edge of the lowermost heater.

Chu et al. (1976) obtained a two-dimensional numerical solution for flow and heat transfer in a rectangular cavity, with a strip heater mounted flush to one vertical wall. The opposing wall was cooled, while the horizontal top and bottom walls were either cooled or adiabatic. With increasing Rayleigh number, the heater location associated with the maximum Nusselt number shifted downward, and, for cooled horizontal walls, the Nusselt number increased with increasing cavity width (decreasing A_z). For chilled horizontal surfaces, the major aspect ratio, A_z, becomes an important parameter, particularly if the opposing vertical wall is adiabatic.

3.4.2 Flush-Mounted Heat Sources with Substrate Conduction

Referring to Figure 3.1, Heindel et al. (1995d) performed two-dimensional conjugate heat transfer calculations for a vertical array of three flush-mounted ($B_h = 0$) strip heaters ($L_{h,x} = W \to \infty$) whose thickness was equated to that of the substrate ($B_{sub} = t_{sub}$). The opposite vertical wall and the horizontal walls of the cavity were assumed to be isothermal and adiabatic, respectively. Although the Prandtl number was fixed at $Pr = 25$, which corresponds approximately to that of FC-77, other studies (Keyhani et al., 1988a; Prasad et al., 1990) have shown Prandtl number effects to be negligible for the range $5 < Pr < 1000$. The prescribed thermal conductivity of the heater corresponded to that of silicon ($k_h = 148$ W/m · K), in which case the heater-to-fluid conductivity ratio was fixed at $R_h = k_h/k_f = 2350$. The cavity, heater, and substrate geometries were also fixed, with $L_{h,z} = 12.7$ mm, $S/L_{h,z} = 1$, $A_z = H/S = 7.5$, $H_1/L_{h,z} = H_3/L_{h,z} = 2$, $P_z/L_{h,z} = 1.25$, and $t_{sub}/S = 0.5$. Calculations were performed for Rayleigh numbers and substrate-to-fluid thermal conductivity ratios in the ranges $10^4 < Ra^*_{L_{h,z}} < 10^9$ and $0.1 < R_{sub} < 1000$. A value of $R_{sub} = 575$ corresponds to an alumina substrate and FC-77, which is a representative combination for liquid immersion cooling of electronic packages.

Dimensionless streamlines ψ^* and isotherms θ^* predicted for a nominal conductivity ratio of $R_{sub} = 10$ and the lower and upper limits of the Rayleigh number range ($Ra^*_{L_{h,z}} = 10^4, 10^9$) are shown in Figure 3.10. For $Ra^*_{L_{h,z}} = 10^4$, thick boundary layers exist on the vertical walls and flow within the cavity is unicellular and weak. With increasing $Ra^*_{L_{h,z}}$, the boundary layers thin, and additional, stronger recirculation zones develop in the cavity. For $Ra^*_{L_{h,z}} = 10^9$, large velocities in the boundary layers are manifested by closely spaced streamlines, which appear as dark bands on the walls. Portions of the fluid descending along the cold wall are swept toward the leading edge of each heater row, while a much weaker flow descends all the way

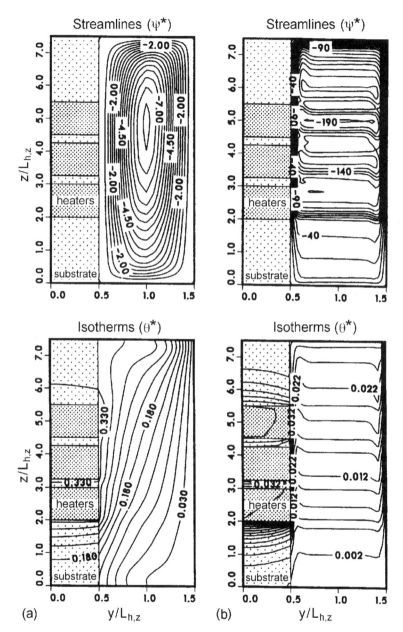

FIGURE 3.10 Dimensionless streamlines and isotherms for a vertical array of strip heaters ($Pr = 25$, $R_h = 2350$, $R_{sub} = 10$): (a) $Ra^*_{L_{h,z}} = 10^4$ and (b) $Ra^*_{L_{h,z}} = 10^9$. Adapted from Heindel et al. (1995d).

to the bottom of the cavity. Core regions of the cellular flows associated with the streamline patterns correspond to zones of nearly stagnant fluid.

The dimensionless isotherms of Figure 3.10 are displayed in both the fluid and the solid (heater/substrate) regions. The nearly vertical isotherms in the fluid for $Ra^*_{L_{h,z}} = 10^4$ indicate the strong influence of conduction on heat transfer across the cavity. However, with increasing Rayleigh number, the core region of the fluid begins to stratify, and, apart from the thin thermal boundary layers on the heated and chilled surfaces, temperature gradients are predominantly in the vertical direction for $Ra^*_{L_{h,z}} = 10^9$. Because of the large value of R_h, the surface of each heater is nearly isothermal, whereas largely vertical temperature gradients exist in the substrate and temperatures differ from heater to heater.

The distribution of the heat flux along the interface between the heater/substrate surfaces and the fluid provides an indication of the significance of conjugate effects. The distribution is normalized with respect to the heat flux applied at the back surfaces of each heater and plotted in Figure 3.11 for different values of the Rayleigh number. The *spike* at the leading edge of each heater, which corresponds to a large value of the local convection coefficient, and its reduction with decreasing row number are due to the effects of boundary layer development in the ascending flow on the heater/substrate surfaces. The spikes in the heat flux ratio at the leading edges of the first and second heater rows are affected by partial dissipation of the thermal bound-

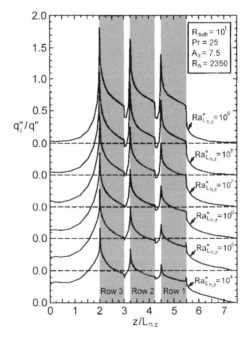

FIGURE 3.11 Effect of Rayleigh number on distribution of the heat flux, q''_i/q'', along the fluid-heater/substrate interface. Adapted from Heindel et al. (1995d).

ary layer in the substrate region between heater rows and renewal of boundary layer development at the leading edge. The smallest value of $Ra^*_{L_{h,z}} = 10^4$ corresponds to the smallest convection coefficients and, hence, the most significant influence of conjugate effects, which are manifested by conduction from the heater to the substrate and convection from the surface of the substrate to the fluid. With increasing $Ra^*_{L_{h,z}}$, fluid motion intensifies, thereby thinning the boundary layer and increasing convection coefficients on the heated wall. The net effect is to increase and decrease convection heat transfer from the heater and substrate surfaces, respectively. The much smaller, secondary maximum in the dimensionless heat flux, which is evident at the trailing edge of heater 1 for smaller values of $Ra^*_{L_{h,z}}$, is due to the effect of streamwise conduction in the fluid.

The foregoing trends are affected by the substrate-to-fluid thermal conductivity ratio R_{sub}. With decreasing R_{sub}, less heat is transferred from the heaters to the substrate, thereby reducing heat transfer by convection from the substrate to the fluid. One consequence of this condition is a reduction in the extent to which fluid below row 3 is heated and, hence, the strength of recirculation in the corresponding portion of the cavity. In the limit as $R_{sub} \to 0$, fluid below row 3 becomes stagnant and isothermal, with motion confined to a region extending from the leading edge of row 3 to the top of the cavity (Heindel et al., 1995d). In contrast, heat transfer to the substrate increases with increasing R_{sub}, thereby increasing convection from the substrate to the fluid and providing a more uniform temperature distribution along the entire solid/fluid interface. These conditions foster a unicellular flow that extends over the entire height of the cavity, and, in the limit as $R_{sub} \to \infty$, the cavity is characterized by isothermal hot and cold walls. The stagnant core associated with the cell is then located at the center ($y/L_{h,z} = 1.0$, $z/L_{h,z} = 3.75$) of the cavity.

The effect of R_{sub} on the dimensionless heat flux distribution is shown in Figure 3.12. With increasing R_{sub}, spikes at the leading (and trailing) edges of the heaters decrease, while convection heat transfer from the substrate increases. The heaters become less discernible, and, at $R_{sub} = 10^3$, the nearly monotonic decay of the heat flux with increasing $z/L_{h,z}$ resembles conditions along an isothermal wall. Because such a condition provides for maximum *spreading* of heat into the substrate and minimum heater temperatures, it is favored in electronic cooling applications. Thermal spreading would, of course, decrease with increasing contact resistance between the heaters and the substrate.

A three-dimensional simulation of conjugate effects has been performed for an experimental test cell used to determine convection coefficients associated with a 3×3 array of square heaters mounted flush ($B_h = 0$) with one wall of a rectangular cavity (Heindel et al., 1995c). Calculations were performed for both water and FC-77, which provided conductivity ratios of $R_h = 650$ and $R_{sub} = 0.48$ and $R_h = 6420$ and $R_{sub} = 4.7$, respectively. The substrate-to-fluid thermal conductivity ratio was found to have a significant effect on the velocity field by progressively muting its three-dimensionality with increasing R_{sub}. For example, if $R_{sub} = 0$, conjugate effects are nonexistent, and, as shown in Figure 3.3a, the spanwise distribution of the vertical velocity component is characterized by pronounced maxima at x locations corresponding to central regions of the heat sources and minima at locations corre-

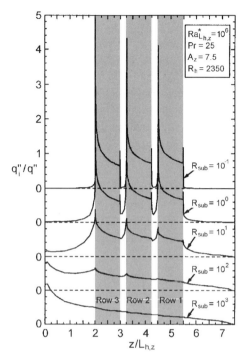

FIGURE 3.12 Effect of substrate-to-fluid thermal conductivity ratio on distribution of the dimensionless heat flux, q_i''/q'', along the fluid-heater/substrate interface. Adapted from Heindel et al. (1995d).

sponding to the unheated substrate between sources. However, with increasing R_{sub}, there is increased heat transfer by conduction from the sources to the substrate and by convection from the substrate to the fluid. Accordingly, the spanwise distribution of the heat flux becomes more uniform, causing the velocity distribution to do likewise. Similarly, just as the previous two-dimensional simulations revealed spikes in the dimensionless heat flux at the leading edge of each row, the three-dimensional calculations revealed spikes at spanwise locations corresponding to interfaces between the sides of the heaters and the substrate. This enhancement of heat transfer along the lateral edges of each heater is due to spanwise diffusion of energy from the warmer fluid passing over the heater to the cooler fluid in adjoining substrate regions. In the limit as $R_{sub} \to \infty$, isothermal conditions are approached in the spanwise, as well as the vertical, direction of the heated wall.

3.4.3 Protruding Heat Sources

Numerical simulations of free convection for protruding heat sources mounted to one wall of a vertical liquid-filled cavity have been performed by several investigators. Sathe and Joshi (1991, 1992) considered a single, protruding heat source mounted at the center of a conducting substrate and an isothermal cold plate for the opposing wall. Results were found to be independent of the Prandtl number over the range $10 < Pr < 1000$, whereas the effects of the substrate were found to increase with decreasing Rayleigh number and/or increasing R_{sub}. Additional calculations performed by Wroblewski and Joshi (1993, 1994) revealed a strong influence of substrate conduction for $R_{sub} > 10$ and the existence of nearly isothermal conditions at the hot wall for $Ra_{L_{h,z}} < 10^3$ and $R_{sub} = 575$. For heater aspect ratios $A_h \equiv L_{h,x}/L_{h,z} > 4$, predictions of the maximum heater temperature based on a two-dimensional approximation were within 10% of results based on a more accurate three-dimensional model. The amount by which the two-dimensional predictions exceeded the three-dimensional results increased with decreasing A_h.

Numerical simulations have also been performed for protruding heat sources mounted to one vertical wall of a rectangular cavity, while the opposing vertical wall was adiabatic and heat was extracted at horizontal top and bottom cold plates. Lee et al. (1987) considered a water-filled enclosure and, for a single heat source, found that heat transfer was enhanced by positioning the heater closer to the top cold plate. Calculations performed by Liu et al. (1987) for a 3×3 array of protruding sources revealed a thermally stratified core region, which caused average heater temperatures to increase with proximity to the top cold plate.

A two-dimensional simulation of flow and heat transfer associated with a vertical array of three protruding heat sources has been performed for representative chip/substrate conditions (Heindel et al., 1996a). Referring to Figure 3.1, fixed values of $B_{sub} = 0$, $B_h = 1$ mm, $L_{h,z} = S = 12.7$ mm, $P_z = 15.9$ mm, $t_{sub} = 3.18$ mm, $H = 95.3$ mm, and $H_1 = H_3 = 25.4$ mm were maintained. The fluid Prandtl number and thermal conductivity were fixed at $Pr = 25$ and $k_f = 0.063$ W/m · K, respectively, corresponding to FC-77, and the heater-to-fluid conductivity ratio was fixed at $R_h = 2350$, which corresponds to silicon chips (148 W/m · K). Calculations were performed for Rayleigh numbers and substrate-to-fluid conductivity ratios in the ranges $10^4 < Ra^*_{L_{h,z}} < 10^9$ and $23.5 < R_{sub} < 2350$ ($1.48 < k_{sub} < 148$ W/m · K), respectively. In addition, a thermal resistance between the chip and the substrate was prescribed, and variations in the range $0.003 < R''_{th,c} < 10$ cm^2 · K/W were considered. Although the two-dimensional approximation overpredicts chip temperatures by neglecting edge effects and thermal spreading in the spanwise direction, which are included in a three-dimensional model, it is known to predict correctly key trends and to provide reasonable accuracy over a significant range of conditions (Heindel et al., 1995a, c).

Dimensionless streamlines and isotherms predicted for a conductivity ratio of $R_{sub} = 575$ ($k_{sub} = 36$ W/m · K), a heater/substrate contact resistance of $R''_{th,c} = 0.003$ cm^2 · K/W, and the lower and upper limits of the Rayleigh number range ($Ra^*_{L_{h,z}} = 10^4, 10^9$) are plotted in Figure 3.13. The values of R_{sub} and $R''_{th,c}$ correspond to an alumina substrate in good thermal contact with a silicon chip.

For $Ra^*_{L_{h,z}} = 10^4$, Figure 3.13a, the flow is weak, with a centrally located stagnant core. The small thermal contact resistance and large substrate conductivity facilitate significant energy transfer through the substrate. The fluid is, therefore, heated along the entire vertical wall, establishing a nearly unicellular flow. The approxi-

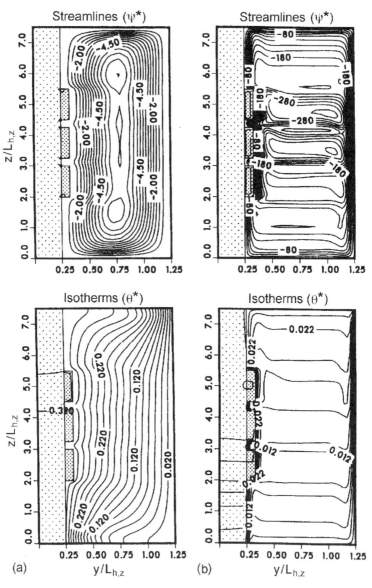

FIGURE 3.13 Dimensionless streamlines and isotherms for a vertical array of protruding heaters ($Pr = 25$, $R_h = 2350$, $R_{sub} = 575$, $R''_{th,c} = 0.003$ cm$^2 \cdot$ K/W): (a) $Ra^*_{L_{h,z}} = 10^4$ and (b) $Ra^*_{L_{h,z}} = 10^9$. Adapted from Heindel et al. (1996a).

mately vertical isotherms in the cavity indicate that energy transport is strongly influenced by conduction, while the paucity of isotherms in the substrate indicates nearly isothermal conditions. Apart from slight *bumps* in the streamlines and isotherms, which follow the protrusion contours along the heated wall, there is little difference between results for protruding (Fig. 3.13a) and flush-mounted (Fig. 3.10a) chips. Moreover, there is negligible fluid penetration between the protrusions.

As for flush-mounted heaters (Fig. 3.10b), with increasing $Ra^*_{L_{h,z}}$, the flow intensifies and becomes multicellular, while fluid in the core region becomes thermally stratified (Fig. 3.13b). For $Ra^*_{L_{h,z}} = 10^9$, conditions along the heated and cooled vertical walls are characterized by thin hydrodynamic and thermal boundary layers. There is some penetration of fluid between the protrusions, with large temperature gradients existing in the fluid.

The distribution of the normalized heat flux along the interface between the heater/substrate surfaces and the fluid is shown in Figure 3.14, where ζ_i measures the distance along the solid/fluid interface from the bottom to the top of the cavity. The distribution is highly nonuniform and strongly influenced by the Rayleigh number.

For $Ra^*_{L_{h,z}} = 10^4$, heat transfer by convection from the heater surface is weak relative to transfer by conduction through the substrate and thence by convection to the fluid. Accordingly, there is significant heat transfer to the fluid from the substrate

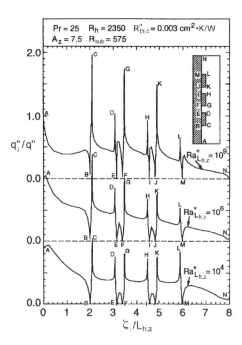

FIGURE 3.14 Effect of Rayleigh number on distribution of the dimensionless heat flux, q''_i/q'', along the fluid-heater/substrate interface for a vertical array of protruding heaters. Adapted from Heindel et al. (1996a).

below the array (region A–B). However, stagnant conditions near the corner of the protrusion (point B) provide a large local resistance to heat transfer by convection, and thermal spreading to lower regions of the substrate is favored. Accordingly, the heat flux decays with increasing ζ_i, approaching zero as point B is approached. Because the flow is stagnant in all of the corners, similar conditions characterize the approaches to points E, F, I, J, and M.

The large increase in heat flux along the bottom surface of the first source (region B–C) is attributed to *thinning* of the boundary layer, and a corresponding increase in the convection heat transfer coefficient, as fluid approaches and turns the corner at point C. This behavior characterizes the leading edges (C, G, K) of each heater, as well as the trailing edges (D, H, L). However, edge effects do not extend to central portions of the heaters (regions C–D, G–H, K–L), where heat transfer is strongly influenced by conduction in the fluid and the heat flux distribution is nearly uniform. Heat transfer at the trailing edges of each source is also influenced by streamwise diffusion, which is most significant for the uppermost heater (point L), because the temperature gradients that sustain diffusion are no longer attenuated by a downstream source. Despite stagnant conditions at the corners, a small amount of energy enters the fluid regions of the substrate between the heaters (regions E–F and I–J).

With increasing $Ra^*_{L_{h,z}}$, fluid circulation intensifies, reducing boundary layer thicknesses and increasing the amount of heat transfer from the protrusions to the fluid. Heat transfer from regions of the substrate between the heaters also increases because of increased flow penetration, which advects cooler, free-stream fluid into the gap. With the increased influence of convection from the protrusions, the heat flux distribution between the leading and trailing edges ceases to be nearly uniform and is, instead, characterized by a monotonic decay resulting from boundary layer growth from the leading edge. However, there is still a reversal of this decay because of subsequent thinning of the boundary layer as the trailing edge is approached.

Heindel et al. (1996a) also predicted the manner in which the dimensionless temperature, $\theta_i^* \equiv (T_i - T_c)/(q''L_{h,z}/k_f)$, varies along the interface between the substrate/protrusions and the fluid. Apart from highly localized minima at corners of the protrusions associated with boundary layer thinning (points C, D, G, H, K, and L), the temperature increases monotonically with increasing distance from the bottom of the cavity. Hence, despite conditions that favor thermal spreading in the substrate, a uniform interface temperature is not maintained.

Relative to the ultimate fate of the heat that is dissipated at the back of each heat source, it is noteworthy that typically less than 6% is transferred to the fluid from the top and bottom horizontal surfaces of a protrusion. Most of the heat is transferred to the fluid from the vertical surface of the protrusion and indirectly via conduction through the substrate. Respectively, these contributions increase and decrease with increasing Rayleigh number, as well as with increasing row number (Heindel et al., 1996a).

Although the thermal contact resistance was varied between 0.003 cm² · K/W (intimate contact) and 10 cm² · K/W (poor contact), calculations performed for $Ra^*_{L_{h,z}} = 10^9$, $R_h = 2350$, and $R_{sub} = 575$ revealed a negligible influence on results for $R''_{th,c} < 1$ cm² · K/W and only a small effect for $R''_{th,c} = 10$ cm² · K/W. The effect is one of marginally increasing the contribution of convection heat transfer from the vertical surface of a protrusion relative to transfer by conduction through the substrate. It follows that, for large Rayleigh numbers, thermal performance would only be affected by an extremely large contact resistance.

Heindel et al. (1996a) assessed the effect of substrate thermal conductivity for fixed values of $Ra^*_{L_{h,z}} = 10^9$ and $R''_{th,c} = 0.003 \text{ cm}^2 \cdot \text{K/W}$. Relative to the results of Figure 3.13b, an increase in R_{sub} from 575 to 2350 intensified fluid circulation, thinning the boundary layers and causing streamlines to follow the protrusion contours closely. For $R_{sub} = 23.5$, thermal spreading through the substrate is diminished, thereby reducing the amount of energy entering the fluid below the heater array and, hence, the strength of the circulation. With decreasing R_{sub}, the assumption of an isothermal substrate worsens, as vertical temperature gradients increase, particularly below row 3.

Irrespective of the value of R_{sub} over the range, $23.5 < R_{sub} < 2350$, the corresponding distribution of the interfacial heat flux is still characterized by local maxima and minima at protrusion corners shared with the fluid and substrate, respectively, as shown in Figure 3.14. However, with decreasing R_{sub}, the attendant reduction in thermal spreading reduces the amount of energy entering the fluid from the substrate below and above the heater array. For $R_{sub} = 23.5$, the corresponding heat flux distributions are characterized by monotonic decays in both the lower (from B to A) and the upper (from M to N) regions of the substrate.

The significant effect of R_{sub} on the dimensionless interface temperature, θ_i^*, is shown in Figure 3.15. Apart from the maxima and minima at protrusion corners associated with stagnant fluid (B, E, F, I, J, M) and boundary layer thinning (C, D, G, H, K, L), for $R_{sub} = 2350$, θ_i^* increases gradually from the bottom of the cavity to the trailing edge of row 1, beyond which it is approximately constant. To a first approximation, isothermal conditions may be assumed to exist over the entire vertical wall. However, with decreasing R_{sub} and thermal spreading, substrate temperatures decrease below and above the heater array, while the heater temperatures, as well as differences between these temperatures, increase. For $R_{sub} = 23.5$, the surface temperature distribution is highly nonuniform and there is a significant increase in the mean heater temperature from row 3 to row 1.

The conductivity ratio also has a significant effect on how energy transfer is partitioned among the four surfaces of a protruding heater. As shown in Figure 3.16, heat transfer from each of the surfaces exposed to fluid (the bottom, right, and top faces) increases with decreasing R_{sub}, while heat transfer to the substrate (from the left face) decreases. The monotonic increase in heat transfer from the right face with increasing row number for $R_{sub} \geq 235$ is due to a corresponding increase in the average convection coefficient associated with each row. However, for $R_{sub} \leq 47$, maximum heat transfer from the right face is associated with row 2. This behavior is attributed to diminished thermal spreading and the role that rows 1 and 3 begin to play as *guard heaters* for row 2, reducing heat transfer from this row to its adjoining substrate. In effect, elevated substrate temperatures maintained in proximity to rows 1 and 3 reduce substrate temperature gradients in proximity to heater 2, thereby reducing conduction in the substrate. In contrast, for $R_{sub} \geq 235$, conduction from

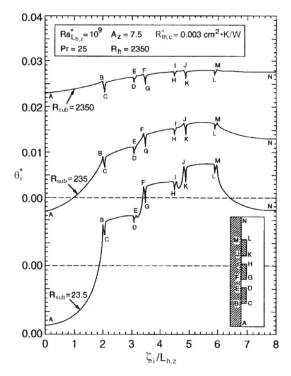

FIGURE 3.15 Effect of substrate-to-fluid thermal conductivity ratio on distribution of the dimensionless temperature, $\theta_i^* = (T_i - T_c)/(q''L_{h,z}/k_f)$, along the fluid-heater/substrate interface for a vertical array of protruding heaters. Adapted from Heindel et al. (1996a).

rows 1 and 3 to upper and lower regions of the substrate, respectively, is sufficiently effective to sustain large substrate temperature gradients and, hence, heat transfer, in proximity to row 2.

3.5 VERTICAL CAVITIES: EXPERIMENTAL RESULTS

3.5.1 Flush-Mounted Heat Sources

Experimental studies of heat transfer from a vertical array of strip heaters ($A_x \gg 1$) mounted flush with one wall of a liquid-filled cavity have been performed by Keyhani and co-workers. Alternating unheated and isoflux surfaces of equivalent height comprised one vertical wall of the cavity, while the opposing wall consisted of a chilled plate. In one case (Keyhani et al., 1988a), experiments were performed for ethylene glycol in a large-aspect-ratio cavity ($A_z = 16.5$) with 11 alternating unheated and flush-mounted sources. Transition to turbulent flow was observed for $9.3 \times 10^{11} < Ra_z^* < 1.9 \times 10^{12}$, where Ra_z^* is based on the distance z from the

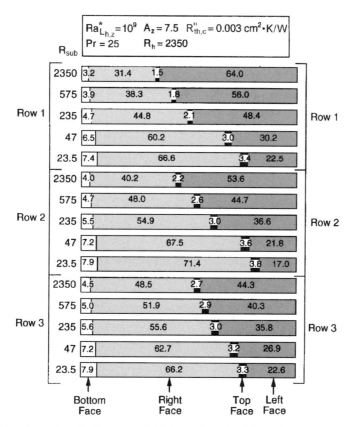

FIGURE 3.16 Effect of substrate-to-fluid thermal conductivity ratio on partition of energy transfer from the four surfaces of each heat source. Adapted from Heindel et al. (1996a).

floor of the cavity. Experiments performed in a smaller cavity ($A_z = 4.5$) with three alternating unheated and isoflux surfaces (Keyhani et al., 1988b) revealed multicellular flow patterns and thermal conditions that were independent of Prandtl number over the range $105 < Pr < 166$. For the top, middle, and bottom heaters (rows 1, 2, and 3, respectively), their heat transfer data were correlated by the following expressions:

$$\text{Row 1:} \quad \overline{Nu}_S = 0.195 Ra_S^{*0.234} \quad (3.9a)$$

$$\text{Row 2:} \quad \overline{Nu}_S = 0.184 Ra_S^{*0.253} \quad (3.9b)$$

$$\text{Row 3:} \quad \overline{Nu}_S = 0.292 Ra_S^{*0.259} \quad (3.9c)$$

where the characteristic length is the distance S between vertical walls. In addition, theoretical considerations revealed that the results could be applied to other liquids such as FC-77 ($Pr \approx 25$).

For three alternating adiabatic and isoflux surfaces, Prasad et al. (1990) performed numerical calculations to determine the effect of aspect ratio and Prandtl number on heat transfer. For $1 \leq A_z \leq 9$ and $10^6 \leq Ra_H^* \leq 10^{10}$, the average Nusselt number of each heater was independent of the Prandtl number over the range $25 \leq Pr \leq 166$ and reduced by approximately 5% for $Pr = 1$. The Nusselt number was independent of A_z for $A_z \geq 3$, and recommended correlations are of the form

Row 1: $\quad \overline{Nu}_H = 0.366 Ra_H^{*0.205}$ (3.10a)

Row 2: $\quad \overline{Nu}_H = 0.713 Ra_H^{*0.191}$ (3.10b)

Row 3: $\quad \overline{Nu}_H = 0.969 Ra_H^{*0.190}$ (3.10c)

where the characteristic length is the cavity height H. Like the cavity spacing S, use of H as a characteristic length masks the fact that heat transfer is more strongly influenced by the heater length $L_{h,z}$.

In contrast to the foregoing studies of two-dimensional conditions associated with strip heaters, Polentini et al. (1993) investigated three-dimensional conditions for a 3×3 array of discrete ($L_{h,z} = L_{h,x} = 12.7$ mm) heat sources mounted flush to one wall of a liquid-filled enclosure. The opposing wall was chilled, and experiments were performed with FC-77 and water. The aspect ratios were varied over the ranges $2.5 \leq A_z < 7.5$ and $1.5 \leq A_x \leq 4.5$ by fixing the cavity height ($H = 95.3$ mm) and width ($W = 57.2$ mm), while performing experiments for wall spacings of $S = 12.7$, 19.0, and 38.1 mm. The heat sources were located 6.35 mm from the side walls and at distances of $H_1 = H_3 = 25.4$ mm from the top and bottom of the cavity, and were closely spaced with a pitch of $P_x = P_z = 15.9$ mm. Although most experiments were performed with adiabatic top and bottom surfaces, the effect of using cold plates at both the top and the side walls was considered.

Heat transfer was independent of aspect ratio and Prandtl number and was invariant for heat sources within a row. However, confirming the predictions described in Section 3.4, maximum and minimum values of $\overline{Nu}_{L_{h,z}}$ corresponded to the bottom and top rows, respectively. Correcting for heat losses to the substrate, which ranged from approximately 5% to 35%, according to heat source location and operating conditions, heat transfer data were correlated by the following expressions:

Row 1: $\quad \overline{Nu}_{L_{h,z}} = 0.242 Ra_{L_{h,z}}^{0.276}$ (3.11a)

Row 2: $\quad \overline{Nu}_{L_{h,z}} = 0.308 Ra_{L_{h,z}}^{0.272}$ (3.11b)

Row 3: $\quad \overline{Nu}_{L_{h,z}} = 0.440 Ra_{L_{h,z}}^{0.270}$ (3.11c)

where $\overline{Nu}_{L_{h,z}} \equiv \overline{h} L_{h,z}/k_f = q_h L_{h,z}/k_f A_{s,h}(\overline{T}_h - T_c)$ and $Ra_{L_{h,z}} = g\beta(\overline{T}_h - T_c)L_{h,z}^3/\nu\alpha_f$ are based on the difference between temperatures of the heated and chilled surfaces. The Nusselt number is based on the heat rate q_h that is transferred directly from the heater surface to the fluid, and all properties are evaluated at the film temperature, $T_f = (\overline{T}_h + T_c)/2$. The average temperature of a heat source was based on two measurements, which were within 0.1°C in all cases. The data for all three rows were also correlated by a single expression of the form

$$\overline{Nu}_{z'} = 0.607 Ra_{z'}^{0.243} \qquad (3.12)$$

where $z' \equiv z - 20.65$ mm is the distance from the center of the heat source to a virtual origin located 4.75 mm below the bottom edge of the bottom row. The fictitious origin provides a measure of the effect of substrate conduction on advancing the onset of thermal boundary layer development. However, because the starting point depends on conjugate effects and, hence, on the conductivity ratio, $R_{sub} = k_{sub}/k_f$, as well as other factors, the correlation should be used with caution. The maximum deviations of Eqs. 3.11 and 3.12 from the data are 10% and 20%, respectively.

Polentini et al. (1993) also studied the effect of heater location on the wall by powering various combinations of one or two heater rows. If only one of the three rows was powered, the largest and smallest Nusselt numbers corresponded to locations closest to the bottom and top of the enclosure, respectively. Cooling the top, as well as the side wall, of the enclosure had a small effect on heat transfer, increasing Nusselt numbers by approximately 10% and 5% for the top and middle rows of the array, respectively. The effect on the bottom row was negligible.

Heindel et al. (1995b) also performed experiments for the 3×3 array and cavity geometry used by Polentini et al. (1993), but for a fixed spacing of $S = 12.7$ mm and, hence, aspect ratios of $A_z = 7.5$ and $A_x = 4.5$. Heat transfer data were again corrected for substrate conduction, and Nusselt numbers were based on energy transferred directly to the fluid. Heat transfer data obtained for FC-77 are shown in Figure 3.17, where different symbols are used to differentiate results for each column of the array. Column-to-column variations are, in fact, small, with slightly larger Nusselt numbers corresponding to the outer (1 and 3) columns. As determined theoretically, the Nusselt number decreases with decreasing row number, which is attributable to boundary layer growth and increasing fluid temperature in the streamwise (upward) direction. Correspondingly, for a fixed heat flux q'', the average heater temperature $\overline{T}_h$ increases with decreasing row number.

If it is assumed that laminar, thermal boundary layer development begins at the lower edge of row 3, the data for this row should be in agreement with Eq. 2.61a, which is expressed as

$$\overline{Nu}_{L_{h,z}} = 0.59 Ra_{L_{h,z}}^{0.25} \qquad (3.13)$$

Overprediction of the data by this correlation (Fig. 3.17) is due to substrate conduction and, hence, initiation of boundary layer growth below row 3. However, the Rayleigh number dependence of the data for each row tracks that of Eq. 3.13, prompting use of the same exponent (0.25) for the correlations of Figure 3.17. Similar results were obtained for water (Heindel et al., 1995b), with better agreement between Eq. 3.13 and the data of row 3 because of the diminished role of conjugate effects.

Fixing the Rayleigh number exponent at 0.25, Heindel et al. (1995b) correlated all of their data ($5 \times 10^4 < Ra_{L_{h,z}} < 10^8, 5 < Pr < 25$) by the following expressions:

116 NATURAL CONVECTION

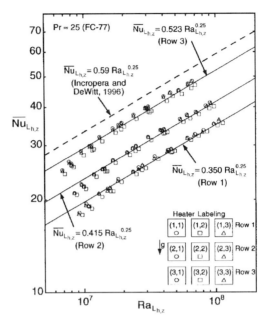

FIGURE 3.17 Effect of Rayleigh number on average Nusselt number for each row of a 3 × 3 array of heat sources mounted flush to the vertical surface of an enclosure, with an opposing chilled surface ($L_{h,z} = L_{h,x} = S = 12.7$ mm, $P_z = P_x = 15.9$ mm, $A_z = 7.5$). Adapted from Heindel et al. (1995b).

$$\text{Row 1:} \quad \overline{Nu}_{L_{h,z}} = 0.348 Ra_{L_{h,z}}^{0.25} \quad (3.14a)$$

$$\text{Row 2:} \quad \overline{Nu}_{L_{h,z}} = 0.415 Ra_{L_{h,z}}^{0.25} \quad (3.14b)$$

$$\text{Row 3:} \quad \overline{Nu}_{L_{h,z}} = 0.530 Ra_{L_{h,z}}^{0.25} \quad (3.14c)$$

The data and the correlation for row 1 are shown in Figure 3.18, along with plots of the correlations of Keyhani et al. (1988b), Eq. 3.9a; Prasad et al. (1990), Eq. 3.10a; and Polentini et al. (1993), Eq. 3.11a. To facilitate the comparison, the modified Rayleigh numbers Ra_S^* and Ra_H^*, used by Keyhani et al. and Prasad et al., respectively, were converted to $Ra_{L_{h,z}}$.

Although the experiments of Heindel et al. (1995b) and Polentini et al. (1993) were performed for equivalent geometries, slight differences in the two correlations are attributed to differences in procedures for obtaining the conjugate correction factors. The procedure used by Polentini et al. (1993) underestimated substrate losses for large values of k_{sub}/k_f, thereby yielding an overprediction of the FC-77 data by up to 15%. The slope of the correlation of Keyhani et al. (1988b) exceeds that of the other correlations, and the correlation itself underpredicts the data. At least in part, differences are attributed to the fact that Keyhani et al. (1988b) evaluated the properties of ethylene glycol, which depend strongly on temperature, at $T_c + 0.25(T_h - T_c)$,

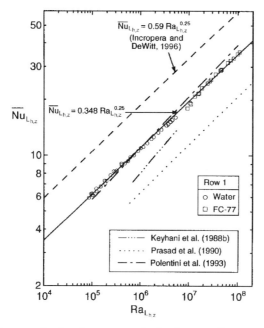

FIGURE 3.18 Heat transfer data for top row of a 3 × 3 array of heat sources mounted flush to the vertical surface of an enclosure, with an opposite chilled surface ($L_{h,z} = L_{h,x} = S = 12.7$ mm, $P_z = P_x = 15.9$ mm, $A_z = 7.5$), and comparison with selected correlations. Adapted from Heindel et al. (1995b).

which differs from the commonly used film temperature. Differences are also due to the fact that row 1 was located adjacent to the top adiabatic surface in the experiments of Keyhani et al. (1988b), which was not the case for the experiments of Heindel et al. (1995b) and, hence, the data presented in Figure 3.18. Underprediction of the data by the correlation of Prasad et al. (1990) is also attributed to the location of row 1 adjacent to the top surface, as well as to the use of a two-dimensional model in their numerical simulation. Agreement between the data and the correlation of Keyhani et al. (1988b) and Prasad et al. (1990) improves with increasing row number (Heindel et al., 1995b).

3.5.2 Protruding Heat Sources

Chen et al. (1988) performed experiments for an array of 10 strip heaters protruding from one vertical wall of a fluid-filled (ethylene glycol) cavity, with an isothermal (chilled) top surface and insulated bottom and opposing surfaces. For $Ra^*_{(S-B_h)} < 6.5 \times 10^6$, where the modified Rayleigh number is based on the separation between the tip of the protrusion and the opposing wall, the largest Nusselt number was associated with the lowermost heater in the array. In contrast, for larger values of $Ra^*_{(S-B_h)}$, the largest Nusselt number was associated with the uppermost heater.

The differences were attributed to the competing effects of increasing fluid temperature and velocity with increasing elevation along the vertical wall, as well as to the proximity of the top heater to the chilled surface. Keyhani et al. (1991) extended the work of Chen et al. (1988) by performing experiments for five protruding strip heaters. The spacing S between vertical walls was varied, providing aspect ratios in the range $3.67 < A_z < 12.22$ and ratios of the cavity width to the heater height in the range $1.5 < (S/B_h) < 5.0$. Combining their data with those of Chen et al. (1988), Keyhani et al. (1991) obtained the following correlation for the Nusselt number associated with up to 10 heaters in an array:

$$\overline{Nu}_z = 0.296 Ra_z^{*0.223} A_z^{-0.53} \qquad (3.15)$$

where the characteristic length is the distance from the bottom of the cavity to the center of each heat source. The correlation also corresponds to data for which $8 < L_{h,z} < 15$ mm, $62 < Pr < 110$, and $8.7 \times 10^4 < Ra_z^* < 8.7 \times 10^{10}$. However, in view of the weak effect of the Prandtl number observed in other studies, the correlation should be applicable for much smaller values of Pr.

Joshi et al. (1990) performed experiments for a 3×3 heat source array mounted to one vertical wall of a cavity, with an insulated opposing wall and chilled top and bottom surfaces. Three dielectric liquids were considered, FC-75, FC-43, and FC-71, which have Prandtl numbers of approximately 25, 43, and 1400, respectively. For a fixed heater power, FC-75 yielded the smallest heater temperatures. Because it also has the largest boiling point (102°C at atmospheric pressure), it was concluded to be the most suitable fluid for cooling by single-phase convection.

3.6 HORIZONTAL AND INCLINED CAVITIES

In their experimental study of heat transfer from a 3×3 array of flush-mounted 12.7 mm $\times$ 12.7 mm heaters on one wall of a cavity, Polentini et al. (1993) also considered the effect of cavity orientation. The cavity was filled with FC-77 ($Pr \approx 25$), and Rayleigh numbers, $Ra_{L_{h,z}}$, ranged from approximately 10^7 to 10^8. The spacing S between the heated and the opposing (chilled) surfaces was varied from 12.7 to 38.1 mm, providing aspect ratios in the range $2.5 < A_z < 7.5$. As in their study of the vertical orientation (Polentini et al., 1993), the heat sources were located 6.35 mm from the side walls and at distances of $H_1 = H_3 = 25.4$ mm from the ends of the cavity. The spacings between chips corresponded to $P_x = P_z = 15.9$ mm.

Referring to Figure 3.1, Polentini et al. (1993) performed experiments for cavity inclinations corresponding to $\phi = 0°, 30°, 45°, 60°$, and $90°$, where ϕ is the angle formed by the orientation of the z axis and a fixed direction that is antiparallel to the gravitational vector. Hence, for the horizontal orientation ($\phi = 90°$), the outward normal from each heater surface opposes the gravitational vector, creating an unstable thermal condition. For small angles of inclination (orientations close to vertical), flow and heat transfer are strongly influenced by the component of gravity that is parallel to the heater surfaces ($g \cos \phi$). However, as $\phi \to 90°$, the normal component ($g \sin \phi$) becomes dominant.

For $\phi = 30°$ and $Ra_{L_{h,z}} \approx 10^7$, the boundary layer was turbulent near the trailing edge of the top row (row 1) of heaters, and plumes of warm fluid were observed to ascend vertically and intermittently from this region. The region was also observed to expand in the negative z direction with increasing ϕ and/or $Ra_{L_{h,z}}$, eventually encompassing row 3 for $\phi = 60°$ and $Ra_{L_{h,z}} \approx 4 \times 10^7$. For $\phi = 90°$ (heating from below), flow was characterized by turbulent plumes ascending from all of the heaters to the cold plate, with the return flow descending principally to regions outside the footprint of the array and subsequently feeding upflow from the array.

For $\phi = 30°$, $45°$, and $60°$, Nusselt numbers, $\overline{Nu}_{L_{h,z}} = q_h L_{h,z} / k_f A_h (T_h - T_c)$, were generally largest for row 3, with the extent and nature of row-to-row variations depending on the values of ϕ and $Ra_{L_{h,z}}$ (Polentini et al., 1993). For the horizontal cavity ($\phi = 90°$), Nusselt numbers were approximately equivalent for all heaters of the array.

The extent to which heat transfer is enhanced by inclining the cavity is shown in Figure 3.19 for each of the three rows. For each angle, enhancement is largest and smallest for rows 1 and 3, respectively. This behavior is attributed to the fact that the largest row-to-row variations correspond to the vertical orientation (the reference

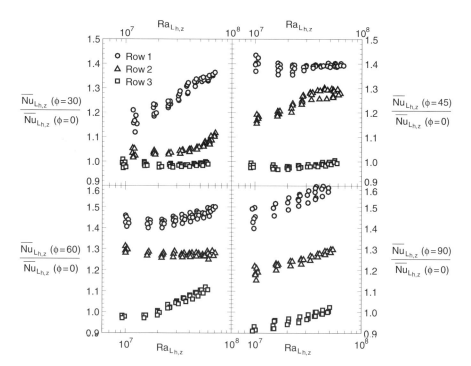

FIGURE 3.19 Effect of row location and Rayleigh number on ratio of Nusselt number for $\phi > 0$ to that for $\phi = 0$. Adapted from Polentini et al. (1993).

condition), for which the smallest and largest Nusselt numbers are associated with rows 1 and 3, respectively. The general trend, which is one of increasing heat transfer and decreasing row-to-row differences with increasing inclination, is attributed to downward propagation and amplification of turbulence with increasing ϕ. The largest enhancement is approximately 60% and is achieved in row 1 for the horizontal orientation. However, for all orientations, negligible or negative enhancements are associated with row 3. This result is due to the thin boundary layer and, hence, large Nusselt numbers associated with the row for the vertical orientation. Nevertheless, in view of the negligible row-to-row differences and overall heat transfer, the horizontal orientation would be best suited for electronic cooling.

For the vertical orientation, Polentini et al. (1993) correlated the average Nusselt number for each row by Eq. 3.12, where $z' \equiv z - 20.65$ mm is the distance from a virtual origin to the center of the row. Similar correlations were developed for the inclined cavities and are of the form

$$\phi = 30°: \quad \overline{Nu}_{z'} = 0.407 Ra_{z'}^{0.272} \qquad (3.16)$$

$$\phi = 45°: \quad \overline{Nu}_{z'} = 0.281 Ra_{z'}^{0.296} \qquad (3.17)$$

$$\phi = 60°: \quad \overline{Nu}_{z'} = 0.318 Ra_{z'}^{0.292} \qquad (3.18)$$

$$\phi = 90°: \quad \overline{Nu}_{z'} = 0.168 Ra_{z'}^{0.325} \qquad (3.19)$$

Although use of z' as the length scale for the horizontal cavity lacks a physical basis, the close proximity of the Rayleigh number exponent (0.325) to 1/3 renders the corresponding convection coefficient approximately independent of length scale.

Example 3.1

A 3×3 array of chips, each 15 mm $\times$ 15 mm on a side, is mounted to a substrate that forms one of the vertical walls of a rectangular cavity. The opposing wall is a cold plate whose temperature is maintained at $T_c = 15°C$, and the cavity is filled with FC-77. If the chip temperature may not exceed $T_h = 85°C$, what is the maximum allowable heat flux that may be dissipated by each chip?

Solution

Known: Maximum operating temperature of nine chips mounted to the vertical wall of a rectangular cavity filled with FC-77. Temperature of the opposing chill plate.

Find: Maximum allowable heat flux for each chip.

Assumptions:

1. Steady-state conditions
2. Each chip is isothermal ($\bar{T}_h = T_h$)

3. Negligible effect of chip protrusion (flush-mounted conditions)
4. Negligible effect of cavity height, H, wall spacing, S, and chip pitches, P_x and P_z, on heat transfer
5. Negligible variations from chip to chip in a row
6. Negligible effect of substrate conduction

Properties: FC-77 ($T_f = 50°C$): From Appendix B: $\rho = 1716$ kg/m^3, $c_{p,f} = 1087$ J/kg·K, $k_f = 0.061$ W/m·K, $\alpha_f = k_f/\rho c_{p,f} = 3.27 \times 10^{-8}$ m^2/s, $\mu = 9.25 \times 10^{-4}$ kg·s/m, $\nu = \mu/\rho = 5.39 \times 10^{-7}$ m^2/s, $Pr = \nu/\alpha_f = 16.5$, and $\beta = 0.00143$ K^{-1}.

Analysis: Subject to the foregoing assumptions, chip heat fluxes will vary from row to row and, as a first approximation, may be computed from Eq. 3.14. Each chip is characterized by the same Rayleigh number,

$$Ra_{L_{h,z}} = \frac{g\beta(\bar{T}_h - T_c)L_{h,z}^3}{\nu\alpha_f}$$

$$= \frac{9.8 \text{ m/s}^2 (0.00143) \text{ K}^{-1}(70 \text{ K})(0.015 \text{ m})^3}{5.39 \times 10^{-7} \text{ m}^2/\text{s} (3.27 \times 10^{-8} \text{ m}^2/\text{s})} = 1.88 \times 10^8$$

Referring to Figure 3.2, it follows that

Row 1: $\overline{Nu}_{L_{h,z}} = 0.348 Ra_{L_{h,z}}^{0.25} = 40.7$

$$\bar{h} = \overline{Nu}_{L_{h,z}} \frac{k_f}{L_{h,z}} = \frac{40.7(0.061 \text{ W/m} \cdot \text{K})}{0.015 \text{ m}} = 166 \text{ W/m}^2 \cdot \text{K}$$

$$q'' = \bar{h}(\bar{T}_h - T_c) = 166 \text{ W/m}^2 \cdot \text{K} (85 - 15) \text{ K}$$
$$= 11{,}600 \text{ W/m}^2 = 1.16 \text{ W/cm}^2$$

Row 2: $\overline{Nu}_{L_{h,z}} = 0.415 Ra_{L_{h,z}}^{0.25} = 48.6$

$$\bar{h} = \overline{Nu}_{L_{h,z}} \frac{k_f}{L_{h,z}} = \frac{48.6(0.061 \text{ W/m} \cdot \text{K})}{0.015 \text{ m}} = 198 \text{ W/m}^2 \cdot \text{K}$$

$$q'' = \bar{h}(\bar{T}_h - T_c) = 198 \text{ W/m}^2 \cdot \text{K} (85 - 15) \text{ K}$$
$$= 13{,}800 \text{ W/m}^2 = 1.38 \text{ W/cm}^2$$

Row 3: $\overline{Nu}_{L_{h,z}} = 0.530 Ra_{L_{h,z}}^{0.25} = 62.1$

$$\bar{h} = \overline{Nu}_{L_{h,z}} \frac{k_f}{L_{h,z}} = \frac{62.1(0.061 \text{ W/m} \cdot \text{K})}{0.015 \text{ m}} = 252 \text{ W/m}^2 \cdot \text{K}$$

$$q'' = \bar{h}(\bar{T}_h - T_c) = 252 \text{ W/m}^2 \cdot \text{K} \, (85 - 15) \text{ K}$$
$$= 17{,}700 \text{ W/m}^2 = 1.77 \text{ W/cm}^2$$

Comments:

1. To the extent that substrate conduction occurs, the foregoing results represent conservative estimates. Hence, if the predicted values of q'' were maintained, the actual values of $\bar{T}_h$ would be less than 85°C; correspondingly, larger values of q'' could be dissipated if $\bar{T}_h$ were maintained at 85°C.
2. With $q = q''A_s$, the heat rates corresponding to the foregoing predictions are 2.61, 3.11, and 3.98 W for rows 1, 2, and 3, respectively.
3. It may be undesirable to have different constraints on the maximum allowable heat rate for the different rows of chips. To remove this constraint, the cavity could be oriented in the horizontal configuration. From Eq. 3.19, the convection coefficient may be expressed as

$$\bar{h} = \frac{\overline{Nu_{z'}} k_f}{z'} = \frac{0.168 Ra_{z'}^{0.325} k_f}{z'} = 0.168 \left[\frac{g\beta(\bar{T}_h - T_c)}{\nu \alpha_f} \right]^{0.325} k_f \frac{(z')^{0.975}}{z'}$$

If it is assumed that $[(z')^{0.975}/z'] = 1$, the average convection coefficient for each row is

$$\bar{h} \approx 0.168 \left(\frac{9.8 \text{ m/s}^2 \, (0.00143 \text{ K}^{-1}) 70 \text{ K}}{5.39 \times 10^{-7} \text{ m}^2/\text{s} \, (3.27 \times 10^{-8} \text{ m}^2/\text{s})} \right)^{0.325} (0.061 \text{ W/m} \cdot \text{K})$$
$$= 301 \text{ W/m}^2 \cdot \text{K}$$

Each chip would then dissipate a heat flux and heat rate of approximately

$$q'' = \bar{h}(\bar{T}_h - T_c) = 301 \text{ W/m}^2 \cdot \text{K} (70 \text{ K}) = 21{,}100 \text{ W/m}^2 = 2.11 \text{ W/cm}^2$$

and

$$q = q''A_s = 2.11 \text{ W/cm}^2 (1.5 \text{ cm})^2 = 4.74 \text{ W}$$

3.7 SUMMARY

Cooling of electronic components by immersion in a dielectric liquid is conveniently implemented if the components are mounted to one wall of a liquid-filled rectangular cavity. A steady-state condition is maintained by transferring heat from the liquid to one or more of the remaining walls that are cooled by an external agent.

In a vertical rectangular cavity, an array of heat-dissipating components is mounted to one of the vertical walls, while, typically, the opposing wall is chilled.

With increasing Rayleigh number, ascending and descending boundary layers begin to develop along the heated and cooled walls, respectively, while multiple flow cells develop in the core of the cavity between the walls. The boundary layer thicknesses decrease with increasing Rayleigh number, while thermal conditions within the core become stratified (temperature increases with increasing vertical distance z). Because the boundary layer thickness on the heated wall and the temperature of the core region increase with increasing z, Nusselt numbers associated with heat sources in a vertical array decrease with increasing z.

The ratio of the substrate-to-fluid thermal conductivity, $R_{sub} = k_s/k_f$, affects the manner in which heat is transferred from the wall-mounted sources to the fluid. For small values of R_{sub}, most of the heat transfer occurs by convection at the heater/fluid interface, and there are pronounced differences between temperatures at the heater and substrate surfaces. For equivalent power dissipation in each heater of a vertical array, the heater temperatures would significantly exceed substrate temperatures and would increase with increasing elevation. However, with increasing R_{sub}, more of the energy dissipated by the sources is transferred first by conduction to the substrate and then by convection from the substrate to the fluid. An attendant effect is to decrease and increase heater and substrate temperatures, respectively, thereby diminishing the discrete nature of the heat sources. For $R_{sub} \approx 1000$, discrete thermal behavior is barely discernible and conditions resemble a differentially heated cavity with uniform hot and cold temperatures at the heated and chilled walls, respectively.

Because of boundary layer development in the vertical direction, the average Nusselt number associated with a heat source in a vertical array decreases with increasing elevation. For an array with three rows, Nusselt numbers may be estimated from the empirical correlations given by Eq. 3.14. The Rayleigh number exponent of 0.25 implies the existence of laminar flow for the conditions associated with the correlation. If there are more than three rows in the array, Eq. 3.12 may be used *as a first approximation* to compute heat transfer for the upper rows. Although the correlations may be used over a wide range of Rayleigh numbers, Prandtl numbers, and aspect ratios, care must be taken in using them much beyond the values of R_{sub} and $(P/L_h)_{x \text{ or } z}$ for which they were obtained. Both the conductivity ratio and the source pitch affect substrate conduction and thermal boundary layer development.

Modern numerical procedures provide powerful tools for modeling flow and heat transfer in a discretely heated cavity, including conjugate effects related to substrate conduction. Two-dimensional models can be used to a good approximation for small-to-moderate values of R_{sub} and moderate-to-large values of the heater aspect ratio A_h. For large values of R_{sub} and small values of A_h, however, spanwise edge effects become significant and a three-dimensional model should be used.

Rectangular cavities may be inclined, and, in terms of achieving uniform thermal conditions among the heat sources and Nusselt numbers comparable to the maximum achievable for a prescribed Rayleigh number, the horizontal orientation is preferred. In this orientation, the heated wall would be below the chilled wall, creating thermally unstable conditions and turbulent flow in the fluid-filled cavity.

CHAPTER 4

CHANNEL FLOWS

4.1 INTRODUCTION

By virtue of its simplicity and adaptability to different packaging schemes, channel flow is well suited for electronic cooling. Electronic devices could, for example, be attached to circuit boards that comprise one or more walls of a rectangular channel. Heat would be transferred by convection from the devices to a coolant that is circulated through the channel and subsequently routed to an external heat sink. One application could involve a planar, multichip array mounted to a substrate forming one wall of a channel through which a dielectric liquid flows. Cooling may also be implemented without direct exposure of the electronic devices to the coolant. In this case, channels could be machined in a substrate, circuit board, or cold plate to which the electronic devices are attached. A special case involves *microchannels* machined within an electronic chip.

In this chapter, emphasis is placed on flow and thermal conditions in rectangular channels. Consideration is given to low-Reynolds-number, mixed-convection flows, as well as to large-Reynolds-number, forced-convection flows for which chip heat fluxes exceeding 100 W/cm^2 can be achieved with a dielectric liquid.

4.2 FORCED CONVECTION FOR DISCRETE, FLUSH-MOUNTED HEAT SOURCES

4.2.1 Theoretical Considerations

The system of interest consists of an array of chips mounted flush to one wall of a rectangular channel (Fig. 4.1). If buoyancy effects are significant, as in mixed convection, flow and thermal conditions would differ according to the orientation of the

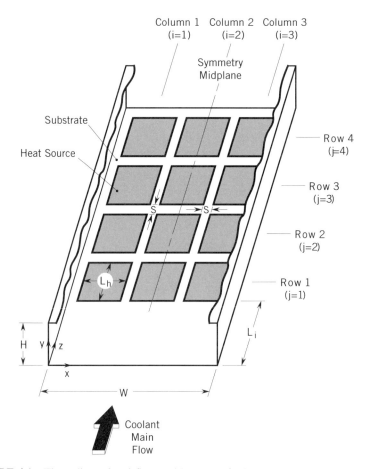

FIGURE 4.1 Three-dimensional flow and heat transfer in a rectangular channel with discrete sources mounted flush in a nonconducting wall.

channel with respect to the gravitational field (horizontal, vertical, or inclined) and the particular wall to which the chips are mounted. However, if buoyancy effects are negligible, as in forced convection, such distinctions are irrelevant. Protrusion of the chips into the flow may also be inconsequential, if the protrusions are slight and have a negligible effect on the flow field.

Fluid is presumed to enter the channel at $z = 0$ with a uniform temperature and velocity. The first row of the array may be preceded by an inlet region of length L_i, over which a laminar or turbulent velocity profile develops. Hence, the velocity profile at $z = L_i$ may be characterized as fully developed ($L_i \geq L_{\text{fd},h}$), developing ($0 < L_i < L_{\text{fd},h}$), or uniform ($L_i = 0$). If the aspect ratio, $A \equiv W/H$, of the channel is large ($A \gg 1$), spanwise (x) variations may be neglected, and, for fully developed flow, the velocity profile could be expressed exclusively in terms of the y coordinate.

Thermal conditions are influenced by the existence of multiple length scales, which can complicate attempts to generalize results. Relevant length scales are associated with the length L_h and spacing S of the heaters, as well as by the channel hydraulic diameter, $D_h = 2(W \cdot H)/(W + H)$. In general, temperature fields in the flow are three dimensional, because intermittent heating is associated with the spanwise and streamwise directions. Moreover, for the array, heat transfer from one chip may be influenced by its neighbors, as when a chip is located in the thermal boundary layer generated by upstream heat sources.

Heat transfer from the chips may also be influenced by the thermal conductivity of the substrate, k_s. Specifically, if k_s is large, the resistance to heat transfer by convection from a chip to the fluid may not be much less than that resulting from conduction from the chip to its substrate. In such cases, chip thermal conditions are determined by both convection to the fluid and conduction to the substrate. Moreover, the two processes are coupled, creating what is known as a *conjugate* heat transfer problem. The implication is that the processes are mutually dependent and, hence, cannot be treated separately.

Three-Dimensional Behavior with Negligible Conjugate Heat Transfer

Mahaney et al. (1989) considered evolution of the three-dimensional temperature field for the channel of Figure 4.1, subject to the assumptions of laminar flow, negligible buoyancy forces, constant properties, a fully developed velocity profile, negligible longitudinal conduction, and negligible substrate conduction. With $u = v = 0$ for a fully developed velocity profile and $\partial^2 T/\partial z^2 = 0$ for negligible longitudinal conduction, Eq. 2.40 reduces to

$$\rho c_{p,f} w \frac{\partial T}{\partial z} = k_f \left(\frac{\partial^2 T}{\partial x^2} + \frac{\partial^2 T}{\partial y^2} \right) \tag{4.1}$$

where the longitudinal velocity w is independent of z. The equation was solved numerically under conditions for which each heater was assumed to dissipate the same power Q and to be characterized by spanwise isothermal conditions, while all other surfaces were assumed to be adiabatic.

The local Nusselt number at any location on the surface of a heater in column i and row j was defined as

$$Nu_{L(i,j)}(x,z) = \frac{q''(x,z)}{T_{h(i,j)}(z) - T_m(z)} \frac{L_h}{k_f} \tag{4.2}$$

where q'' is the local heat flux, $T_{h(i,j)}(z)$ is the spanwise uniform temperature of the heater, and $T_m(z)$ is the corresponding mixed mean temperature of the fluid. An average Nusselt number for the three heaters in a row was defined as

$$\overline{Nu}_{L,j} = \frac{Q/A_h}{T_{h,j} - T_{m,j}} \frac{L_h}{k_f} \tag{4.3}$$

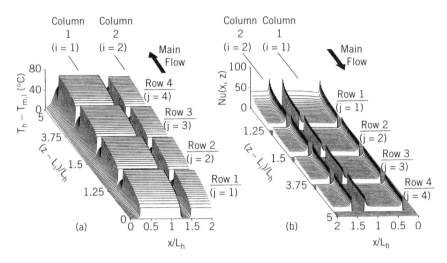

FIGURE 4.2 Two-dimensional bottom surface (*a*) temperature and (*b*) Nusselt number distributions corresponding to an array of flush-mounted heat sources for laminar, forced convection in a rectangular channel ($Re_{D_h} = 500$, $Pr = 7$). Adapted from Mahaney et al. (1989).

where $\overline{T}_{h,j}$ is the average temperature of the three heaters in row j and $T_{m,j}$ is the mixed mean fluid temperature at the leading edge of the row. Although calculations were performed for specific multichip module conditions of $A = 4$, $L_h = H = 4S = 12.7$ mm, and $D_h = 20.3$ mm, the results are applicable to any geometrically similar arrangement. The Reynolds and Prandtl number ranges of the calculations were $200 \leq Re_{D_h} \leq 4000$ ($125 \leq Re_{L_h} \leq 2500$) and $0.7 \leq Pr \leq 30$.

Surface temperature and Nusselt number distributions for $Re_{D_h} = 500$ and $Pr = 7$ are shown in Figure 4.2, where symmetry about the vertical midplane at $x/L_h = 2$ allows representation of results for one half of the channel. From Figure 4.2*a*, the temperature on each heater is seen to increase monotonically in the streamwise direction, whereas the temperature decreases significantly at intervening locations on the substrate surface. In Figure 4.2*b*, coordinate directions have been reversed to illustrate better the streamwise decay in the Nusselt number that occurs on the heaters in a particular row, as well as from row to row. The decay is due to progressive development of the thermal boundary layer. The major implication of these results is that, subject to the limitations of forced convection, equivalent power dissipation in each heater will yield higher temperatures for heaters in downstream rows. The spanwise distribution of the heater Nusselt number is nearly uniform, except near the edges where there is a substantial enhancement resulting from heat exchange with cooler fluid in the adjoining, unheated regions. From Eq. 4.2 and specification of an adiabatic surface condition, the Nusselt number is zero at all unheated (substrate) portions of the surface.

Two-Dimensional Approximation with Conjugate Heat Transfer If the aspect ratio of the channel is large ($A \gg 1$), spanwise (x) variations in the velocity field may be neglected. If spanwise variations in the temperature field are also neglected, a two-dimensional model may be used to determine thermal and flow conditions in the channel. With the numerical simplifications afforded by a two-dimensional approximation, substrate conduction may readily be considered, thereby simulating the coupled effects of heat dissipation by convection to the fluid and conduction to the substrate. For two-dimensional, steady-state conduction in a substrate with no thermal energy generation and a constant thermal conductivity, Eq. 2.1 yields an energy equation of the form

$$\frac{\partial^2 T_s}{\partial y^2} + \frac{\partial^2 T_s}{\partial z^2} = 0 \tag{4.4}$$

where the subscript s distinguishes conditions in the substrate from those in the fluid.

The physical implications of coupling between substrate conduction and convection to a fluid have been delineated by Ramadhyani et al. (1985) and Moffatt et al. (1986) for laminar and turbulent flow, respectively. The numerical simulations considered isothermal, discrete heat sources embedded in a conducting substrate (Fig. 4.3). In principle, a two-dimensional model implies infinite extent in the spanwise (x) direction; in practice, the approximation would be reasonable for aspect ratios as low as $A \approx 2$. The simulations also assumed fully developed flow throughout the channel, rendering the longitudinal velocity component w independent of z, and adiabatic conditions for the ends and bottom of the substrate. Hence, energy that is transferred by conduction from a source to the substrate must leave the substrate by convection from its surface to the coolant. In the analysis, the source-to-substrate thermal contact resistance is neglected.

The extent to which heat transfer from the sources is more strongly coupled to the fluid or the substrate depends on the nature of the flow, the fluid and substrate thermophysical properties, and the spacing between sources. The relative contribution of convection to total heat transfer from a source increases with increasing Reynolds and Prandtl numbers and is larger for turbulent than for laminar flow. Conversely, the

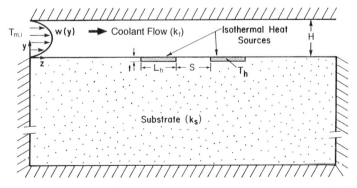

FIGURE 4.3 Two-dimensional flow and heat transfer in a rectangular channel with discrete sources mounted flush in a conducting wall. Adapted from Ramadhyani et al. (1985).

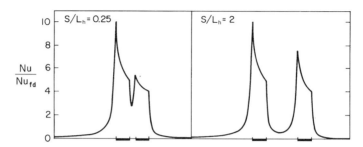

FIGURE 4.4 Variation of normalized local Nusselt number along the substrate and heater surfaces for laminar flow with adjoining heaters mounted flush to one wall of a parallel-plate channel ($Pe_{D_h} = 10^4$, $k_s/k_f = 10$).

relative contribution of substrate conduction increases with increasing values of the conductivity ratio, k_s/k_f, and the dimensionless spacing, S/L_h.

For laminar flow in the channel of Figure 4.3, Ramadhyani et al. (1985) solved Eqs. 4.1 and 4.4 to determine the temperature fields in the channel and the substrate, from which the distribution of the local Nusselt number Nu could be computed, where

$$Nu = \frac{hD_h}{k_f} = \frac{q''(z)}{T_h - T_{m,i}} \times \frac{2H}{k_f} \qquad (4.5)$$

The effect of the heater spacing on the Nusselt number distribution is shown in Figure 4.4, where $Nu_{fd} = 4.86$ is the Nusselt number for fully developed flow and heat transfer in a channel with one wall isothermal and the other adiabatic (Shah and London, 1978). If there were no energy transfer from the source to the substrate, the Nusselt number would be zero at all locations on the substrate surface and would be characterized by a sharp (spike) increase at the leading edge of each heater. However, with energy transfer by conduction from the source into the substrate, there is also energy transfer by convection from the substrate to the fluid and, hence, the nonzero values of the Nusselt number on the substrate surface. The Nusselt number distribution is still sharply peaked at the leading edge of the first heater and decays with thermal boundary layer development, as the trailing edge is approached.

For closely spaced heaters ($S/L_h = 0.25$), the Nusselt number at the leading edge of the second heat source is greatly reduced because of preheating of the fluid by the upstream source and substrate. However, with increasing S/L_h, the preheating effect is diminished and the Nusselt number at the leading edge of the second source recovers, approaching that of the first peak. Values of the Nusselt number on the substrate surfaces upstream and downstream of, as well as between, the heaters decrease with increasing Péclet number, $Pe_{D_h} = Re_{D_h} Pr$, and decreasing thermal conductivity ratio, k_s/k_f. Hence, the ratio of heat transfer by convection from the substrate to the total amount of heat transfer from the sources decreases with increasing Pe_{D_h} and decreasing k_s/k_f, ranging from approximately 0.4 for $Pe_{D_h} = 10^4$ and $k_s/k_f = 10$

to approximately 0.01 for $Pe_{D_h} = 10^6$ and $k_s/k_f = 0.5$ (Ramadhyani et al., 1985). For equivalent values of Pe_{D_h} and k_s/k_f, the ratio is also smaller for turbulent than for laminar flow (Moffatt et al., 1986).

For turbulent flow, Eq. 4.1 was modified to account for the effects of turbulent mixing (Moffatt, 1985), and parametric calculations were performed for a single heat source ($S \to \infty$) in a conducting substrate. The calculations were performed over the range, $4000 < Re_{D_h} < 10^6$, $0.6 < Pr < 25$, $0.1 < L_h/H < 10$, and $0.5 < k_s/k_f < 50$. The predictions were correlated to within ±5% by the expression

$$\overline{Nu}_L = 0.037 Re_{D_h}^{0.75} Pr^{0.35} \left(\frac{L_h}{H}\right)^{0.85} \left(\frac{k_f}{k_s}\right)^{0.02} \tag{4.6a}$$

where $\overline{Nu}_L = \overline{h}L/k_f$ and $Re_{D_h} = w_m D_h/\nu$. A much weaker dependence on (L_h/H) is obtained if the Reynolds number is based on the heater length, $Re_{L_h} = w_m L_h/\nu$, from which it follows that

$$\overline{Nu}_L = 0.062 Re_{L_h}^{0.75} Pr^{0.35} \left(\frac{L_h}{H}\right)^{0.1} \left(\frac{k_f}{k_s}\right)^{0.02} \tag{4.6b}$$

The weak dependence on (k_f/k_s) implies negligible substrate conduction and is a consequence of the increased levels of forced convection associated with turbulent flow.

Two-dimensional, turbulent flow and heat transfer from discrete, flush-mounted heat sources in a liquid-cooled channel has also been modeled by Xu et al. (1998). Conjugate effects were neglected, but an improved, two-equation (k–ε) turbulence model was used in the simulation. For a single heat source, trends were consistent with those of Moffatt et al. (1986), although Nusselt number predictions were larger by approximately 30%. A slightly larger, albeit still weak dependence, $\overline{Nu}_L \sim (L_h/H)^{0.14}$, was predicted for the effect of channel height.

4.2.2 Experimental Results

Early experimental studies of forced-convection heat transfer from flush-mounted sources considered silicone oil and R-113 in parallel flow over small, chiplike heaters (Baker, 1972, 1973). Single-phase forced-convection coefficients increased significantly with decreasing surface area from 200 to 1 mm^2 and exceeded results associated with nucleate pool boiling. For the smallest heaters, the data also exceeded predictions based on two-dimensional boundary layer theory, and differences were attributed to the three-dimensionality of the flow associated with small heating elements. In a related study of heat transfer from a single 0.25-mm-long-by-2.0-mm-wide source mounted flush with one wall of a rectangular duct (Samant and Simon, 1986), results obtained for FC-72 and R-113 in the Reynolds number range $7000 \leq Re_H \leq 1.5 \times 10^5$ were correlated by the expression

$$\overline{Nu}_H = 0.47 Re_H^{0.58} Pr^{1/2} \tag{4.7}$$

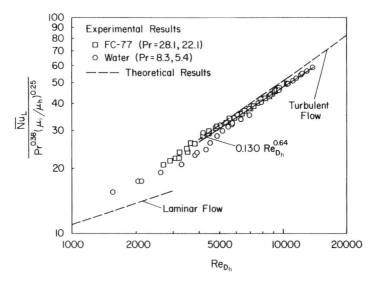

FIGURE 4.5 Comparison of measured and predicted results for a square ($L_h = 12.7$ mm) heat source mounted flush in one wall of a rectangular channel ($D_h = 19.3$ mm). Adapted from Incropera et al. (1986).

In R-113, a peak heat flux of 200 W/cm^2 was achieved without nucleation. Although it is suitable for characterizing hydrodynamic conditions and, hence, as a length scale in the Reynolds number, the channel height is inappropriate for representing chip thermal conditions and should not be used as a length scale in the Nusselt number.

Experimental results obtained by Incropera et al. (1986) for a single, flush-mounted, 12.7 mm × 12.7 mm heater are plotted in Figure 4.5, along with numerical predictions based on two-dimensional, conjugate, forced-convection models for laminar and turbulent flow (Moffatt, 1985). With use of a Prandtl number exponent of 0.38 and a ratio of viscosities evaluated at the channel inlet and heater surface temperatures, the FC-77 and water data are in excellent agreement, and collectively both data sets are correlated to within ±6% by the following equation for $5000 < Re_{D_h} < 14{,}000$:

$$\overline{Nu}_L = 0.13 Re_{D_h}^{0.64} Pr^{0.38} \left(\frac{\mu_i}{\mu_h}\right)^{0.25} \tag{4.8}$$

For $Re_{D_h} > 5000$, the data agree well with predictions based on the two-dimensional, turbulent-flow model, but, for $Re_{D_h} < 3000$, the data are significantly underpredicted by the laminar model. This discrepancy was attributed to the influence of buoyancy-induced flows on the low-Reynolds-number data and the neglect of buoyancy effects in the forced-convection model. Additional experiments were performed by Maddox and Mudawar (1988) for a single, flush-mounted, 12.7 mm × 12.7 mm heater, but with FC-72 and over a larger Reynolds number range. Although their

Reynolds number dependence is consistent with that of Eq. 4.6, their Nusselt numbers are approximately 37% larger in the overlapping Reynolds number range.

Studies related to heat transfer from an array of discrete (12.7 mm × 12.7 mm) isothermal sources mounted flush to one wall of a rectangular channel ($W = 50.8$ mm, $H = 11.9$ mm) have also been performed (Incropera et al., 1986). The array (Fig. 4.1) consisted of four equally spaced rows, with three heaters in each row and $S_L = S_T = 3.18$ mm. Experiments were performed with water and FC-77 for Reynolds numbers in the range $1000 < Re_{D_h} < 14,000$, and results were compared with predictions based on two-dimensional, conjugate forced-convection models for laminar and turbulent flow. The data were significantly underpredicted in laminar flow, and differences were again attributed to the effects of buoyancy on the experimental results. In turbulent flow, however, agreement between the predicted and measured results was good, suggesting that three-dimensional boundary layer effects were negligible. Convection from a single source was quantified in terms of an average Nusselt number, $\overline{Nu}_L = \overline{h}L_h/k_f = q_h'' L_h/(T_h - T_{m,i})k_f$, and variations between sources of a row were negligible. As shown in Figure 4.6, however, average Nusselt numbers in the second row were approximately 25% less than those of the first row, while Nusselt numbers in the third row were approximately 10% less than those of the second row. This reduction in $\overline{Nu}_L$ with increasing row number was attributed to the effects of upstream thermal boundary layer development. However, the decrease between the third and fourth rows was only 3%, suggesting that a fully

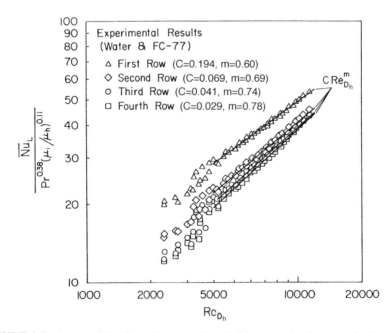

FIGURE 4.6 Average Nusselt numbers associated with convection heat transfer from each row of an in-line array of square ($L_h = 12.7$ mm) heat sources. Adapted from Incropera et al. (1986).

developed condition is approached for which thermal boundary layer development on one row is balanced by boundary layer dissipation in the unheated region between rows. For each row, the data were correlated by an expression of the form

$$\frac{\overline{Nu}_L}{Pr^{0.38}(\mu_i/\mu_h)^{0.11}} = C\,Re_{D_h}^m \tag{4.9a}$$

where C and m decreased and increased, respectively, with increasing row number. All properties were evaluated at the fluid inlet temperature, except μ_h, which was evaluated at the heater temperature.

Because parametric calculations performed by Moffatt (1985) indicate a strong dependence of $\overline{Nu}_L$ on (D_h/L_h) when the Reynolds number is defined in terms of D_h, but a much weaker dependence when it is defined in terms of L_h (see Section 4.2.1), it is preferable to cast Eq. 4.9a in terms of Re_{L_h}. Because the hydraulic diameter and heat source length of the experiments differed by a constant factor $(D_h/L_h = 1.52)$, Eq. 4.9a may also be expressed as

$$\frac{\overline{Nu}_L}{Pr^{0.38}(\mu_i/\mu_h)^{0.11}} = C\,Re_{L_h}^m \tag{4.9b}$$

where the values of m are equivalent to those prescribed in Figure 4.6, but the values of C are 0.250, 0.092, 0.056, and 0.040 for rows 1 through 4, respectively. As suggested by Eq. 4.6b, which was determined from calculations performed for $0.2 < D_h/L_h < 20$, $\overline{Nu}_L \sim (L_h/D_h)^{0.1}$ when $\overline{Nu}_L$ is correlated in terms of Re_{L_h}. Hence, Eq. 4.9b may be used to a good approximation over a wide range of D_h and L_h. Good agreement between Eq. 4.9b and predictions based on a two-dimensional model of turbulent flow over the heater array (Moffatt et al., 1986) was obtained for each of the four rows, suggesting that three-dimensional boundary layer edge effects are negligible.

Somewhat different behavior was observed by Gersey and Mudawar (1992), who performed experiments for FC-72 and a linear array of nine flush-mounted heat sources, each of width $L_h = 10$ mm and spacing $S = 10$ mm. The channel width and height were fixed at $W = 20$ mm and $H = 5$ mm, respectively, and the flow rate was varied to provide turbulent flow in the Reynolds number range, $3000 < Re_{L_h} < 2 \times 10^5$. The average Nusselt number was defined as $\overline{Nu}_L = q_h'' L_h/(T_h - T_{m,i})k_f$, where $T_{m,i}$ is the channel inlet temperature, at which all fluid properties were evaluated. In contrast to the results of Incropera et al. (1986), which indicated a reduction in $\overline{Nu}_L$ from the first to the third rows of the array, Gersey and Mudawar (1992) found that the data for all nine heat sources were correlated to within $\pm 5\%$ by the following expression:

$$\frac{\overline{Nu}_L}{Pr^{1/3}} = 0.362\,Re_{L_h}^{0.614} \tag{4.10}$$

where all fluid properties were evaluated at $T_{m,i}$. This behavior was confirmed by the two-dimensional simulations of Xu et al. (1998), who considered an array of four heat sources for which $S/L_h = 1$. The existence of equivalent Nusselt numbers for each chip in an array for which the dimensionless heater spacing is $S/L_h = 1$ suggests that thermal boundary layer dissipation between heaters is nearly complete and boundary layer development is reinitiated at the leading edge of each heater. However, with decreasing S/L_h, the effects of upstream boundary layer development on heat transfer from downstream sources becomes more pronounced, and one should expect a reduction in the average Nusselt number with increasing row number, particularly for the first few rows of the array.

4.3 MIXED CONVECTION FOR DISCRETE, FLUSH-MOUNTED HEAT SOURCES

4.3.1 Physical Features

When a liquid is heated, its density decreases and attendant buoyancy forces may induce motion (free convection) or influence a preexisting forced flow (mixed convection). If the preexisting flow is laminar, the effect can be significant and its physical manifestations depend on the orientation of the gravitational vector relative to the flow. If a channel is oriented vertically and the flow is upward, buoyancy forces induced by heating the fluid enhance (accelerate) fluid motion. Conversely, if the flow is downward, fluid motion would be retarded by heating. For a horizontal channel, there is no effect on fluid motion if heating occurs at the top surface. In this case, warm fluid would overlie cool fluid and the corresponding stable vertical density variation would preclude development of a buoyancy-driven flow. However, if heating occurred at the side walls or the bottom of the channel, secondary, buoyancy-driven flows would be superimposed on the main (preexisting) flow and convection heat transfer coefficients would be enhanced. Heat transfer enhancement is most pronounced for heating from below.

The effect of buoyancy on convection heat transfer in laminar, horizontal channel flows is well known for continuously heated surfaces. Bottom heating induces a thermal instability that is manifested by ascending plumes of warm fluid and the development of longitudinal vortices. In this mixed convection regime, Nusselt numbers are as much as six times larger than those corresponding to pure forced convection (Osborne and Incropera, 1985; Incropera et al., 1987).

The fact that mixed-convection flows offer the potential for significant heat transfer enhancement with only a modest penalty in pressure loss suggests that such flow conditions may be well suited to applications involving multichip arrays. Such an application has been considered both theoretically and experimentally for flush-mounted chips.

Theoretical Considerations The effects of buoyancy on the flow and heat transfer associated with an array of discrete heat sources mounted flush with the bottom wall of a horizontal, rectangular channel have been considered by Mahaney et al. (1989) for laminar flow and the geometry prescribed in Figure 4.1. In particular, Eqs. 2.36 to 2.40 were solved for $g_x = g_z = 0$ and $g_y = g$.

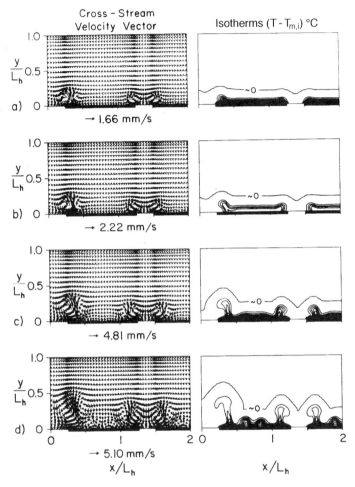

FIGURE 4.7 Cross-stream velocity vectors and isotherms for laminar, mixed-convection flow in a horizontal channel with an array of discrete heat sources mounted flush with the bottom surface ($Re_{D_h} = 500$, $Ra^*_{L_h} = 2.5 \times 10^7$, $Pr = 7$): (a) $(z - L_i)/L_h = 1.0$, (b) $(z - L_i)/L_h = 1.25$, (c) $(z - L_i)/L_h = 2.25$, and (d) $(z - L_i)/L_h = 3.5$. Adapted from Mahaney et al. (1989).

Evolution of the buoyancy-driven secondary flow is revealed in Figure 4.7 by representative results for the cross-stream (x, y) velocity vectors and isotherms at different locations in the longitudinal (z) direction. Once again, symmetry about the longitudinal midplane of the channel permits representation of results for one half of the channel. The modified Rayleigh number, $Ra^*_{L_h} \equiv g\beta q''_h L_h^4 / k_f \alpha_f \nu$, provides a measure of the buoyancy force induced by each heat source, and the isotherms provide the amount by which the local fluid temperature exceeds the inlet temperature,

$T - T_{m,i}$. The temperature difference between isotherms is 5°C, and the scale of the velocity vectors is indicated below each plot.

At the trailing edge of the first heater row, $(z - L_i)/L_h = 1$, cooler, more dense fluid descends along the vertical wall ($x/L_h \approx 0$) and in the region between heaters ($1.25 \leq x/L_h \leq 1.5$). This motion is accompanied by the ascension of adjacent warmer fluid, which is cooled and ultimately feeds the descending fluid. Although the unstable condition of warmer, less dense fluid beneath cooler, more dense fluid occurs across the entire bottom surface, a pronounced spanwise temperature gradient exists at the heater edges. This gradient provides a region where the buoyancy force varies with x as well as with y, thereby providing a source of vorticity. Thus, secondary flows initially develop at spanwise locations corresponding to discontinuities in the thermal boundary condition.

At the trailing edge of the nonheated region beyond the first heater row, $(z - L_i)/L_h = 1.25$, the bottom surface temperature has substantially decreased. However, as shown by the velocity scale, which corresponds to the maximum cross-stream velocity at the designated longitudinal position, cross-stream velocities have increased, even though the heating has been interrupted. At the trailing edge of the second heater row, $(z - L_i)/L_h = 2.25$, the three zones of ascending fluid have bifurcated and are feeding the two original downflows, as well as three new zones of downward-moving fluid ($x/L_h \approx 0.5, 1.0, 1.75$). These new downflows act to compress isotherms near the heated surface. At the trailing edge of the third row, $(z - L_i)/L_h = 3.5$, the three zones of ascending fluid have created two additional regions of downflow at $x/L_h \approx 0.75$ and 2.0, which also act to compress the isotherms locally. At this point, thermal plumes are prominent near the heater edges, with each plume bifurcating to form counterrotating vortex pairs that significantly enhance heat transfer. As indicated by the incipience of additional plumes at intermediate locations on the heat sources (Fig. 4.7d), the buoyancy-driven secondary flow continues to evolve as the fourth heater row is approached. Disruption of thermal boundary layer development by the secondary flow is the mechanism by which heat transfer is enhanced.

Effects of the buoyancy-induced flows on surface temperature and Nusselt number distributions may be discerned by contrasting the results for which buoyancy was neglected (Fig. 4.2) with those for which it was considered (Fig. 4.8). Compared with the forced-convection results, Figure 4.8a reveals substantially lower temperatures at the heated surfaces and the intervening unheated sections. Moreover, heat source temperatures do not increase monotonically for rows 2, 3, and 4.

As for pure forced convection, the local Nusselt number distribution (Fig. 4.8b) reveals spanwise peaks. However, by $(z - L_i)/L_h \approx 0.35$, the magnitudes of these peaks begin to increase, rather than decrease, with longitudinal position. This increase, along with reductions in the Nusselt number at locations adjacent to the peaks, is due to development of the buoyancy-induced secondary flow that causes unheated fluid to descend to the heated surface near the locations of the peaks. This downward motion is sustained by ascending, relatively warmer fluid at locations where the Nusselt number is depressed. As additional heaters are encountered in the longitudinal direction, multiple circulation cells develop, yielding additional peaks

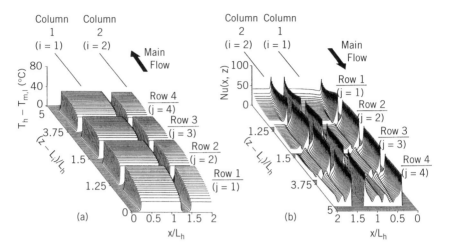

FIGURE 4.8 Two-dimensional, bottom surface (a) temperature and (b) Nusselt number distributions corresponding to an array of flush-mounted heat sources for laminar, mixed convection in a horizontal, rectangular channel ($Re_{D_h} = 500$, $Ra^*_{L_h} = 2.5 \times 10^7$, $Pr = 7$). Adapted from Mahaney et al. (1989).

and depressions in the local Nusselt number distribution, but with an overall enhancement in convection heat transfer.

Predictions of the row-averaged Nusselt number, Eq. 4.3, are plotted as a function of the Reynolds number in Figure 4.9. At large values of Re_{L_h}, conditions are characteristic of forced convection, and the form of the Reynolds number dependence, $\overline{Nu}_{L,j} \sim Re_{L_h}^{1/3}$, is consistent with that obtained by solving the integral form of the boundary layer energy equation for flow between parallel plates with a fully developed laminar velocity profile and a cubic temperature profile (Mahaney, 1989). However, with decreasing Reynolds number, there is a transition from forced to mixed convection, as manifested by heat transfer enhancement above the forced convection limit. The transition first occurs in the last row and results from development of the secondary flow because of heating in the upstream rows. The Nusselt number continues to decay until a minimum is reached, at which point the reduction in $\overline{Nu}_{L,4}$ resulting from diminishing forced-convection effects is balanced by an increase caused by the buoyancy-driven secondary flow. With a further reduction in Re_{L_h}, the secondary flow intensifies and the corresponding enhancement more than offsets the reduction due to the weakened forced flow. Similar behavior characterizes the upstream rows, although reduced preheating of the fluid necessitates a lower Reynolds number for onset of the transition and the effect is not significant for the first row.

4.3.2 Experimental Results

Experiments have been performed for a configuration like that shown in Figure 4.1 (Mahaney et al., 1990b). Specifically, an in-line array of four rows of discrete heat

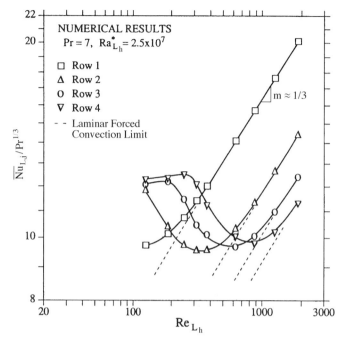

FIGURE 4.9 Numerical prediction of the effect of Reynolds number on the row-averaged Nusselt number for laminar, mixed-convection flow over an array of heat sources mounted flush with the bottom surface of a horizontal, rectangular channel ($Ra^*_{L_h} = 2.5 \times 10^7$, $Pr = 7$). Adapted from Mahaney et al. (1990a).

sources, with three sources per row, was mounted flush to the bottom wall of a horizontal rectangular channel of width $W = 50.8$ mm and height $H = 11.9$ mm ($D_h = 19.1$ mm). Each heater measured 12.7 mm on a side, and the spacing between heaters was $S = 3.18$ mm.

Measurements of the row-averaged Nusselt number obtained for water and a power dissipation of $Q = 7.5$W per heater are plotted in Figure 4.10. In reducing the data, all properties were evaluated at the mixed mean temperature of the fluid, T_m, except μ_h, which was evaluated at the heater surface temperature, T_h. The dashed lines correspond to forced convection limits for which the Reynolds number dependence is characterized by an exponent of $m = 1/3$.

The experimental trends of Figure 4.10 are consistent with the numerical results of Figure 4.9. At large Reynolds numbers, conditions correspond to forced convection, and the reduction in Nusselt number with increasing row number results from thermal boundary layer development caused by upstream heating. The departure from the laminar forced-convection limit with increasing Re_{L_h} was attributed to the onset of transition to turbulence. However, with decreasing Re_{L_h}, there is also heat transfer enhancement above the forced-convection limit that initially occurs for the fourth heater row. As indicated by the numerical predictions, this behavior results

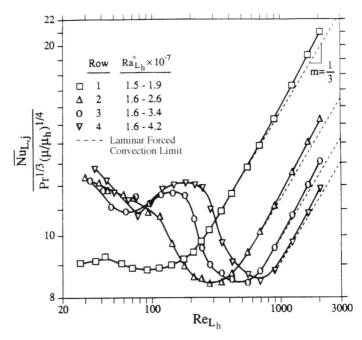

FIGURE 4.10 Experimentally determined row-averaged Nusselt numbers for water flow over an in-line array of heat sources mounted flush to the bottom wall of a rectangular channel. Adapted from Mahaney et al. (1990b).

from the onset of thermal instabilities and the development of secondary flow resulting from heating at the three upstream rows. The Nusselt number continues to decay with decreasing Reynolds number, until a minimum is reached, where the reduction in $\overline{Nu}_{L,4}$ resulting from forced-convection effects is balanced by an increase caused by the buoyancy-driven secondary flow. With further reduction in Re_{L_h}, the strength of the buoyancy-induced secondary flow increases and the corresponding enhancement more than offsets the decrease resulting from a reduction in the forced flow. Having achieved conditions dominated by mixed convection, a further decrease in Reynolds number results in decay of the forced-convection component, transition to pure natural convection, and a decreasing Nusselt number. The subsequent increase in $\overline{Nu}_{L,4}$ with decreasing Reynolds number for $Re_L \leq 75$ is attributed to substantial increases in the mean fluid temperature and variations of fluid properties. Although the data appear to be strongly Reynolds number dependent, the results correspond to pure natural convection and the increase in Nusselt number is due to substantial increases in Rayleigh number. The wide Rayleigh number variation resulting from changes in fluid temperature is indicated for each heater row.

The transitions from pure forced convection to mixed convection to pure natural convection occur in a similar manner for the upstream rows. Because transition from forced to mixed convection is caused by the onset of thermal instabilities, which is influenced by the cumulative effect of heating at all upstream rows, this transition occurs for each successive upstream row at a smaller Reynolds number. Because of the absence of heat transfer upstream of the first row, transition from forced to mixed convection for the first row results solely from thermal instabilities induced by heating in this row. Thus, for the prescribed Rayleigh number range, significant enhancement above the minimum heat transfer rate does not occur with decreasing Reynolds number for the first row. However, with decreasing Reynolds number, substantial enhancement above the forced-convection limits occurs for every row, including the first. The fact that the variation of the row-averaged Nusselt number with Reynolds number exhibits a minimum for each row suggests that heat transfer may be enhanced because of buoyancy-induced flow by reducing the flow rate and, hence, the pumping power requirements. The data associated with the smallest value of Re_{L_h} correspond to a flow rate of only 5.2 L/h (1.4 gal/hr).

Mahaney et al. (1990b) also performed experiments for FC-77 and were able to correlate data for the two fluids by an expression of the form

$$(\overline{Nu_L})^n = (\overline{Nu_{L,F}})^n + (\overline{Nu_{L,N}})^n \tag{4.11}$$

where the subscripts F and N refer to pure forced and natural convection, respectively. An expression for the forced-convection Nusselt number was obtained by solving the integral form of the boundary layer energy equation for laminar flow between parallel plates, with a fully developed velocity profile and an assumed cubic temperature profile (Mahaney, 1989). The solution is of the form

$$\overline{Nu_{L,F}} = 1.448 Pe_{L_h}^{1/3} \left(\frac{L_h}{H} \right)^{1/3} \tag{4.12a}$$

which, for fixed values of the heater length and channel height, may be expressed as

$$\overline{Nu_{L,F}} = C_1 Pe_{L_h}^{1/3} \tag{4.12b}$$

The natural-convection Nusselt number for a flat, horizontal plate with a heated upper surface may be correlated by an expression of the form (Incropera and DeWitt, 1996)

$$\overline{Nu_{L,N}} = C_2 Ra_{L_h}^{*m} \tag{4.13a}$$

where $1/5 \leq m \leq 1/4$. The best correlation of the data of Mahaney et al. (1990b) was achieved with an intermediate value of $m = 2/9$, in which case,

$$\overline{Nu_{L,N}} = C_2 Ra_{L_h}^{*2/9} \tag{4.13b}$$

Substituting Eqs. 4.12b and 4.13b into Eq. 4.11, the mixed-convection correlation becomes

$$\left[\frac{\overline{Nu_L}}{Pe_{L_h}^{1/3}} \right]^n = C_1^n + \left[\frac{C_2 Ra_{L_h}^{*2/9}}{Pe_{L_h}^{1/3}} \right]^n \tag{4.14}$$

Thus, the ratio $\overline{Nu_L}/Pe_{L_h}^{1/3}$ is solely a function of $Ra_{L_h}^{*2/9}/Pe_{L_h}^{1/3}$, or, alternatively, by raising the mixed-convection parameter to the $-9/2$ power, $\overline{Nu_L}/Pe_{L_h}^{1/3}$ is solely a function of $Pe_{L_h}^{3/2}/Ra_{L_h}^*$.

Extensive data obtained by Mahaney et al. (1990b) for water and FC-77 were plotted in terms of the foregoing scaling parameters, and the results are shown in Figure 4.11. Except for regions identified as *outliers*, all of the data fell within the solid lines shown for each of the rows. For small values of $Pe_{L_h}^{3/2}/Ra_{L_h}^*$, the results are asymptotic to a slope of $m = -2/9$, which is consistent with the natural-convection correlation and for which the Nusselt number is independent of the Péclet number. If a viscosity ratio of the form $(\mu/\mu_h)^{0.25}$ is included in Eq. 4.12b, the corresponding values of C_2 for rows 1 to 4 are $C_2 \approx 0.47, 0.46, 0.44$, and 0.46, respectively. Thus, as expected, heat transfer in the natural-convection regime is relatively insensitive to row number. For large values of $Pe_{L_h}^{3/2}/Ra_{L_h}^*$, the results should collapse to horizontal lines that correspond to pure forced convection. The ordinate values of the horizontal asymptotes correspond to the constant C_1 of Eq. 4.12b, and, if a viscosity ratio is included, values of $C_1 \approx 1.56, 1.14, 0.99$, and 0.91 fit the data for rows 1 to 4, respectively. The outliers are associated with transition to turbulence, which occurs for larger values of $Pe_{L_h}^{3/2}/Ra_{L_h}^*$. Transition depends principally on the Reynolds number and not on the Rayleigh number. Hence, for increasing values of $Q(Ra_{L_h}^*)$, transition occurs for decreasing values of $Pe_{L_h}^{3/2}/Ra_{L_h}^*$.

The results of Figure 4.11 can be used to identify regions of significant heat transfer enhancement. If the onset of enhancement resulting from mixed convection is assumed to occur when the parameter $\overline{Nu_L}/Pe_{L_h}^{1/3}(\mu/\mu_h)^{0.25}$ exceeds the forced-convection limit of a row by 10%, the results of Figure 4.11 provide corresponding values of $Pe_{L_h}^{3/2}/Ra_{L_h}^* = 0.0022, 0.0065, 0.012$, and 0.019 for rows 1 to 4, respectively. Figure 4.11 reveals increasing heat transfer enhancement above the forced-convection limit with decreasing $Pe_{L_h}^{3/2}/Ra_{L_h}^*$. For the smallest mixed-convection parameter values of the study, enhancement above the forced-convection limit exceeds 300%.

Example 4.1

A 5×5 array of chips, each of length $L_h = 10$ mm, is mounted to a substrate that forms the bottom wall of a horizontal channel. The array is characterized by longitudinal and transverse pitches of $P_L = P_T = 12.5$ mm, and the channel width and height are $W = 65$ mm and $H = 15$ mm, respectively. Coolant (FC-77) enters the channel at a mean velocity and temperature of $w_m = 0.5$ m/s and $T_{m,i} = 15°C$, respectively. If the maximum allowable chip temperature is $T_h = 85°C$, what is the maximum allowable chip heat rate q?

Solution

Known: Dimensions of chip array and rectangular channel. Maximum allowable chip temperature and coolant inlet conditions.

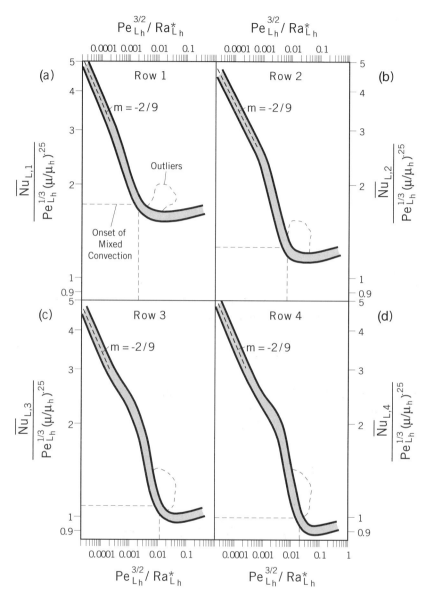

FIGURE 4.11 Correlation of mixed-convection heat transfer data obtained for an in-line, 4×3 array of discrete heat sources with water ($1 \leq Q \leq 7.5\,\text{W}, 0.2\times 10^7 \leq Ra^*_{L_h} \leq 4.2\times 10^7$) and FC-77 ($0.5 \leq Q \leq 2.5\,\text{W}, 0.5 \times 10^9 \leq Ra^*_{L_h} \leq 2.9 \times 10^9$). Adapted from Mahaney et al. (1990b).

144 CHANNEL FLOWS

Find: Maximum allowable chip heat rate.

Assumptions:

1. Steady-state conditions.
2. Isothermal chips.
3. Negligible effect of chip protrusion and channel height on heat transfer.
4. If flow is turbulent, equivalent thermal conditions exist on the fourth and fifth rows.

Properties: FC-77. Appendix B. $T_h = 85°C$: $\mu_h = 6.30 \times 10^{-4}$ kg/s·m. $T_{m,i} = 15°C$: $k_f = 0.0639$ W/m·K, $\alpha_f = 3.44 \times 10^{-8}$ m²/s, $\nu = 9.54 \times 10^{-7}$ m²/s, $\mu = 1.72 \times 10^{-3}$ kg/s·h, $\beta = 0.00136$ K^{-1}, and $Pr = 27.8$.

Analysis: With $D_h = 4A_c/P = 4 \times (65 \times 15)$ mm²/2(65 + 15) mm = 24.4 mm, the Reynolds number based on hydraulic diameter is $Re_{D_h} = w_m D_h/\nu = 0.5$ m/s(0.0244 m)/9.54×10^{-7} m²/s = 12,800. Hence, flow in the channel is turbulent, and with $P_L/S = P_T/S = 1.25$, the average convection coefficient for each row may be determined from Eq. 4.9b:

$$\overline{Nu}_L = C\, Re_{L_h}^m\, Pr^{0.38} \left(\frac{\mu}{\mu_h}\right)^{0.11}$$

where $C = 0.250, 0.092, 0.056, 0.040,$ and 0.040 and $m = 0.60, 0.69, 0.74, 0.78,$ and 0.78 for rows 1 to 5, respectively. Hence, with $Re_{L_h} = Re_{D_h}(L_h/D_h) = 12,800(10$ mm/24.4 mm$) = 5250$,

$$\overline{Nu}_L = C(5250)^m (27.8)^{0.38} \left(\frac{1.72 \times 10^{-3}\ \text{kg/s·m}}{6.30 \times 10^{-4}\ \text{kg/s·m}}\right)^{0.11}$$

$$= 3.95C(5250)^m$$

$$\bar{h} = \frac{k_f}{L_h}\overline{Nu}_L = \frac{0.0639\ \text{W/m·K}}{0.01\ \text{m}}(3.95C)(5250)^m$$

$$= 25.2C(5250)^m$$

The heat rate for a single chip is

$$q = \bar{h} A_s (T_h - T_{m,i}) = \bar{h}(0.01\ \text{m})^2 (85 - 15)°C = 0.007\bar{h}$$

It follows that, for each of the five rows,

Row 1: $\bar{h} = 1080$ W/m²·K, $q = 7.5$ W.
Row 2: $\bar{h} = 860$ W/m²·K, $q = 6.0$ W.
Row 3: $\bar{h} = 800$ W/m²·K, $q = 5.6$ W.

Row 4: $\bar{h} = 800$ W/m$^2 \cdot$ K, $q = 5.6$ W.
Row 5: $\bar{h} = 800$ W/m$^2 \cdot$ K, $q = 5.6$ W.

The maximum allowable heat rate of $q = 7.5$ W is associated with chips in the first row.

Comments:

1. For the largest value of $q = 7.5$ W, $q'' = q/A_s = 7.5 \times 10^4$ W/m^2, and the modified Rayleigh number is $Ra^*_{L_h} = g\beta q'' L_h^4/\alpha_f \nu = 9.8$ m/s^2 (0.00136 K^{-1}) 7.5 $\times 10^4$ W/m^2(0.01 m)^4/(3.44 $\times 10^{-8}$ m^2/s)(9.54 $\times 10^{-7}$ m^2/s) $= 3.05 \times 10^8$. With $Pe_{L_h} = Re_{L_h} Pr = 5250 \times 27.8 = 1.46 \times 10^5$, the mixed-convection parameter $Pe_{L_h}^{3/2}/Ra^*_{L_h}$ has a value of 0.183, which is well in excess of the lower limit for forced convection. Mixed-convection effects are, therefore, negligible.

2. To circumvent the effect of row number on $\bar{h}$ and q for the first three rows, the dimensionless longitudinal pitch would have to be increased. The results of Gersey and Mudawar (1992) indicate that heat transfer is approximately independent of row number for $P_L/L_h = 2.0$ and is correlated by Eq. 4.10. It would follow that, for each source in the array,

$$\overline{Nu_L} = 0.362 Re_{L_h}^{0.614} Pr^{1/3} = 0.362(5250)^{0.614}(27.8)^{1/3} = 211$$

$$\bar{h} = \frac{k_f}{L_h} \overline{Nu_L} = 1350 \text{ W/m}^2 \cdot \text{K}$$

$$q = \bar{h} A_s (T_h - T_{m,i}) = 1350 \text{ W/m}^2 \cdot \text{K}(0.01 \text{ m})^2 (70°\text{C}) = 9.4 \text{ W}$$

This result exceeds that obtained from Eq. 4.9b for the first row by approximately 25%.

4.4 PROTRUDING HEAT SOURCES

4.4.1 Forced Convection

Although heat transfer from protruding sources in a rectangular channel has been studied extensively for air flow, comparatively few studies have been performed for liquids. As shown in Figure 4.12, the sources are typically in the form of rectangular blocks of height B_h and planform lengths $L_{h,L}$ and $L_{h,T}$ in the longitudinal and transverse directions, respectively. The longitudinal and transverse spacings, S_L and S_T, may differ, and the sources may be arranged as in-line (Fig. 4.12) or staggered arrays.

The effect of a slight protrusion ($B_h = 0.5$ mm) was considered experimentally for a single heat source, and the results were compared with those for a flush-mounted source (Incropera et al., 1986). When the comparison was based on the heat

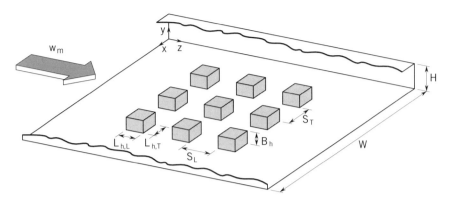

FIGURE 4.12 Channel flow with an in-line array of protruding heat sources.

rate per exposed surface area, which included the sides of the protruding source, heat transfer enhancement resulting from the protrusion was less than approximately 5%.

Garimella and Eibeck (1990, 1991) performed experiments for water flow in a channel with six rows of protruding heaters and five heaters per row. Both staggered and in-line arrays were considered. With element dimensions fixed at $L_{h,L} = L_{h,T} = 25.4$ mm and $B_h = 10$ mm, measurements were performed for Reynolds numbers up to $Re_{L_h} \approx 9000$. Row-independent Nusselt numbers were obtained beyond the first three rows, and heat transfer was enhanced by staggering the elements, as well as by decreasing the channel height and/or by increasing the longitudinal and transverse spacings between elements.

Gersey and Mudawar (1993) performed experiments for FC-72 and a linear array of nine heat sources for which $L_{h,L} = L_{h,T} = 10$ mm, $B_h = 1$ mm, and $S_L = 10$ mm. The channel width was fixed at $W = 20$ mm, and two heights ($H = 2$ and 5 mm) were considered. Unlike results for the flush-mounted heaters (Gersey and Mudawar, 1992), Nusselt numbers for the first two heaters in the array were slightly larger than those for the downstream heaters. Adjusting the definition of the Nusselt number to account for the increased surface area resulting from the protrusion, $\overline{Nu_L} = QL_h/A_{ex}k_f(T_h - T_{m,i})$, data for the downstream heaters were well correlated by Eq. 4.10, thereby implying a negligible effect of the protrusion on heat transfer. The effect of channel height was also found to be negligible.

Gudapati (1993) and Garimella and Schlitz (1995) performed experiments for water and FC-77, using an in-line array of 15 rows with 5 elements per row. The channel width and element planform dimensions were fixed at $W = 110$ mm and $L_{h,L} = L_{h,T} = 10.2$ mm, whereas the channel and element heights were varied over the ranges $1.9 < H < 40$ mm and $0.64 < B_h < 15.0$ mm. The spacing between elements was fixed at $S_L = S_T = 6.1$ mm. In their experiments, all elements were unheated except that occupying the middle column of the 10th row.

The data acquired by Gudapati (1993) and Garimella and Schlitz (1995), as well as results obtained by Garimella and Eibeck (1990) for water and other investigators for air, were collected by Morris and Garimella (1996) and correlated by an expression of the form

$$\overline{Nu}_{L,\text{ad}} = 0.158 Re_{L_h}^{0.655} Pr^{0.5} \left(\frac{\mu_i}{\mu_h}\right)^{0.13} \left(\frac{S_L}{H}\right)^{0.125} \left(\frac{H-B_h}{L_{h,L}}\right)^{0.05} \quad (4.15)$$

where the ranges of the data are $610 < Re_{L_h} < 69{,}600$, $0.71 < Pr < 25.2$, $0.025 < S_L/H < 3.29$, and $0.06 < (H - B_h)/L_{h,L} < 3.87$. In this expression, the Reynolds number is based on the mean velocity of the fluid upstream of the array, and the Nusselt number is defined in terms of an *adiabatic heat transfer coefficient*:

$$\overline{h}_{\text{ad}} = \frac{Q}{A_{\text{ex}}(T_h - T_{\text{ad}})} \quad (4.16)$$

where the reference temperature, T_{ad}, is the temperature that the element would assume if it were unheated, while all other elements were heated. For a designated element, it, therefore, represents the effect of heating resulting from the *thermal wakes* generated by upstream elements. In the absence of specific data for T_{ad}, it may be approximated by the mixed mean temperature of the fluid just upstream of the designated element or by the channel inlet temperature, if heating of the fluid by upstream elements is negligible.

Data for water and FC-77 were correlated by Eq. 4.15 to within $\pm 6.8\%$, and data for air, as well as the liquids, were correlated to within $\pm 9.4\%$. Although the Prandtl number exponent of 0.5 exceeds customary values ($0.33 < n < 0.40$), a larger exponent is often associated with heat transfer from roughened surfaces. The Nusselt number increases modestly with an increase in the longitudinal spacing relative to the channel height, as well as to a very small degree with an increase in the channel-to-element clearance relative to the heater length.

Tou et al. (1998) performed experiments for FC-72 flowing in a vertical channel of width $W = 20$ mm, with four in-line simulated chips of length $L_h = 10$ mm and spacing $S_L = 10$ mm. Protrusion and channel heights of $B_h = 0$, $0.2L_h$ (0, 2 mm), and $H = L_h$ (10 mm) were considered for Reynolds numbers in the range $2200 < Re_{L_h} < 6.3 \times 10^4$. The average Nusselt number was defined as $\overline{Nu}_L = QL_h/A_{\text{ex}}(T_h - T_{m,i})$, and the mean velocity was determined from the actual cross-sectional area of the flow, $w_m = \dot{m}/\rho(WH - B_h L_h)$. For the first chip ($j = 1$) and $B_h = 0$, the data were correlated by

$$\overline{Nu}_L = C\, Re_{L_h}^m\, Pr^{0.33} \quad (4.17)$$

with $C = 0.397$ and $m = 0.617$. The correlation overpredicts the results of Maddox and Mudawar (1988) by approximately 10%. The Nusselt number decreased from the first to the second row, with a nearly fully developed condition being reached after the second row and $\overline{Nu}_L$ for the first row exceeding that of the last row by approximately 22%. For $B_h = 2$ mm, $\overline{Nu}_L$ decreased by approximately 40% from the first to the second row, and results for the second through the fourth row were correlated by Eq. 4.17, but with $C = 0.367$ and $m = 0.609$. The correlation is in good agreement with that of Gersey and Mudawar (1993). For the downstream rows, there was little difference between results for $B_h = 0$ and $B_h = 2$ mm. For the first row, results for $B_h = 2$ mm exceeded those for $B_h = 0$ by less than 20% for $Re_{L_h} < 10^4$, but the difference increased with increasing Reynolds number.

4.4.2 Mixed Convection

A detailed study of mixed-convection heat transfer was performed by Heindel et al. (1992a) for a linear array of 10 heat sources mounted flush to substrates protruding from the bottom wall of a horizontal channel (Fig. 4.13). The heat sources of length $L_h = 6.35$ mm simulated electronic chips, whereas the protrusions of length $L_c = 25.4$ mm and spacing $S_L = 6.35$ mm represented chip carriers. The height of the protrusion and the channel width were fixed at $B_h = 2$ mm and $W = 38.1$ mm, respectively, whereas channel heights of $H = 3.58$ mm and 6.96 mm were considered. Experiments were performed with water and FC-77 for Reynolds and modified Rayleigh numbers in the ranges $50 < Re_{L_h} < 13{,}200$ and $4.5 \times 10^5 < Ra^*_{L_h} < 6.4 \times 10^8$, respectively. The mean velocity used in the Reynolds number accounts for the effect of the protrusion on the cross-sectional area of the flow. With the test section located 600 mm from the channel entrance, fully developed or nearly fully developed conditions were achieved at its leading edge for the entire range of Reynolds numbers.

Representative results for the dependence of the Nusselt number on the Reynolds number are shown in Figure 4.14. The Nusselt number is defined as $\overline{Nu_L} = \overline{h}L_h/(T_h - T_{m,j})k_f$, where $T_{m,j}$ is the mixed mean temperature of the fluid immediately above the center of the jth heater. Five distinct heat transfer regimes may be delineated from the results. For very low Reynolds numbers at and below point A, heat transfer is dominated by free convection and the effect of the main flow,

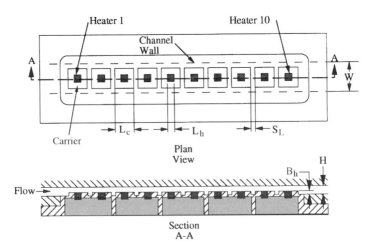

FIGURE 4.13 Linear array of 10 heaters mounted flush to protruding carriers at the bottom surface of a horizontal channel. Adapted from Heindel et al. (1992a).

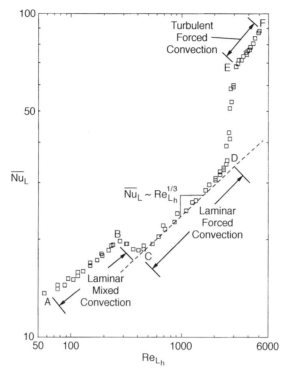

FIGURE 4.14 Flow regimes associated with heat transfer from the 10th source of a linear array to water ($Q = 2$W, $H = 6.96$ mm). Adapted from Heindel et al. (1992a).

and, hence, Re_{L_h}, is negligible. As the Reynolds number is increased from A to B, forced convection effects intensify, decreasing the thermal boundary layer thickness and increasing heat transfer resulting from mixed convection. However, as the Reynolds number is increased from B to C, there is a transition from mixed convection to laminar forced convection and a corresponding reduction in the Nusselt number. Similar behavior was numerically predicted and experimentally observed by Mahaney et al. (1989, 1990b) for flush-mounted sources and was attributed to elimination of thermal instabilities and secondary flows resulting from thinning of the thermal boundary layer with increasing Reynolds number. From C to D, the Nusselt number varies with the Reynolds number raised to the 1/3 power, which is consistent with the integral solution for laminar forced convection on a heated surface in parallel flow (Mahaney et al., 1989). At point D, there is an abrupt increase in $\overline{Nu}_L$ because of a transition from laminar forced convection to turbulent forced convection, which extends from points E to F and beyond.

Similar trends were observed for the other heaters, although the transition Reynolds numbers differed. The Reynolds number for transition to laminar forced convection increased with increasing heater number, and the effect was attributed to upstream heating of the fluid and the attendant enhancement of buoyancy effects.

Because of disturbances induced by upstream protrusions, transition to turbulence occurred at smaller Reynolds numbers for downstream heaters. Hence, transition from laminar to turbulent forced convection occurred first at heater 10 and last at heater 1. With increasing Reynolds number, transition propagated upstream until, eventually, heater 1 experienced fully turbulent forced convection. Hence, the Reynolds number range associated with laminar forced convection decreases with increasing heater number.

As shown in Figure 4.15, the mixed-convection region is extended to larger Reynolds numbers as the heater power, and, hence, the temperature difference between the surface and the bulk fluid, is increased. The correspondingly stronger buoyancy forces delay the transition to laminar forced convection. Transition to turbulence is also delayed, but only slightly, as the heater power is increased. This transition occurs over a very small Reynolds number range for heater 1 ($3700 < Re_{L_h} < 4000$), but the range increases with increasing heater number, occurring over the interval $2500 < Re_{L_h} < 3500$ for heater 10.

The laminar forced-convection correlation for heater 10 is shown by the dashed line in Figure 4.15. The correlation coefficient ranged from 1.89 for heater 1 to 1.32 for heater 10. The coefficient decreased monotonically with increasing heater number and approached a constant value, suggesting that, for downstream heaters, the Nusselt number is approximately independent of position, which is consistent with the results of other studies of in-line protruding and flush-mounted sources. Similar behavior was observed for FC-77, although with some differences. The Reynolds number range of the laminar forced-convection region decreased with increasing power and, in some cases, was not observed at downstream heaters because of a combination of upstream heating and the strong buoyancy forces associated with

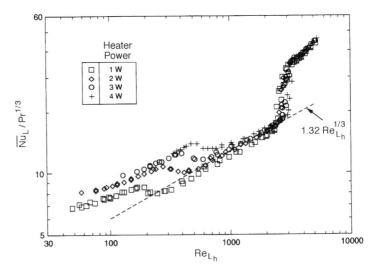

FIGURE 4.15 Effect of power dissipation on heat transfer from the 10th source of a linear array to water ($H = 6.96$ mm). Adapted from Heindel et al. (1992a).

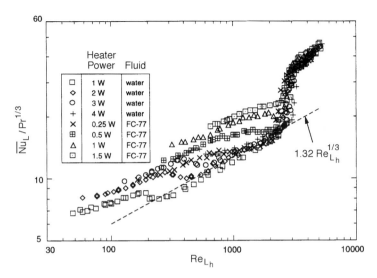

FIGURE 4.16 Comparison of water and FC-77 heat transfer data for the 10th source of a linear array ($H = 6.96$ mm). Adapted from Heindel et al. (1992a).

FC-77. In such cases, mixed convection exists over an extended Reynolds number range, until the transition to turbulence. Such behavior is shown by the FC-77 data of Figure 4.16, for which the laminar forced-convection region is nonexistent at heater powers of 1.0 and 1.5 W.

When the channel height was reduced from 6.96 to 3.58 mm, experiments performed with water revealed a significant suppression of buoyancy effects, including elimination of the mixed-convection regime. The reduced channel height restricts thermal boundary layer development, thereby suppressing the onset of thermal instability and the development of buoyancy-driven flows that would enhance heat transfer.

4.5 MICROCHANNELS

4.5.1 General Considerations

Using microfabrication techniques developed by the electronics industry, it is possible to manufacture three-dimensional structures with length scales as small as 0.1 μm. In turn, this capability permits the fabrication of microchannel heat sinks that are characterized by extremely large surface areas per unit volume ($\sim 10^5$ m^2/m^3) and volumetric heat transfer coefficients (~ 300 $MW/m^3 \cdot K$). Microchannels and microfins can be fabricated in different materials, such as metals, plastics, and silicon, as well as by different methods, such as photo- or X-ray lithography, electroplating, chemical etching, and precision machining (sawing or milling). Although departures from purely rectangular shapes are inherent in the manufacturing process and are manifested by conditions such as rounded corners or tapered fins, tolerances may typically be maintained to within 10% of the characteristic length scales (Phillips, 1988).

Two options exist for using microchannels to cool integrated circuits. In one case, channels may be machined in the chip itself; in the other, they may be machined in a substrate or heat sink to which a chip or an array of chips is attached.

The notion of forming cooling channels in silicon wafers was first advanced by Tuckerman and Pease (1981, 1982). As shown in Figure 4.17, grooves are etched on one side of the wafer and a cover plate is used to form rectangular channels through which the coolant is pumped. Heat dissipated by IC elements on the circuit side of the chip is transferred by conduction through the base of the wafer, as well as through the walls of the channel, which act as continuous, longitudinal fins. From the surfaces of the channel, heat is transferred by convection to the liquid coolant. Alternatively, micro pin fin arrays fabricated with semiconductor dicing saws may be used in lieu of channels (Tuckerman, 1984); different flow cross sections may be created (Hoopman, 1990); or machined wafers may be stacked to create parallel-, counter-, or cross-flow arrangements (Bier et al., 1990). Pin fin arrangements allow for transverse mixing, thereby alleviating the adverse effects associated with possible closure of a channel.

Tuckerman and Pease (1981, 1982) considered laminar, water flow through microchannels machined in a silicon wafer measuring 10 mm × 10 mm on a side, with channel width and height variations of $55 \leq W \leq 60$ μm and $287 \leq H \leq 376$ μm, respectively. They experimentally determined unit thermal resistances less than 0.1 cm$^2 \cdot$ K/W and dissipated heat fluxes as large as 1000 W/cm^2, while maintaining the chip temperature below 110°C and the pressure drop below 345 kPa (50 psi). Much of this superior thermal performance is attributable to the large convection coefficients associated with small channel dimensions. Consider, for ex-

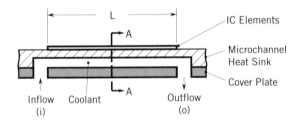

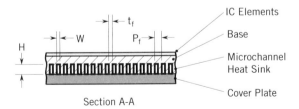

FIGURE 4.17 Schematic of microchannel heat sink.

ample, representative microchannel dimensions of $W = 60$ μm and $H = 300$ μm ($A = 0.2$, $D_h = 100$ μm) and a uniform wall temperature T_s. Even for fully developed laminar flow ($Nu_{D_h} = 4.80$ from Table 2.1) and a dielectric liquid ($k_f \approx 0.06$ W/m·K), a large convection coefficient of $h = (k_f/D_h)Nu_{D_h} = (0.06$ W/m·K/100×10^{-6}m$)4.80 = 2880$ W/m^2·K would be maintained. Mahalingam (1985) also performed experiments for a water-cooled silicon wafer, but under conditions for which the wafer was 38 mm × 38 mm on a side and the channel width and height were $W = 200$ μm and $H = 1700$ μm, respectively. For these conditions, a unit thermal resistance of $R''_{th} \approx 0.3$ cm^2·K/W was recorded.

With respect to the second cooling option, namely, that for which the cooling channels are formed in a substrate to which the chips are mounted, Sasaki and Kishimoto (1986) performed experiments for a 2 × 2 array of four 8 mm × 8 mm chips soldered to a silicon substrate within which microchannels of fixed height $H = 900$ μm and length $L = 24$ mm, but of variable width, $W = t_f = 70$, 140, and 340 μm, were machined. Experiments were also performed for a single large channel (no fins) of width $W = 20$ mm. Subject to the constraint of a maximum allowable pressure drop of $\Delta p = 19.6$ kPa for water flow through the channels, the optimum channel width that minimized the thermal resistance was estimated to be $W_{op} \approx 250$ μm. The value of W_{op} was found to increase with decreasing pressure drop and, hence, with decreasing coolant flow rate. Similar studies have been performed by Hwang et al. (1987) and Nayak et al. (1987), who considered arrays of 10 mm × 10 mm chips mounted on silicon-carbide and copper modules, respectively, in which large-aspect-ratio channels ($A \sim 5, 50$) were machined. Channel widths were approximately 5 mm, with channel heights of 100 and 1000 μm providing for laminar and turbulent flow of water, respectively.

4.5.2 Channel Pressure Drop and Heat Transfer

A focal point of studies on friction and heat transfer in microchannel flows has dealt with the extent to which well-established correlations for macrochannels are applicable. The results of these studies suggest departures from traditional behavior but fail to reveal consistent trends.

Although Tuckerman and Pease (1981) and Wu and Little (1983) report friction factors that exceed classical results for both laminar and turbulent flows, respectively, the opposite behavior was observed by Pfahler et al. (1991) and Choi et al. (1991). Moreover, in addition to reporting values of $f \cdot Re_{D_h}$ that are less than classical results, Pfahler et al. (1991) report an increase in $f \cdot Re_{D_h}$ with increasing Re_{D_h} for small Reynolds numbers.

Peng et al. (1994a) measured pressure losses for water flow through rectangular ducts for which $133 \leq D_h \leq 343$ μm and $1 \leq A \leq 3$. The product $f \cdot Re_{D_h}$ was not constant but decreased significantly with increasing Re_{D_h} and varied widely with A and D_h. Moreover, depending on the particular channel geometry, $f \cdot Re_{D_h}$ was under- or overpredicted by the classical correlations, which is consistent with the inconclusive trends of other investigators. However, more recently, Kendall and Rao (1997) obtained agreement to within 8% between pressure drop measurements for low-Reynolds-number ($10 < Re_{D_h} < 100$) flow of water through microchannels and numerical predictions based on the assumption of two-dimensional laminar, continuum flow. The favorable comparison suggests that classical, continuum behavior is preserved in microchannels, at least for the 30- and 55-μm channel heights considered in their study.

The experimental results of Peng et al. (1994a) also indicate transition Reynolds numbers well below the commonly accepted value of $Re_{D_h} \approx 2000$ for channel flows. The critical Reynolds number corresponding to the onset of transition to turbulence, $Re_{D_{h,c}}$, exhibited a strong dependence on the hydraulic diameter, decreasing from approximately 700 for $267 < D_h < 333$ μm to approximately 200 for $D_h < 200$ μm, while Reynolds numbers associated with the establishment of fully turbulent conditions ranged from approximately 2000 to 300. These results are consistent with those of Wu and Little (1983), who reported transition Reynolds numbers ranging from approximately 400 to 900.

Peng et al. (1994b) obtained heat transfer data for the same geometries considered in their earlier paper (1994a), but with three sides heated and the other (the cover plate) insulated. Their convection coefficient, $h \equiv q''/(T_{s,o} - T_{m,i})$, was defined in terms of an average heat flux, the surface temperature at the channel outlet, and the fluid inlet temperature. The corresponding Nusselt numbers based on the hydraulic diameter were contrasted with predictions obtained from Eqs. 2.154 and 2.156 for laminar and turbulent flow, respectively. For laminar flow ($Re_{D_{h,c}} < 700$), their data were significantly overpredicted by Eq. 2.154. For turbulent flow, their data were correlated by a Colburn-like equation (Incropera and DeWitt, 1996), $Nu_{D_h} = C\, Re_{D_h}^{4/5} Pr^{1/3}$, although the coefficient C depended on the channel geometry (D_h, A) and was well below that of the Colburn correlation (0.023).

Wang and Peng (1994) performed heat transfer measurements for channels of fixed height ($H = 700$ μm) and length ($L = 45$ mm), but of variable width ($200 < W < 800$ μm). The channels were heated on three sides, and the convection coefficient was defined in the manner of Peng et al. (1994b). The data suggested the existence of fully developed turbulent flow for $Re_{D_h} > 1500$, and corresponding Nusselt numbers were correlated by the expression $Nu_{D_h} = 0.00805 Re_{D_h}^{4/5} Pr^{1/3}$, where the coefficient is nearly a factor of 3 smaller than that associated with the Colburn correlation. For laminar and transitional flows, the data were subject to considerable scatter and were not correlated.

Collectively, the foregoing results suggest that, relative to macrochannels, transition to turbulence occurs at much smaller Reynolds numbers in microchannels and conventional correlations for friction and heat transfer may not be applicable. However, a consistent pattern of differences between hydrodynamic and thermal conditions in micro- and macrochannels has yet to emerge and the physical basis for such differences is uncertain. Hence, in modeling the performance of cooling systems involving microchannels, there is currently some uncertainty associated with treating flow and thermal conditions. Accordingly, in implementing such models, use continues to be made of standard results for macrochannels (Copeland, 1995; Goodson et al., 1997; Harms et al., 1997).

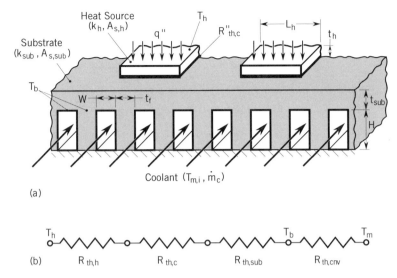

FIGURE 4.18 Chip array mounted on substrate with microchannel cooling: (*a*) system schematic and (*b*) equivalent thermal circuit.

4.5.3 System Performance

Figure 4.18*a* illustrates a cooling scheme for which microchannels are machined in a substrate with an attached chip array. Heat is dissipated by IC devices on the top surface of the chips and is transferred to the coolant through a series of thermal resistances (Fig. 4.18*b*).

Although there is a *spreading resistance* associated with heat transfer from each IC device, the resistance is difficult to quantify and it is common to assume a uniform heat flux just below the top surface. Resistances to this heat flow are associated with conduction through the chip ($R_{th,h} = t_h/k_h A_{s,h}$), a contact resistance at the chip/substrate interface ($R_{th,c} = R''_{th,c}/A_{s,h}$), a three-dimensional conduction resistance associated with heat transfer from the surface of the substrate to the base of the cooling channel ($R_{th,sub}$), and a convection resistance ($R_{th,cnv}$) associated with heat transfer to the coolant from the base of the channel, as well as from the side walls, which act as fins.

For the discrete sources shown in the figure, there is a *spreading* effect associated with three-dimensional conduction in the substrate beneath the heat sources, and $R_{th,sub}$ increases with decreased spacing between the heat sources. Moreover, because the thermal resistance resulting from conduction in the channel side walls, which act as continuous longitudinal fins, typically differs from that resulting from convection at the base of the channel, there is a *constriction* effect associated with conduction in the substrate and a diversion of heat flow lines toward the smallest resistance. Typically, $\varepsilon_f > 1$ and there is preferential direction of heat transfer to the fins. The maximum value that $R_{th,sub}$ could assume corresponds to a continuous

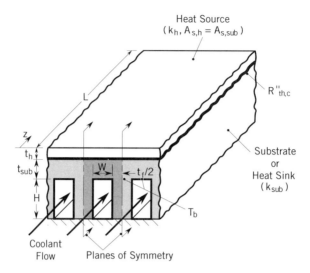

FIGURE 4.19 Microchannel heat sink attached to heat source of equivalent area.

heat source for which $A_{s,sub} = A_{s,h}$ and conditions for which the *constriction* effect is negligible. Conduction through the base of the substrate would then be one dimensional and the resistance would correspond to $R_{th,sub} = (t/kA_s)_{sub}$.

Phillips (1988, 1990) developed a detailed thermal/fluid model for conditions corresponding to equivalent heater/substrate surface areas (Fig. 4.19), a special case of which involves microchannels that are integrally machined with the chip (Fig. 4.17). Heat was assumed to be dissipated uniformly over the surface of the heat source, and constant properties were assumed for the heat source (k_h), substrate or heat sink (k_{sub}), and the fluid (ρ_f, ν, $c_{p,f}$, k_f, Pr), with properties evaluated at appropriate average temperatures. A *thin fin* approximation was made for the channel side walls, with equivalent temperatures T_b assumed for the base of the fins and the channels. Consistent with the assumption of an adiabatic cover plate, the fin tip condition was treated as adiabatic.

Phillips (1988, 1990) also assumed a uniform convection coefficient over the perimeter of the channel. Although variations in h are, in fact, significant for *laminar flow*, with $h \rightarrow 0$ in the corners, accurate estimates of the overall thermal performance may still be made by using average values of h over the perimeter. However, although such averages are typically known for the conditions of uniform heat flux or temperature over the four surfaces of the channel (Shah and London, 1978), such is not the case for the current application, where one of the surfaces is adiabatic. Correction factors may be used to account for departures from completely uniform thermal conditions over the channel perimeter (Phillips, 1988, 1990), but the effect becomes negligible when the width of the adiabatic surface is much smaller than the channel height ($A \ll 1$). Heat transfer correlations for fully developed laminar flow in rectangular channels with isothermal or longitudinally isoflux side walls are provided in Section 2.8.4. For *turbulent flow*, mixing reduces the influence of an adiabatic surface, as well as that of the corners, and, to a good approximation, heat transfer may be modeled by using circular tube correlations based on the channel hydraulic diameter, D_h. Moreover, because the entry length for turbulent flow is small, it is often reasonable to approximate the average Nusselt number over the length of the channel as the local Nusselt number for fully developed flow.

As shown in Figure 4.19, Phillips (1988, 1990) bases his model on the surface area associated with a unit channel/wall configuration, which is replicated by however many channels are machined in the heat sink. A unit total thermal resistance is defined as

$$R''_{th,tot} \equiv \frac{L(W + t_f)\Delta T_{max}}{q_{c,1}} \quad (4.18)$$

where $q_{c,1}$ is the heat rate associated with a unit surface area, $L(W + t_f)$, and $\Delta T_{max} \equiv T_{h,max} - T_{m,i}$. The analysis acknowledges the fact that, as the coolant temperature increases in the flow direction, the heater temperature will do likewise, and the driving potential for heat transfer is defined in terms of the maximum possible heater-to-fluid temperature difference, where $T_{h,max} = T_h(z = L)$.

Phillips divides the total unit resistance into six components, which include a *spreading resistance*, $R''_{th,spr}$, associated with discrete IC devices on the top surface of the heat source; a cumulative one-dimensional *conduction resistance* associated with the heater and substrate, $R''_{th,cnd} = (t_h/k_h) + (t_{sub}/k_{sub})$, a *contact resistance* between the two solids, $R''_{th,c}$; a *constriction resistance*, $R''_{th,cns}$, associated with two-dimensional conduction in proximity to the fin/channel base; a *convection resistance*, $R''_{th,cnv}$, associated with heat transfer to the coolant from both the base and the side walls of the channel; and a *bulk resistance*, $R''_{th,bulk}$, which is attributed to the temperature rise of the coolant as it flows through a channel.

The spreading resistance, which depends on the size and shape of the semiconductor devices (Joy and Schlig, 1970; Phillips, 1988), decreases with increasing device concentration (decreasing feature size) and, therefore, can be expected to make a progressively smaller contribution to the total resistance with increasing circuit integration. If the heat source is attached to a heat sink, such as a cold plate, the value of $R''_{th,c}$ would depend on the specific interface, which could be metallurgical, as in a soldered or brazed joint, gas filled, evacuated, or a thermal grease. There is no contact resistance, $R''_{th,c} = 0$, if the heat sink is an integral part of the heat source, as, for example, when channels are machined in the chip itself.

Phillips (1988, 1990) uses the following approximation suggested by Kraus and Bar-Cohen (1983) to determine the constriction resistance at the fin/channel base:

$$R''_{th,cns} = \frac{(W + t_f)}{\pi k_{sub}} \ln\left\{\frac{1}{\sin[0.5\pi t_f/(W + t_f)]}\right\} \quad (4.19)$$

By assuming that all of the heat is transferred to the side walls (fins), before being transferred by convection to the coolant, the expression provides a conservative (upper) estimate of the resistance. Consistent with this assumption, the resistance approaches zero as the channel width approaches zero and infinity as the fin thickness approaches zero, a condition for which use of the expression is inappropriate.

The convection resistance is determined by assuming resistances resulting from heat flow from the base of the channel and through the fin to act in parallel. From Eqs. 2.13 and 2.25, respectively, these resistances may be expressed as $R_{\text{th,cnv,b}} = 1/(\bar{h}LW)$ and $R_{\text{th,f}} = 1/[\eta_f \bar{h}(2LH)]$. With $(R_{\text{th,cnv}})^{-1} = (R_{\text{th,cnv,b}})^{-1} + (R_{\text{th,f}})^{-1}$, $R''_{\text{th,cnv}} = [L(W+t_f)]R_{\text{th,cnv}}$, and use of Eq. 2.21 for the efficiency of a fin with an adiabatic tip, it follows that

$$R''_{\text{th,cnv}} = \frac{W+t_f}{\bar{h}W + 2\bar{h}H\eta_f} \tag{4.20}$$

where $\eta_f = (\tanh mH)/mH$ and $m = (2\bar{h}/k_{\text{sub}}t_f)^{1/2}$. The bulk resistance, which is introduced to account for the increase in the mixed mean temperature of the coolant as it flows through a channel, is expressed as

$$R''_{\text{th,bulk}} = \frac{L(W+t_f)}{\dot{m}_{c,1}c_{p,f}} \tag{4.21}$$

where the mass flow rate of coolant through a single channel is $\dot{m}_{c,1} = \rho_f w_m(HW)$.

Focusing on VLSI chips with integral microchannels, for which $R''_{\text{th,spr}}$ may be neglected and $R''_{\text{th,c}} = 0$, Phillips (1988, 1990) expresses the total unit resistance as

$$R''_{\text{th,tot}} = R''_{\text{th,cnd}} + R''_{\text{th,cns}} + R''_{\text{th,cnv}} + R''_{\text{th,bulk}} \tag{4.22}$$

where the conduction resistance may be expressed as $R''_{\text{th,cnd}} = t_{\text{sub}}/k_{\text{sub}}$ and the constriction, convection and bulk resistances are given by Eqs. 4.19, 4.20, and 4.21, respectively. The foregoing expression may be used with Eq. 4.18 to determine the maximum temperature rise:

$$\Delta T_{\max} = T_{h,\max} - T_{m,i} = R''_{\text{th,tot}} q''_{c,1} + \Delta T_{\text{vis}} \tag{4.23}$$

where $q''_{c,1} = q_{c,1}/[L(W+t_f)]$ and ΔT_{vis} is added to account for the temperature rise of the coolant resulting from viscous heating in the channel. The effect is related to pressure losses, $\Delta p = p_i - p_o$, between the channel inlet and outlet (Fig. 4.17) and the attendant conversion from mechanical to thermal energy. With the energy production rate corresponding to $q_{\text{vis}} = \dot{m}_{c,1}\Delta p/\rho_f$ and the corresponding temperature rise obtained from the energy balance, $q_{\text{vis}} = \dot{m}_{c,1}c_{p,f}\Delta T_{\text{vis}}$, it follows that

$$\Delta T_{\text{vis}} = \frac{\Delta p}{\rho_f c_{p,f}} \tag{4.24}$$

The pressure drop between the inlet and outlet plenums of the heat sink is evaluated from the following expression:

$$\Delta p = \frac{\rho_f w_m^2}{2}(2K_{90}) + \frac{\rho_f w_m^2}{2}\left[K_c + K_e + 4f_{F,\text{app}}\left(\frac{L}{D_h}\right)\right] \quad (4.25)$$

This expression presumes the existence of 90° bends between the channel and its inlet and outlet plenums and accounts for channel entrance and exit losses, as well as friction losses within the channel. The loss coefficient associated with a 90° bend was estimated as $K_{90} \approx 1.2$, whereas the entrance and exit loss coefficients, K_c and K_e, were approximated by the results of Kays and London (1984) for flow in circular tubes. However, as discussed in Section 2.8.3, loss coefficients depend strongly on specific manifold conditions, and good engineering judgment must be exercised to obtain reasonable approximations to actual behavior.

It should be noted that the artifice of using a bulk resistance in the foregoing model may be circumvented. One option would be to replace Eq. 4.18 by the following expression:

$$R''_{\text{th,tot}} \equiv \frac{L(W + t_f)\Delta T_{\text{lm}}}{q_{c,1}} \quad (4.26)$$

where the total resistance is now expressed as

$$R''_{\text{th,tot}} = R''_{\text{th,cnd}} + R''_{\text{th,cns}} + R''_{\text{th,cnv}} \quad (4.27)$$

and ΔT_{lm} is a log-mean temperature difference defined in terms of the *average* heater temperature, $\overline{T}_h$:

$$\Delta T_{\text{lm}} \equiv \frac{T_{m,o} - T_{m,i}}{\ln[(\overline{T}_h - T_{m,i})/(\overline{T}_h - T_{m,o})]} \quad (4.28)$$

With the heat rate also expressed as

$$q_{c,1} = \dot{m}_{c,1} c_{p,f}(T_{m,o} - T_{m,i}) \quad (4.29)$$

the foregoing equations may be used to compute $\overline{T}_h$ and $T_{m,o}$.

Phillips (1988) presents iterative procedures and a computer code for using the foregoing model to predict thermal and hydrodynamic conditions for liquid-cooled, microchannel heat sinks. The procedures vary according to the type of constraint imposed on operating conditions. In one case, the designer may choose to fix the maximum allowable pressure drop Δp on the basis of limitations associated with the coolant pump or the structural integrity of the heat sink. Alternatively, the unit volumetric flow rate, $\dot{V}_{c,1}$, may be limited, again by the pump, or the objective may be to limit the unit pump power requirement, $\Delta p \dot{V}_{c,1}/\eta_p$, where η_p is the pump efficiency. In each case, the mean velocity w_m of the channel flow may be expressed in terms of the corresponding constraint. With specification of other system parameters, such as the coolant inlet temperature and properties, the heat dissipation, and the channel and plenum geometries, the solution procedure may be implemented and representative parametric calculations have been performed (Phillips, 1988, 1990; Phillips et al., 1988). A significant conclusion is that improved thermal performance and reduced pump power requirements may be achieved by maintaining turbulent, rather than laminar, flow within the microchannels of a chip. Turbulent flow may readily be maintained for channel widths in the range $200 < W < 300$ μm, which exceed those ($W \approx 60$ μm) originally considered by Tuckerman and Pease (1981, 1982) and for which laminar flow is maintained. Moreover, if one or more chips are joined to a liquid-cooled heat sink, turbulent flow should be maintained in larger micro- or millichannels to achieve improved thermal performance and reduced costs by using conventional machining techniques. In a study by Harms et al. (1997), it was found that improved thermal performance may be achieved by maximizing the depth H of the channels, while the opposite effect was determined by Copeland et al. (1995). Copeland (1995) and Copeland et al. (1995) also found that the thermal resistance may be reduced by using multiply sequenced inlet and outlet manifolds (Harpole and Eninger, 1991), while Goodson et al. (1997) concluded that a significant reduction in thermal resistance may be achieved by using a microchannel heat sink made of chemical-vapor-deposited diamond instead of silicon.

It should be noted that more sophisticated modeling approaches may be used. For example, finite-difference or finite-element numerical procedures may be used to predict detailed temperature fields in a chip and/or substrate; such procedures could also be used to predict flow conditions within a channel. For example, Weisberg et al. (1992) performed a two-dimensional conjugate analysis to solve concurrently for thermal conditions in the solid and fluid domains associated with laminar water flow through microchannels machined in a silicon wafer. Results were used to determine resistances, which, in turn, were used to assess the accuracy of simpler models.

Example 4.2

A chip that is 1 mm thick and $L = 10$ mm on a side is cooled by pumping water at a mean velocity of $w_m = 5$ m/s through microchannels machined in the chip (Fig. 4.17). Each channel is of width $W = 250$ μm, wall thickness $t_f = 150$ μm, and height $H = 600$ μm, and the substrate thickness and thermal conductivity are $t_{sub} = 400$ μm and $k_{sub} = 150$ W/m · K, respectively. If the inlet temperature of the water is $T_{m,i} = 15°C$ and the maximum allowable chip temperature is $T_{h,max} = 85°C$, what is the maximum heat rate that may be dissipated by the chip? What is the pressure drop and pump power requirement associated with flow through the channels?

Solution

Known: Dimensions of chip and microchannels. Inlet temperature and mean velocity of coolant. Maximum allowable chip temperature.

Find: Maximum heat dissipation by the chip. Pressure loss and pump power requirement associated with coolant flow.

Assumptions:

1. Steady-state conditions.
2. Turbulent flow.
3. Smooth channel walls.
4. Applicability of macro flow and heat transfer correlations.
5. Negligible viscous dissipation.
6. Adiabatic surface at base of channels.
7. Properties may be evaluated at the inlet temperature.

Properties: Appendix B ($T_{m,i}$ = 15°C). Water: ρ = 999 kg/m³, $c_{p,f}$ = 4188 J/kg · K, k_f = 0.588 W/m · K, μ = 0.00114 N · s/m², and Pr = 8.12.

Analysis: The heat dissipation for the chip is

$$q = N_c q_{c,1}$$

where the number of channels is $N_c = L/(W + t_f) = 0.01 \text{ m}/(250 + 150) \times 10^{-6}$ m = 25 and, from Eq. 4.18, the heat rate per channel is

$$q_{c,1} = \frac{L(W + t_f)\Delta T_{max}}{R''_{th,tot}}$$

From Eq. 4.22, the total unit thermal resistance is

$$R''_{th,tot} = R''_{th,cnd} + R''_{th,cns} + R''_{th,cnv} + R''_{th,bulk}$$

where

$$R''_{th,cnd} = \frac{t_{sub}}{k_{sub}} = \frac{400 \times 10^{-6} \text{ m}}{150 \text{ W/m} \cdot \text{K}} = 2.67 \times 10^{-6} \text{ m}^2 \cdot \text{K/W}$$

and, from Eq. 4.19, the constriction resistance is

$$R''_{th,cns} = \frac{(W + t_f)}{\pi k_{sub}} \ln\left\{\frac{1}{\sin[0.5\pi t_f/(W + t_f)]}\right\}$$

$$= \frac{(250 + 150) \times 10^{-6} \text{ m}}{\pi(150 \text{ W/m} \cdot \text{K})} \ln\left\{\frac{1}{\sin[0.5\pi(150/400)]}\right\}$$

$$= 4.99 \times 10^{-7} \text{ m}^2 \cdot \text{K/W}$$

From Eq. 4.20, the convection resistance is

$$R''_{th,cnv} = \frac{W + t_f}{\bar{h}(W + 2H\eta_f)}$$

where $\bar{h} = \overline{Nu}_{D_h} k_f / D_h$, $\eta_f = (\tanh mH)/mH$, and $m = (2\bar{h}/k_{sub}t_f)^{1/2}$. With $D_h = 4A_c/P = 4(HW)/2(H+W) = 2(600 \times 250)\mu m/(600+250)\mu m = 353\ \mu m$,

$$Re_{D_h} = \frac{\rho w_m D_h}{\mu} = \frac{999 \text{kg/m}^3 (5 \text{ m/s}) 353 \times 10^{-6}\text{ m}}{0.00114 \text{ N} \cdot \text{s/m}^2} = 1547$$

With transition known to occur at smaller Reynolds numbers for microchannels, it is reasonable to assume turbulent flow for the foregoing value of Re_{D_h}. Of the internal, turbulent flow correlations provided in Chapter 2, Eq. 2.157 is known to be most applicable at low Reynolds numbers. Hence, with $f_{F,fd} = (1.58 \ln Re_{D_h} - 3.28)^{-2} = (1.58 \ln 1547 - 3.28)^{-2} = 0.0144$ and Eq. 2.159 used to account for entrance effects,

$$\overline{Nu}_{D_h} = \frac{(f_{F,fd}/2)(Re_{D_h} - 1000) Pr}{1 + 12.7(f_{F,fd}/2)^{1/2}(Pr^{2/3} - 1)} \left[1 + \left(\frac{D_h}{L}\right)^{2/3} \right]$$

$$= \frac{0.0072(547)8.12}{1 + 12.7(0.0072)^{1/2}(4.04 - 1)} \left[1 + \left(\frac{353\ \mu m}{10^4\ \mu m}\right)^{2/3} \right] = 8.28$$

Hence, $\bar{h} = \overline{Nu}_{D_h} k_f / D_h = 8.28(0.588 \text{ W/m} \cdot \text{K})/(353 \times 10^{-6}\text{ m}) = 13{,}800 \text{ W/m}^2 \cdot \text{K}$, $m = (2\bar{h}/k_{sub}t_f)^{1/2} = (2 \times 13{,}800 \text{ W/m}^2 \cdot \text{K}/150 \text{ W/m} \cdot \text{K} \times 150 \times 10^{-6}\text{ m})^{1/2} = 1108 \text{ m}^{-1}$, $mH = 0.665$, $\tanh mH = 0.582$, $\eta_f = 0.876$, and

$$R''_{th,cnv} = \frac{(250 + 150) \times 10^{-6}\text{ m}}{13{,}800 \text{ W/m}^2 \cdot \text{K}(250 + 2 \times 600 \times 0.876) \times 10^{-6}\text{ m}}$$

$$= 2.23 \times 10^{-5}\text{ m}^2 \cdot \text{K/W}$$

With $\dot{m}_{c,1} = \rho w_m A_c = 999 \text{ kg/m}^3 (5 \text{ m/s})(1.5 \times 10^{-7}\text{ m}^2) = 7.49 \times 10^{-4}\text{ kg/s}$,

$$R''_{th,bulk} = \frac{L(W + t_f)}{\dot{m}_{c,1} c_{p,f}} = \frac{0.01 \text{ m}(250 + 150) \times 10^{-6}\text{ m}}{7.49 \times 10^{-4}\text{ kg/s}(4188 \text{ J/kg} \cdot \text{K})}$$

$$= 1.27 \times 10^{-6}\text{ m}^2 \cdot \text{K/W}$$

and the total resistance is

$$R''_{tot} = (2.67 \times 10^{-6} + 4.99 \times 10^{-7} + 2.23 \times 10^{-5} + 1.27 \times 10^{-6})\text{ m}^2 \cdot \text{K/W}$$

$$= 2.67 \times 10^{-5}\text{ m}^2 \cdot \text{K/W}$$

Hence, the heat rate per channel is

$$q_{c,1} = \frac{0.01 \text{ m}(250 + 150) \times 10^{-6}\text{ m}(85 - 15)°\text{C}}{2.67 \times 10^{-5}\text{ m}^2 \cdot \text{K/W}} = 10.5 \text{ W}$$

and the heat rate for the chip is

$$q = N_c q_{c,1} = 25 \times 10.5 \text{ W} = 263 \text{ W}$$

From Eq. 4.25, the pressure drop is

$$\Delta p = \frac{\rho w_m^2}{2} \left[2K_{90} + K_c + K_e + 4 f_{F,\text{app}} \left(\frac{L}{D_h} \right) \right]$$

where $f_{F,\text{app}}$ is obtained from Eq. 2.147. With $A = 0.0929 + 1.0161/(L/D_h) = 0.129$, $B = -0.2680 - 0.3193/(L/D_h) = -0.279$, an aspect ratio of $A = W/H = 0.417$ and $D_e/D_h = [(2/3)+(11/24)A(2-A)] = 0.969$ from Eq. 2.144, and $Re_{D_e} = Re_{D_h}(D_e/D_h) = 1547(0.969) = 1500$, it follows that

$$f_{F,\text{app}} = A\,Re_{D_e}^B = 0.129(1500)^{-0.279} = 0.0168.$$

Hence, with $K_{90} \approx 1.2$, $K_c \approx 0.5$, and $K_e \approx 1.0$ (see Section 2.8.3),

$$\Delta p \approx \frac{999 \text{ kg/m}^3 (5 \text{ m/s})^2}{2} \left[2.4 + 0.5 + 1.0 + 4(0.0168)\frac{10 \text{ mm}}{0.353 \text{ mm}} \right]$$

$$\approx 7.25 \times 10^4 \text{ N/m}^2 = 72.5 \text{ kPa}$$

With $\dot{\forall} = N_c \dot{\forall}_1 = N_c w_m A_c = 25(5 \text{ m/s})(250 \times 600)10^{-12} \text{ m}^2 = 1.88 \times 10^{-5} \text{ m}^3/\text{s}$ and a pump efficiency of 100%, the power requirement is

$$P = \dot{\forall} \Delta p = 1.88 \times 10^{-5} \text{ m}^3/\text{s}(7.25 \times 10^4 \text{ N/m}^2) = 1.36 \text{ W}$$

Comments:

1. From Eq. 4.29, the outlet temperature of the coolant is $T_{m,o} = T_{m,i} + q_{c,1}/\dot{m}_{c,1} c_{p,f} = 15°C + 10.5 \text{ W}/(7.49 \times 10^{-4} \text{ kg/s} \times 4188 \text{ J/kg} \cdot \text{K}) = 18.35°C$. Hence, evaluation of the fluid properties at $T_{m,i} = 15°C$ is appropriate.
2. With $\Delta T_{\text{vis}} = \Delta p/\rho c_{p,f} = 7.25 \times 10^4 \text{ N/m}^2/(999 \text{ kg/m}^3 \times 4188 \text{ J/kg} \cdot \text{K}) = 0.017°C$, it is certainly reasonable to have neglected viscous heating effects.
3. The system could be optimized by selecting values of W and t_f that maximize q.

4.6 SUMMARY

Attractive options for dissipating chip heat fluxes by means of single-phase, liquid immersion cooling involve the use of channel flow conditions. The substrate to which the chips are mounted could comprise one or more walls of a channel through which the liquid coolant is forced, or microchannels could be machined within the chips. The results discussed in this chapter provide a useful, albeit far from complete, knowledge base for assessing the merits of such options. A complicating feature relates to the numerous length scales that can characterize specific operating conditions and the attendant difficulties in obtaining generalizations or identifying optimal conditions. Nevertheless, channel flows are highly compatible with chip packaging requirements and, as for air cooling, represent a promising vehicle for implementing liquid immersion cooling.

CHAPTER 5

JET IMPINGEMENT COOLING

5.1 INTRODUCTION

One of the most promising options for dissipating large heat fluxes, while maintaining acceptable surface temperatures, involves cooling by liquid jet impingement. Upon impingement, the jet forms very thin hydrodynamic and thermal boundary layers (δ, $\delta_{th} \sim 0.01$ to 0.1 mm), for which convection heat transfer coefficients may be well in excess of 10^4 W/m^2·K and corresponding surface-to-coolant thermal resistances are well below 1 cm^2·K/W. Such conditions may be achieved with a dielectric liquid, relatively simple manifold and nozzle designs, and comparatively low flow rates.

In this chapter, the literature associated with liquid jet impingement heat transfer is reviewed, with the goal of providing a suitable knowledge base for electronic cooling system design. The review considers both *circular* and *rectangular (planar or slot)* jets operating under *free-surface* or *submerged* conditions, as well as *unconfined* or *semiconfined* jets. *Cross-flow* effects associated with *multiple* jets are also considered, as are the effects of jet intermittency and liquid dispersion in the form of sprays. Related hydrodynamic conditions are described in Section 2.7.

5.2 CIRCULAR, UNCONFINED, FREE-SURFACE JETS

5.2.1 Stagnation Zone

As discussed in Section 2.7.2, as a free-surface jet approaches an impingement surface, a *stagnation zone* is created in which the jet is decelerated and accelerated in directions normal and parallel to the surface, respectively. Within this zone (Fig.

2.16a), the favorable pressure gradient that accelerates flow in the streamwise direction also acts to suppress turbulence associated with the preimpingement jet. Within the zone, boundary layer thicknesses are uniform and thin, thereby providing a uniform and large convection coefficient across the zone.

Theoretical Results for Laminar Flow For a circular jet, laminar boundary layer flow and heat transfer in the stagnation region are governed by the following forms of the continuity, momentum, and energy equations, respectively:

$$\frac{1}{r}\frac{\partial (ru)}{\partial r} + \frac{\partial v}{\partial y} = 0 \tag{5.1}$$

$$u\frac{\partial u}{\partial r} + v\frac{\partial u}{\partial y} = u_\infty(r)\frac{du_\infty(r)}{dr} + v\frac{\partial^2 u}{\partial y^2} \tag{5.2}$$

$$u\frac{\partial T}{\partial r} + v\frac{\partial T}{\partial y} = \alpha_f \frac{\partial^2 T}{\partial y^2} \tag{5.3}$$

where $u_\infty(r) = C_r r$ characterizes radial flow in the inviscid region outside the boundary layer (Section 2.7.2).

Expressing the velocity components in terms of the stream function, $u = (1/r)(\partial \psi / \partial y)$ and $v = -(1/r)(\partial \psi / \partial r)$, a similarity solution may be obtained by introducing a transformation of the form (Falkner and Skan, 1931; Mangler, 1948)

$$\psi \equiv r \left(\frac{rvu_\infty}{2}\right)^{1/2} f(\eta) \tag{5.4}$$

where the similarity parameter η is defined as

$$\eta \equiv y \left(\frac{2u_\infty}{vr}\right)^{1/2} \tag{5.5}$$

The transformed momentum equation is then

$$f''' + ff'' = \left(\frac{1}{2}\right)(f'^2 - 1) \tag{5.6}$$

where primes signify the order of differentiation with respect to η and $f'(\eta) = u/u_\infty$. Similarly, for an isothermal surface and introduction of the dimensionless temperature difference,

$$\theta(\eta) \equiv \frac{T_s - T}{T_s - T_i} \tag{5.7}$$

the transformed energy equation becomes

$$\theta'' + Pr\, f\theta' = 0 \tag{5.8}$$

The momentum and energy equations, Eqs. 5.6 and 5.8, have been solved numerically for $f(\eta)$ and $\theta(\eta)$, where knowledge of $\theta(\eta)$ may be used to determine the stagnation zone Nusselt number. With the convection coefficient given by $h_o = -k_f \partial T/\partial y|_{y=0}/(T_s - T_i) = k_f(2u_\infty/\nu r)^{1/2}\theta'(0)$, the corresponding expression for the Nusselt number is

$$Nu_{o,r} \equiv \frac{h_o r}{k_f} = (2)^{1/2}[\theta'(0)]Re_r^{1/2} \qquad (5.9)$$

where $Re_r = u_\infty r/\nu$. However, because u_∞ is linearly proportional to r in the stagnation zone, $u_\infty = C_r r$, $Nu_{o,r}$ is proportional to r and h_o is a constant, independent of r. The function $\theta'(0)$ is obtained from the numerical solution, and, for liquids (Pr > 3), it may be approximated as (Lienhard, 1995)

$$\theta'(0) = 0.601 Pr^{1/3} - 0.0508 \qquad (5.10)$$

Substituting from Eq. 2.112, the Nusselt number may be expressed in terms of the dimensionless velocity gradient G_r:

$$Nu_{o,r} = \frac{h_o r}{k_f} = (2)^{1/2}[\theta'(0)]\left(\frac{V_i r^2 G_r}{\nu D_i}\right)^{1/2} \qquad (5.11)$$

Canceling the radial coordinate and multiplying both sides of the equation by the diameter of the impinging jet, it follows that

$$Nu_{o,D_i} = \frac{h_o D_i}{k_f} = (2)^{1/2}[\theta'(0)]Re_{D_i}^{1/2} G_r^{1/2} \qquad (5.12)$$

where $Re_{D_i} \equiv V_i D_i/\nu$. Note the strong dependence of h_o on the jet diameter and velocity, as well as on the velocity gradient, which, in turn, depends strongly on nozzle discharge conditions. Note also that the combination of a uniform surface temperature and a uniform convection coefficient corresponds to the existence of a uniform heat flux, where $q''_{s,o} = h_o(T_s - T_i)$. Treating the thermal boundary layer as a planar conduction layer of thickness δ_{th} and invoking Fourier's law, $q''_{s,o} = k_f(T_s - T_i)/\delta_{th,o}$, it follows that

$$\delta_{th,o} = \frac{k_f}{h_o} = \frac{D_i}{Nu_{o,D_i}} \qquad (5.13)$$

The expression for the Nusselt number may be completed by substituting for G_r and $\theta'(0)$. For a laminar jet with a *uniform velocity profile*, $G_r = 0.916$ (Section 2.7.2), in which case, Eqs. 5.10 and 5.12 yield

$$Nu_{o,D_i} = 1.353(0.601 Pr^{1/3} - 0.0508) Re_{D_i}^{1/2} \qquad (5.14)$$

or, neglecting the constant of 0.0508,

$$Nu_{o,D_i} = 0.813 Re_{D_i}^{1/2} Pr^{1/3} \qquad (5.15)$$

This result is approximately 9% larger than the following empirical correlation by Liu et al. (1993):

$$Nu_{o,D_i} = 0.745 Re_{D_i}^{1/2} Pr^{1/3} \quad (5.16)$$

which was developed from data obtained using sharp-edged orifice nozzles, with $Re_{D_i} \leq 130,000$, $Pr \geq 3$, and $r/D_i \leq 0.7$. For $r/D_i \geq 0.7$, the convection coefficient decays rapidly with increasing r.

For a parabolic velocity profile, with V_i defined as the average jet velocity, $G_r = 4.646$ and, neglecting the constant of 0.0508, Eqs. 5.10 and 5.12 yield

$$Nu_{o,D_i} = 1.832 Re_{D_i}^{1/2} Pr^{1/3} \quad (5.17)$$

This result is approximately 11% larger than the following correlation proposed by Scholtz and Trass (1970):

$$Nu_{o,D_i} = 1.648 Re_{D_i}^{1/2} Pr^{1/3} \quad (5.18)$$

which is based on data obtained for jets emerging from tubes long enough to yield a parabolic velocity profile. Hence, use of the correlation is limited to $Re_{D_n} < 2300$. Results for the uniform and parabolic velocity profiles provide lower and upper bounds, respectively, for h_o.

The foregoing results apply under conditions for which the effects of surface tension and turbulence may be neglected. For $Re_{D_n} < 2300$, surface tension may act to increase the jet diameter at the nozzle exit, thereby reducing the Nusselt number (Webb and Ma, 1995). At intermediate to large Reynolds numbers, jet turbulence intensifies and may influence heat transfer in the stagnation zone.

Effects of Turbulence Turbulence may influence boundary layer development on an impingement surface by providing disturbances in the free-stream fluid outside the boundary layer. In the absence of a pressure gradient, increased turbulence in the free stream has no other effect than to advance the transition from a laminar to a turbulent boundary layer. However, with a favorable pressure gradient, increasing free-stream turbulence can enhance heat transfer across a laminar boundary layer by as much as a factor of 2 (Kestin, 1966).

Sutera et al. (1963) proposed that the mechanism for enhancement of stagnation flow heat transfer is the *amplification of vorticity* associated with free-stream turbulence. Vorticity, which is oriented in the same direction as flow acceleration along a surface, is stretched, with the attendant amplification creating a system of counterrotating longitudinal vortices just outside the laminar stagnation boundary layer. VanFossen and Simoneau (1987) have shown that flow in the boundary layer between the longitudinal vortices and the wall is three dimensional and that heat transfer is enhanced at locations where vortex motion is toward the surface. However, although *vortex amplification* provides a physical basis for explaining the relationship of jet turbulence to heat transfer enhancement in the impingement zone, it has yet to be used as basis for predicting enhancement.

Difficulties in accounting for the effects of jet turbulence relate to the fact that turbulence intensity measurements are sparse and that few investigators have attempted to isolate the separate effects of jet turbulence and the streamwise velocity gradient on stagnation heat transfer. Pan et al. (1992) attempted to resolve these effects for the experimental conditions of Stevens et al. (1992) and concluded that stagnation heat transfer was determined principally by the velocity gradient and was influenced only weakly by turbulence. They proposed the following correlation:

$$Nu_{o,D_n} = 0.69 Re_{D_n}^{1/2} Pr^{0.4} G_r^{1/2} \tag{5.19}$$

where $G_r \approx 1.2$, 1.8, and 2.2 for a convergent nozzle, tubular nozzle, and sharp-edged orifice, respectively. The correlation applies for $15{,}000 \leq Re_{D_n} \leq 48{,}000$ and $r/D_n \leq 0.7$.

For fully developed turbulent flow exiting a tubular nozzle ($G_r = 1.8$), Eq. 5.19 reduces to

$$Nu_{o,D_n} = 0.93 Re_{D_n}^{1/2} Pr^{0.4} \tag{5.20}$$

which is in good agreement with correlations developed by Faggiani and Grassi (1990) and DiMarco et al. (1994), but exceeds the laminar flow correlation, Eq. 5.16, of Liu et al. (1993) by approximately 25%. For the converging nozzle, which reduces the velocity gradient and the turbulence, $Gr \approx 1.2$ and Eq. 5.19 reduces to

$$Nu_{o,D_n} = 0.76 Re_{D_n}^{1/2} Pr^{0.4} \tag{5.21}$$

which is in good agreement with Eq. 5.16.

The effects of free-stream turbulence increase with increasing Reynolds number and are manifested by a stronger dependence on this parameter (Faggiani and Grassi, 1990; Gabour and Lienhard, 1994). For $25{,}000 < Re_{D_n} < 85{,}000$, Gabour and Lienhard (1994) recommend a correlation of the form

$$Nu_{o,D_n} = 0.278 Re_{D_n}^{0.633} Pr^{1/3} \tag{5.22}$$

Alternatively, Faggiani and Grassi (1990) represented the Reynolds number dependence in terms of two exponents. For jets discharged from tubular nozzles providing fully developed turbulent flow, data corresponding to a uniform surface heat flux and $S/D_n = 5$ yielded correlations of the form

$$Nu_{o,D_n} = 1.10 Re_{D_n}^{0.473} Pr^{0.4} \qquad Re_{D_n} < 77{,}000 \tag{5.23a}$$

$$Nu_{o,D_n} = 0.229 Re_{D_n}^{0.615} Pr^{0.4} \qquad Re_{D_n} > 77{,}000 \tag{5.23b}$$

Equations 5.16 and 5.18, which represent lower and upper limits for laminar flow, as well as Eqs. 5.20, 5.22, and 5.23b, which are representative of moderately to highly turbulent flows, are plotted in Figure 5.1. Although laminar flow may be extended to large Reynolds numbers ($Re_{D_n} \to 10^5$) a sharp-edged orifice nozzle,

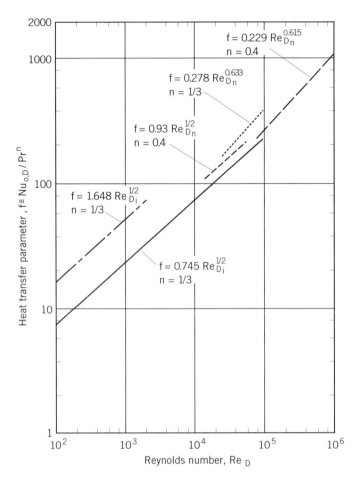

FIGURE 5.1 Nusselt number correlations for the stagnation zone associated with impingement of a circular, free-surface jet.

larger Nusselt numbers are associated with maintaining turbulent flow for an equivalent Reynolds number.

Because the effects of nozzle-to-surface spacing fall within the uncertainty of the foregoing correlations for $1 \leq S/D_n \leq 30$ (Lienhard, 1995; Webb and Ma, 1995), they may be neglected.

5.2.2 Boundary Layer Development Regions

For a laminar impinging jet, Figure 2.17, hydrodynamic and thermal boundary layers develop downstream of the *stagnation zone* ($r > r_s$), eventually reaching the free surface and undergoing transition to turbulence. As the velocity boundary layer grows, the radial velocity gradient, G_r, decreases, while $u_\infty(r) = V_s$ increases to V_i at $r/D_i \approx 2.23$.

Assuming a uniform surface heat flux, $Pr > 1$, and negligible evaporation from the free surface, Liu and Lienhard (1989) and Liu et al. (1991) used integral solution techniques to determine heat transfer in the downstream regions. For the boundary layer region ($r_s/D_i < r/D_i \leq r_v/D_i$), the following correlation was recommended:

$$Nu_{D_i} = 0.632 Re_{D_i}^{1/2} Pr^{1/3} \left(\frac{D_i}{r}\right)^{1/2} \qquad r_s < r < r_v \qquad (5.24)$$

For $r_v < r < r_{th}$, the velocity boundary layer has reached the free surface and V_s begins to decrease, while the thermal boundary layer continues to grow and temperatures remain at T_i for $y > \delta_{th}$. However, for $Pr > 5.23$, which is typical of dielectric liquids, thermal boundary layer development is insufficient for δ_{th} to reach the film thickness t before the onset of transition to turbulence and $r_c < r_{th}$ (Lienhard, 1995). Hence, the following results obtained for $r_v < r < r_{th}$ may be used for $r_v < r < r_c$.

$r_v < r \leq r_c$:

$$Nu_{D_i} = \frac{0.407(Re_{D_i} Pr)^{1/3}(D_i/r)^{2/3}}{[0.171(D_i/r)^2 + 5.147(r/D_i)/Re_{D_i}]^{2/3}[(r/D_i)^2/2 + C_3]^{1/3}} \qquad (5.25a)$$

where

$$C_3 = \frac{0.267(D_i/r_v)^{1/2} Re_{D_i}^{-1/2}}{[0.171(D_i/r_v)^2 + 5.147(r_v/D_i)/Re_{D_i}]^2} - \frac{1}{2}\left(\frac{r_v}{D_i}\right)^2 \qquad (5.25b)$$

For $Pr < 5.23$ and $r_c > r_{th}$, the results obtained for $r_{th} < r < r_c$ (Liu et al., 1991) are within a few percent of predictions based on Eqs. 5.25, in which case, Eqs. 5.25a and 5.25b can be used to a good approximation for $Pr > 1$.

Beyond the location for which the turbulence becomes fully developed ($r > r_t$), the following result may be used to determine the local Nusselt number:

$r \geq r_t$

$$Nu_{D_i} = \frac{8 Re_{D_i} Pr \, f(C_f, Pr)}{49(t/D_i)(r/D_i) + 28(r/D_i)^2 f(C_f, Pr)} \qquad (5.26a)$$

where

$$f(C_f, Pr) = \frac{C_f/2}{1.07 + 12.7(Pr^{2/3} - 1)(C_f/2)^{1/2}} \qquad (5.26b)$$

$$C_f = 0.073 \left[\frac{(r/D_i)}{Re_{D_i}}\right]^{1/4} \qquad (5.26c)$$

and the film thickness t is given by Eqs. 2.121a and 2.121b. Assuming that the Nusselt number increases linearly with r in the region extending from the onset of turbulence (r_c) to fully developed turbulence (r_t), the following result may be used to predict the local Nusselt number:

$r_c < r < r_t$:

$$Nu_{D_i} = Nu_{D_i}(r_c) + [Nu_{D_i}(r_t) - Nu_{D_i}(r_c)]\frac{r - r_c}{r_t - r_c} \quad (5.27)$$

The Nusselt numbers at r_c and r_t are obtained from Eqs. 5.25 and 5.26, respectively.

Predictions based on the foregoing expressions are in good agreement with measurements made using a sharp-edged orifice nozzle. In the boundary layer and similarity regions, Nu_{D_i} decreases with increasing r, and the decay is most pronounced just beyond the stagnation zone. The Nusselt number ceases to decrease or increases with the onset of transition to turbulence and decays when the turbulence becomes fully developed. These effects are illustrated in Figure 5.2, which was generated using Eqs. 5.16 and 5.24 to 5.27 for a representative dielectric liquid ($Pr = 20$).

Effects of Turbulence The foregoing results apply for impinging jets that are laminar. In the case of turbulent jets, the laminar model can significantly underpredict heat transfer in the wall jet region and the transition to turbulent flow can be advanced to much smaller values of r_c.

For fully turbulent jets issuing from tubular nozzles, Stevens and Webb (1991) recommend the following correlation:

$$\frac{Nu_{D_i}}{Nu_{o,D_i}} = \left\{1 + \left[f\left(\frac{r}{D_i}\right)\right]^{-9}\right\}^{-1/9} \quad (5.28a)$$

where

$$f\left(\frac{r}{D_i}\right) = a \exp\left[b\left(\frac{r}{D_i}\right)\right] \quad (5.28b)$$

and the coefficients (a, b), which depend on the jet diameter, are listed in Table 5.1. The stagnation Nusselt number may be obtained from an appropriate correlation, such as that given by Eq. 5.20, 5.22, or 5.23.

The foregoing correlation applies for $(r/D_i) < (r_c/D_i) \approx 3$, beyond which transition to turbulence in the wall jet region causes $Nu_{D_i}/Nu_{o,D_i}$ to increase and subsequently to decay with increasing r/D_i. Lienhard et al. (1992) developed a model that predicts the effect of splattering on local heat transfer in the posttransition region ($r > r_c$).

5.2.3 Average Heat Transfer Coefficients

In electronic cooling applications, it is often the *average* convection coefficient, $\overline{h}$, that is desired. From knowledge of this coefficient, the total rate of heat transfer by

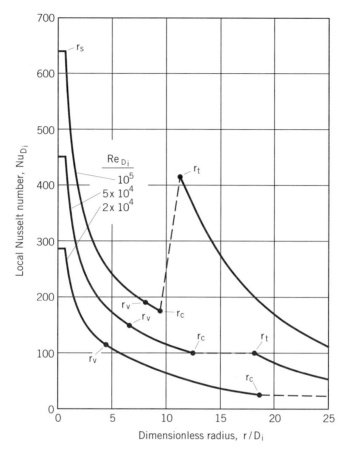

FIGURE 5.2 Local Nusselt number as a function of radius for laminar jet impingement of a representative dielectric liquid ($Pr = 20$).

convection from the surface may be determined:

$$q = \overline{h} A_s (\overline{T}_s - T_i) \qquad (5.29)$$

where $\overline{T}_s$ and T_i are the average surface temperature and the jet impingement temperature, respectively.

TABLE 5.1 Coefficients of Equation 5.28b

D_i (mm)	2.2	4.1	5.8	8.9
a	1.15	1.34	1.48	1.57
b	−0.23	−0.41	−0.56	−0.70

Metzger et al. (1974) measured the average convection coefficients for water and oil jets impinging on an isothermal circular heater for $2200 < Re_{D_n} < 140{,}000$, $3 < Pr < 150$, $1.75 < D_h/D_n < 25$, and $1 < S/D_n < 3$, where D_h is the heater diameter. The results showed a strong dependence on the relative size of the heater, D_h/D_n, and no dependence on the nozzle-to-surface spacing, S/D_n. The first result is a consequence of the large variation in the local convection coefficient with radial location from the stagnation point. The second result is not surprising, as jet velocities were sufficiently large to preclude an effect of gravitational acceleration between the nozzle exit and the heater surface.

For conditions closer to those of electronic cooling, Jiji and Dagan (1988) measured average convection coefficients for a square, heated surface of width $L_h = 12.7$ mm and nozzle diameters of 0.5 and 1 mm. Results for water and FC-77 were correlated as a function of Re_{D_n}, Pr, and L/D_n over the range $2800 < Re_{D_n} < 20{,}000$.

Womac et al. (1993) also considered free-surface water ($Pr \approx 7$) and FC-77 ($Pr \approx 25$) jets impinging on a square, nearly isothermal heater of width $L_h = 12.7$ mm. Nozzle diameters D_n ranged from 0.978 to 6.55 mm, while nozzle-to-surface spacings varied from 3.5 to 10. As shown in Figure 5.3, good agreement between data for the two fluids is obtained by plotting results in terms of the parameter $\overline{Nu}_{D_n}/Pr^{0.4}$. Although the Nusselt number is only weakly dependent on S/D_n, because the nozzle diameter is the length scale used in the Nusselt number, $\overline{Nu}_{D_n}$ in-

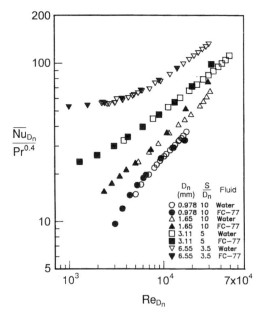

FIGURE 5.3 Average Nusselt number for impingement of a free-surface jet on a square (12.7 mm × 12.7 mm) heater as a function of Reynolds number, nozzle diameter, and nozzle-to-heater separation. Adapted from Womac et al. (1993).

creases with increasing D_n. Although an increase in D_n for fixed Re_{D_n} corresponds to a reduction in V_n and, hence, a corresponding reduction in $\bar{h}$, use of D_n as a characteristic length in the Nusselt number more than compensates for the reduction in $\bar{h}$.

On the logarithmic scale of Figure 5.3, the data only partly fall along straight lines, with significant deviations existing in the lower-Reynolds-number range for nozzle diameters of 0.978 and 6.55 mm. For $D_n = 0.978$ mm, the slope is initially large but diminishes with increasing Re_{D_n}. In contrast, the data for $D_n = 6.55$ mm exhibit slopes that are initially small but increase with increasing flow rate. These opposing trends are the result of separate physical phenomena. At very low flow rates with small-diameter jets, the thickness of the thermal boundary layer forming on the heater surface is comparable to that of the thin film. If the developing thermal boundary layer reaches the free surface, significant downstream attenuation of heat transfer may be expected because of bulk warming of the fluid. Hence, the Nusselt number would diminish rapidly with decreasing Reynolds number for small-diameter jets at low flow rates. Estimates of the thermal boundary layer thickness at low Reynolds numbers suggest that the 0.978-mm-diameter data are affected by this phenomenon.

Data for the 6.55-mm-diameter nozzle deviate from linearity at low Reynolds numbers because of acceleration of the jet by gravity. With $V_i = (V_n^2 + 2gS)^{1/2}$ from Eq. 2.108, the impingement velocity for $S/D_n = 3.5$ exceeds V_n by factors of 4.6 for $Re_{D_n} = 1000$ and 1.28 for $Re_{D_n} = 3000$. If the Nusselt number for the 6.55-mm nozzle is replotted against a Reynolds number based on the impingement velocity, the data no longer deviate from linearity at low flow rates (Womac et al., 1990).

At high flow rates with small diameter jets of FC-77, visual observation revealed that fluid splattered from the free surface in the vicinity of the impingement region, thereby reducing the thickness of the liquid film and allowing the thermal boundary layer to reach the free surface. The expected reduction in heat transfer becomes noticeable for the 0.978-mm jet of FC-77 at high Reynolds numbers.

Womac et al. (1993) correlated their data by using an area-weighted average of standard correlations for the impingement and wall jet regions. The impingement region correlation is of the form

$$\frac{\overline{Nu}_{D_i}}{Pr^{0.4}} = C_1 Re_{D_i}^m \tag{5.30}$$

with the Reynolds number defined in terms of the impingement velocity V_i and diameter $D_i = (V_n/V_i)^{1/2} D_n$. As indicated by Eqs. 5.17 and 5.18, a value of $m = 0.5$ was selected on the basis of the known behavior for local heat transfer at the stagnation point. The wall jet region correlation is of the form

$$\frac{\overline{Nu}_{L^*}}{Pr^{0.4}} = C_2 Re_{L^*}^n \tag{5.31}$$

where the Reynolds number is defined in terms of the impingement velocity and the average length L^* of the wall jet region for a square heater

$$L^* = \frac{0.5(\sqrt{2}L_h - D_i) + 0.5(L_h - D_i)}{2} \quad (5.32)$$

An area-weighted combination of the foregoing correlations yields

$$\frac{\overline{Nu_L}}{Pr^{0.4}} = C_1 Re_{D_i}^m \frac{L_h}{D_i} A_r + C_2 Re_{L^*}^n \frac{L_h}{L^*}(1 - A_r) \quad (5.33)$$

where $A_r = \pi D_i^2 / 4L_h^2$. With $m = 0.5$ and properties evaluated at an arithmetic mean (film) temperature of $T_f = (\overline{T}_s + T_i)/2$, optimal values of $C_1 = 0.516$, $C_2 = 0.491$, and $n = 0.532$ were determined using a least-squares technique to correlate water and FC-77 data obtained with the three larger nozzles for the Reynolds number range $1000 < Re_{D_n} < 51{,}000$. Because data for the 0.978-mm nozzle were affected by bulk warming, they were not utilized in developing the correlation. The correlation is compared with all of the data in Figure 5.4, and apart from a few conditions for $D_n = 0.978$ mm, agreement is within ±15%. Womac et al. (1993) also found good agreement between the correlation and data obtained by Stevens and Webb (1991) for water.

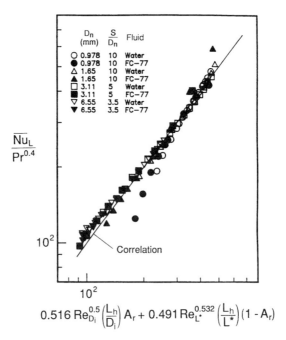

FIGURE 5.4 Correlation of data for a free-surface jet impinging on a square (12.7 mm × 12.7 mm) heater. Adapted from Womac et al. (1993).

By integrating the local heat transfer coefficients associated with Eq. 5.28, Stevens and Webb (1991) obtained the following correlation for fully turbulent jets issuing from tubular nozzles:

$$\frac{\overline{Nu_{D_i}}}{Nu_{o,D_i}} = \left\{1 + \left[g\left(\frac{r}{D_i}\right)\right]^{-7}\right\}^{-1/7} \quad (5.34a)$$

where

$$g\left(\frac{r}{D_i}\right) = \frac{2}{(r/D_i)^2}\left[\frac{f(r/D_i)}{b}\left(\frac{r}{D_i} - \frac{1}{b}\right) + \frac{a}{b^2}\right] \quad (5.34b)$$

The function $f(r/D_i)$ is given by Eq. 5.28b, and the coefficients (a, b) are obtained from Table 5.1. As a first approximation, the correlation may be applied to square heaters of width L by replacing r with an average heater radius, $r = 0.5(\sqrt{2}L+L)/2$.

5.3 RECTANGULAR, UNCONFINED, FREE-SURFACE JETS

5.3.1 Stagnation Zone

As for the circular jet, Figure 2.16a, a *stagnation zone* is created when a rectangular jet approaches an impingement surface, Figure 2.16b, and the favorable pressure gradient $(dp/dx < 0)$ that accelerates flow in the streamwise direction suppresses turbulence associated with the preimpingement jet. Within the zone, thin boundary layers of uniform thickness provide a large, uniform convection coefficient across the zone.

Theoretical Results for Laminar Flow   Flow and heat transfer in the stagnation region are governed by the following forms of the continuity, momentum, and energy equations, respectively:

$$\frac{\partial u}{\partial x} + \frac{\partial v}{\partial y} = 0 \quad (5.35)$$

$$u\frac{\partial u}{\partial x} + v\frac{\partial u}{\partial y} = u_\infty(x)\frac{du_\infty(x)}{dx} + \nu\frac{\partial^2 u}{\partial y^2} \quad (5.36)$$

$$u\frac{\partial T}{\partial x} + v\frac{\partial T}{\partial y} = \alpha_f\frac{\partial^2 T}{\partial y^2} \quad (5.37)$$

where $u_\infty(x) = C_x x$ characterizes longitudinal flow in the inviscid region outside the boundary layer (Section 2.7). As for the circular jet, a similarity solution may be obtained by introducing the following definition of the stream function:

$$\psi \equiv (\nu x u_\infty)^{1/2} f(\eta) \quad (5.38)$$

where

$$\eta \equiv y \left(\frac{u_\infty}{\nu x}\right)^{1/2} \tag{5.39}$$

With $u = (\partial \psi / \partial y)$ and $v = -(\partial \psi / \partial x)$, the transformed momentum equation is

$$f''' + ff'' = f'^2 - 1 \tag{5.40}$$

and, for an isothermal surface, the energy equation is equivalent to that for the circular jet, Eq. 5.8, where the dimensionless temperature difference θ is defined by Eq. 5.7.

Equations 5.40 and 5.8 may be solved numerically (Falkner and Skan, 1931) to obtain $\theta(\eta)$, which may, in turn, be differentiated and evaluated at $\eta = 0$ to obtain $\theta'(0)$. With the stagnation zone convection coefficient defined as $h_o = -k_f \partial T / \partial y|_{y=0} / (T_s - T_i) = k_f (u_\infty / \nu x)^{1/2} \theta'(0)$, the corresponding expression for the Nusselt number is

$$Nu_{o,x} = \frac{h_o x}{k_f} = [\theta'(0)] Re_x^{1/2} \tag{5.41}$$

where $Re_x = u_\infty x / \nu$. However, because u_∞ is linearly proportional to x in the stagnation zone, $u_\infty = C_x x$, $Nu_{o,x}$ is proportional to x and h_o is a constant, independent of x. Hence, because $q_s'' = h(T_s - T_i)$, uniform surface temperature and heat flux are coexisting conditions.

Substituting for $u_\infty(x)$ from Eq. 2.113, the Nusselt number may be expressed as

$$Nu_{o,x} \equiv \frac{h_o x}{k_f} = [\theta'(0)] \left[\frac{V_i x^2 G_x}{\nu w_i}\right]^{1/2} \tag{5.42}$$

Canceling the rectangular coordinate and multiplying both sides of the equation by the width of the impinging jet, it follows that

$$Nu_{o,w_i} \equiv \frac{h_o w_i}{k_f} = [\theta'(0)] Re_{w_i}^{1/2} G_x^{1/2} \tag{5.43}$$

where $Re_{w_i} \equiv V_i w_i / \nu$ and, for liquid jets ($Pr > 3$), the function $\theta'(0)$ may be expressed as (Lienhard, 1995)

$$\theta'(0) = 0.661 Pr^{1/3} - 0.0765 \tag{5.44}$$

From Eq. 5.43, note the strong dependence of h_o on the impingement jet width and velocity, as well as on the velocity gradient, which, in turn, depends strongly on nozzle discharge conditions. As for the circular jet, the thermal boundary layer thickness is uniform and may be approximated as $\delta_{th,o} = k_f / h_o = w_i / Nu_{o,w_i}$.

The expression for the Nusselt number may be completed by substituting for G_x and $\theta'(0)$. Neglecting the constant in Eq. 5.44 and, from Eq. 2.114, selecting $G_x =$

0.785 for a *laminar jet* with a *uniform velocity profile*, it follows that

$$Nu_{o,w_i} = 0.586 Re_{w_i}^{1/2} Pr^{1/3} \qquad (5.45)$$

Alternatively, based on the Falkner–Skan similarity solution for stagnation flow, the results of Levy (1952) and Evans (1962) may be expressed as

$$Nu_{o,w_i} = 0.570 G_x^{1/2} Re_{w_i}^{1/2} Pr^{0.375} \qquad (5.46)$$

or, with $G_x = 0.785$,

$$Nu_{o,w_i} = 0.505 Re_{w_i}^{1/2} Pr^{0.375} \qquad (5.47)$$

This result agrees well with data obtained by Inada et al. (1981) for $Re_{w_i} = 940$ and $S/w_n > 1.5$.

For a *laminar jet* with a *parabolic velocity profile*, use of $G_x = 3$ with Eq. 5.43 yields

$$Nu_{o,w_i} = 1.145 Re_{w_i}^{1/2} Pr^{1/3} \qquad (5.48)$$

Alternatively, with $G_x = 3.2$, Sparrow and Lee (1975) obtained the following result:

$$Nu_{o,w_i} = 1.02 Re_{w_i}^{1/2} Pr^{0.375} \qquad (5.49)$$

Effects of Turbulence As for circular jets, turbulence may significantly increase heat transfer relative to predictions tied to laminar models. McMurray et al. (1966) report data that exceed predictions based on Eq. 5.47 by approximately 40% for $69,000 < Re_{w_i} < 137,000$. Miyasaka and Inada (1980) report data that exceed predictions by approximately 80% for $10,500 < Re_{w_i} < 157,000$.

Using a highly convergent nozzle to reduce turbulence and to create a nearly uniform velocity profile, Vader et al. (1991) correlated stagnation zone $(x/w_i < 1)$ data by an expression of the form

$$Nu_{o,w_i} = 0.28 Re_{w_i}^{0.58} Pr^{0.4} \qquad (5.50)$$

which exceeds predictions based on Eq. 5.47 by up to 50%. In contrast, for $17,000 < Re_{w_i} < 79,000$ and $x/w_i < 0.5$, Wolf et al. (1990) obtained data for fully developed turbulent jets emerging from a long parallel-plate nozzle and correlated their results by an expression of the form

$$Nu_{o,w_i} = 0.116 Re_{w_i}^{0.71} Pr^{0.4} \qquad (5.51)$$

Predictions based on this correlation exceed those of Eq. 5.50 by as much as 80%, which is attributed to increased levels of turbulence and nonuniformity of the velocity profile.

Correlations for a laminar jet, Eqs. 5.47 and 5.48, a weakly turbulent jet, Eq. 5.50, and a fully turbulent jet, Eq. 5.51, are compared in Figure 5.5. Predictions for the

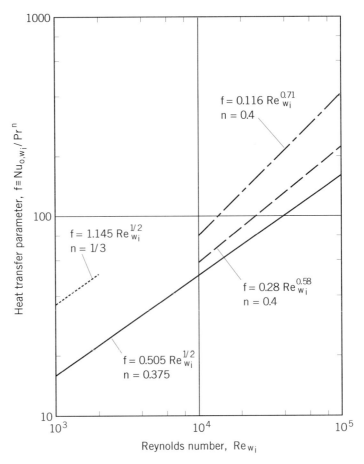

FIGURE 5.5 Nusselt number correlations for the stagnation zone associated with impingement of a planar, free-surface jet.

laminar jet with a uniform velocity profile provide a lower limit for the heat transfer parameter, while the upper limit is associated with a laminar jet of parabolic velocity profile or a fully turbulent jet. However, although laminar jets may be maintained at large Reynolds numbers through the use of a sharp-edged orifice, the corresponding velocity is more nearly uniform. A laminar jet with a parabolic profile can only be discharged from a parallel-plate channel, in which case, the Reynolds number range is restricted to $Re_{w_i} < 2000$.

Wolf et al. (1995b, c) investigated concurrent influence of the free-stream velocity gradient G_x and free-stream turbulence intensity, $\tilde{v} = v'/\bar{V}$, on heat transfer in the stagnation zone for converging and parallel-plate nozzles. Data were obtained in the ranges $29,900 \leq Re_{w_i} \leq 61,200$, $0.79 < G_x < 1.13$, and $0.0116 < \tilde{v} < 0.0547$ and were obtained with and without placement of wire screens and grids at the nozzle exit. Although data were correlated to within ±26% by an expression of the form

$$Nu_{o,w_i} = 0.202 Re_{w_i}^{0.620} Pr^{0.4} \quad (5.52a)$$

inclusion of the velocity gradient and turbulence intensity yielded the following correlation:

$$Nu_{o,w_i} = 0.849 Re_{w_i}^{0.584} G_x^{0.360} \tilde{v}^{0.263} Pr^{0.4} \quad (5.52b)$$

and agreement to within 7% of the data. Because the dependence on turbulence intensity is approximately of the form $Nu_{o,w_i} \sim \tilde{v}^{1/4}$, a twofold increase in the intensity would only increase heat transfer by approximately 20%, suggesting little justification for the use of special turbulence generators.

Because the foregoing results are based on the width w_i and mass average velocity V_i of the *impinging* jet, they may be used over a wide range of nozzle-to-surface spacings.

5.3.2 Boundary Layer Development Regions

For weakly turbulent jets discharged from a converging nozzle, Vader et al. (1991) measured the local convection coefficient as a function of distance x from the stagnation line of an isoflux surface, Figure 2.16*b*, and representative results are shown in Figure 5.6. As the thermal boundary layer thickens, the resistance to heat transfer increases and the local coefficient decays from the maximum associated with the

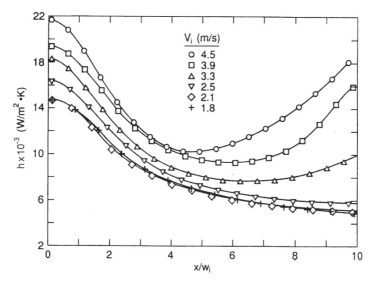

FIGURE 5.6 Longitudinal distribution of the local Nusselt number for a planar, free-surface jet impinging on a heated surface. Adapted from Vader et al. (1991).

stagnation line. However, for $V_i > 2.5$ m/s, a local minimum results from transition to turbulence and the coefficient increases, approaching a maximum comparable to that associated with the stagnation zone. Although not shown in Figure 5.6, the coefficient subsequently declines with increasing x because of the development of a fully turbulent boundary layer.

Adopting the scaling parameters suggested by McMurray et al. (1966), Vader et al. (1991) correlated their pretransition data by an expression of the form

$$Nu_x = 0.89 Re_x^{0.48} Pr^{0.4} \tag{5.53}$$

where $Nu_x = hx/k$ is the local Nusselt number and $Re_x = u_\infty x/\nu$ is the corresponding Reynolds number. Properties are evaluated at the film temperature, $T_f = (T_s + T_\infty)/2$, and the local free-stream velocity $u_\infty(x)$ is evaluated from Eqs. 2.113, 2.114, and 2.122. The correlation is recommended for $100 < Re_x < Re_{x,c}$, where the critical Reynolds number for transition to turbulence was found to be $Re_{x,c} = 3.6 \times 10^5$. For $Re_x < 100$, Eq. 5.50 was recommended. In the posttransition region corresponding to $6 \times 10^5 < Re_x < 2.5 \times 10^6$, McMurray et al. (1966) recommended the following correlation:

$$Nu_x = 0.037 Re_x^{4/5} Pr^{1/3} \tag{5.54}$$

Note that the critical Reynolds number, $Re_{x,c}$, associated with transition on the heated surface decreases with increasing jet turbulence, and values as low as 1.5×10^5 have been reported (Wolf et al., 1995c).

Wolf et al. (1990) obtained heat transfer data for fully developed turbulent jets impinging on an isoflux surface and correlated their results by an expression of the form

$$Nu_{w_i} = f\left(\frac{x}{w_i}\right) Re_{w_i}^{0.71} Pr^{0.4} \tag{5.55a}$$

where

$$f\left(\frac{x}{w_i}\right) = 0.116 + \left(\frac{x}{w_i}\right)^2 \left[0.00404 \left(\frac{x}{w_i}\right)^2 - 0.00187 \left(\frac{x}{w_i}\right) - 0.0199\right] \tag{5.55b}$$

for $0 < (x/w_i) < 1.6$, or

$$f\left(\frac{x}{w_i}\right) = 0.111 - 0.020 \left(\frac{x}{w_i}\right) + 0.00193 \left(\frac{x}{w_i}\right)^2 \tag{5.55c}$$

for $1.6 < (x/w_i) < 6$. As for the stagnation zone, predictions based on this correlation exceed those obtained by Vader et al. (1991), which are associated with jets of lower turbulence intensity and a more uniform velocity profile. The correlation applies in the wall jet region upstream of the transition to turbulence and for Reynolds numbers in the range $1.7 \times 10^4 < Re_{w_i} < 7.9 \times 10^4$.

Average heat transfer coefficients may be obtained by integrating the appropriate correlation for the local convection coefficient over the heater surface.

5.4 CIRCULAR, UNCONFINED, SUBMERGED JETS

5.4.1 Stagnation Zone

As with the free-surface jet, heat transfer for a submerged jet depends on its turbulence intensity and the radial velocity gradient in the stagnation zone. In turn, both quantities depend on the nozzle geometry and the nozzle-to-surface spacing.

If the impingement surface is within the potential core of the jet ($S < L_{pc}$), the Nusselt number has been found to increase slightly with increasing S (Gardon and Akfirat, 1965; Nakatogawa et al., 1970; Sun et al., 1993). With increasing S, the centerline velocity of the jet is preserved, whereas the turbulence intensity of the jet increases because of mixing with the adjoining fluid. Hence, the Nusselt number should increase with increasing S, achieving a maximum at $S \approx L_{pc}$. For $S > L_{pc}$, the turbulence intensity continues to increase with increasing S, but its effect on heat transfer is less than that of the decaying centerline velocity. Hence, the Nusselt number decreases.

Ma and Bergles (1988), Ma et al. (1990), and Sun et al. (1993) obtained stagnation point heat transfer data for submerged liquid jets issuing from a tubular nozzle of $D_n \approx 1.0$ mm. For the Reynolds number range $130 < Re_{D_n} < 29{,}000$ and $S < L_{pc}$, the Nusselt number was assumed to be independent of S and correlated by an expression of the form

$$Nu_{o,D_n} = Nu_{o,D_n,\max} = 1.29 Re_{D_n}^{1/2} Pr^{1/3} \qquad S < L_{pc} \qquad (5.56)$$

For $S > L_{pc}$, the decay in the Nusselt number was correlated by an expression of the form

$$\frac{Nu_{o,D_n}}{Nu_{o,D_n,\max}} = \left[\frac{L_{pc}/D_n}{S/D_n}\right]^{1/2} \qquad (5.57)$$

where the value of L_{pc}/D_n was found to range from 4.5 to 8.0, depending on the Reynolds number and the nature of the liquid. For $Re_{D_n} > 4000$, the length of the potential core may be approximated as $L_{pc}/D_n \approx 6$.

5.4.2 Boundary Layer Development Regions

As fluid flows radially from the stagnation point, the thermal boundary layer thickens and the convection coefficient decreases accordingly. However, depending on the

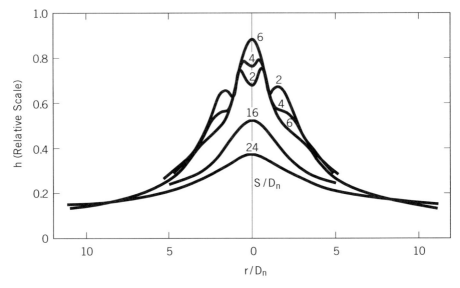

FIGURE 5.7 Effect of separation distance on radial distribution of the local heat transfer coefficient for a turbulent submerged jet of fixed Reynolds number. Adapted from Gardon and Cobonpue (1961).

value of S/D_n, a secondary peak may occur in the local value of h. As shown in Figure 5.7, for $S/D_n > 6$, the radial distribution of the local convection coefficient is bell shaped. However, with decreasing S/D_n, secondary peaks begin to form, along with a local minimum at the stagnation point. The secondary peak at $r/D_n \approx 0.6$ may be due to a local minimum in the boundary layer thickness, whereas the peak at $r/D_n \approx 1.8$ is commonly associated with a transition to turbulence resulting from loss of the favorable pressure gradient. The peak becomes more prominent with increasing Re_{D_n} (Martin, 1977). Local heat transfer measurements for submerged water jets emerging from a tubular nozzle ($D_n = 0.99$ mm) in fully developed turbulent flow (Sun et al., 1993) also revealed the existence of a slight secondary maximum at $r/D_n = 1.9$. The maximum was only observed for $Re_{D_n} > 20,000$ and was attributed to transition from laminar to turbulent flow.

Although secondary maxima in the radial distribution have been observed, correlations of such distributions have not been reported in the literature. Working with data obtained for water emerging from a 0.99-mm-diameter tube in fully developed turbulent flow, Sun et al. (1993) proposed a correlation of the form

$$\frac{Nu_{D_n}}{Nu_{o,D_n}} = \left\{ \left[\left(\frac{\tanh(0.88 r/D_n)}{(r/D_n)} \right)^{1/2} \right]^{-17} + \left[1.69 \left(\frac{r}{D_n} \right)^{-1.07} \right]^{-17} \right\}^{-1/17}$$

(5.58)

However, it is only by deleting the factor of 0.88 in the hyperbolic tangent that the correct limit is obtained as $r/D_n \to 0$. Hence, it may be necessary to delete this factor from the correlation. The expression applies for impingement within the potential core ($S/D_n < 6$). With increasing $S/D_n > 6$, radial variations become less pronounced, with a reduction in h at all locations.

5.4.3 Average Heat Transfer Coefficients

Because total heat transfer from an impingement surface is a manifestation of conditions in the stagnation and boundary layer development regions, Sitharamayya and Raju (1969) chose to correlate data obtained for submerged water jets impinging on a circular heater by superposing contributions for the two regions. For parameter values in the ranges $2000 < Re_{D_n} < 40{,}000$, $1.74 < D_n < 12.7$ mm, $8 < D_h/D_n < 58$, and $1 < S/D_n < 7$, their correlation is of the form

$$\overline{Nu}_{D_n} = \left[32.5 Re_{D_n}^{0.523} + 0.266 \left(\frac{D_h}{D_n} - 8\right) Re_{D_n}^{0.828}\right] \left(\frac{D_n}{D_h}\right)^2 Pr^{1/3} \qquad (5.59)$$

Although the average convection coefficient is independent of the nozzle separation distance, it decreases significantly with increasing heater diameter.

Womac et al. (1993) performed experiments for submerged jets of water and FC-77 impinging on a square (12.7 mm × 12.7 mm) heater for nozzle diameters and spacings in the range $0.978 < D_n < 6.55$ mm and $1 < S/D_n < 14.5$. The data are approximately independent of S/D_n for $1 < S/D_n < 4$, which is within the jet's potential core. As they did for the free-surface jet, the authors correlated their data by using an area-weighted average of correlations for the stagnation and boundary layer development regions. The stagnation zone was assumed to extend to $r = 1.9 D_n$ and to be followed by transition to a turbulent wall jet. The area-weighted combination of correlations for the two regions was expressed as

$$\frac{\overline{Nu}_L}{Pr^{0.4}} = C_1 Re_{D_n}^m \frac{L_h}{D_n} A_r + C_2 Re_{L^*}^n \frac{L_h}{L^*}(1 - A_r) \qquad (5.60)$$

where $A_r = \pi(1.9 D_n)^2 / L_h^2$ and the average length of the wall jet region is

$$L^* = \frac{(0.5\sqrt{2} L_h - 1.9 D_n) + (0.5 L_h - 1.9 D_n)}{2} \qquad (5.61)$$

The values of m and n were set equal to 0.5 and 0.8, respectively, and optimal values of C_1 and C_2 were obtained by a least-squares technique using only the data obtained for separation distances in the range $1.5 < S/D_n \leq 4$. The values were determined to be $C_1 = 0.785$ and $C_2 = 0.0257$ and are close to the values of 0.76 proposed by Nakatogawa et al. (1970) for stagnation point heat transfer and 0.0296 for turbulent, parallel flow over a flat plate (Incropera and DeWitt, 1996). The correlation was determined for both water ($Pr \approx 7$) and FC-77 ($Pr \approx 25$) data over parameter

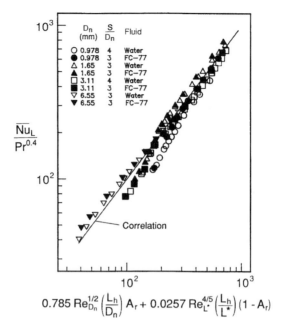

FIGURE 5.8 Correlation of data for a submerged jet impinging on a square (12.7 mm × 12.7 mm) heater. Adapted from Womac et al. (1993).

ranges corresponding to $1.65 \leq D_n \leq 6.55$ mm and $4500 < Re_{D_n} < 51{,}000$. A comparison of the correlation with all of the data is shown in Figure 5.8.

Womac et al. (1993) also compared their water and FC-77 data to a correlation developed for submerged, gas-jet impingement on a circular heater over the parameter ranges $2 < S/D_n < 12$, $5 < D_h/D_n < 15$, and $2000 < Re_{D_n} < 4 \times 10^5$ (Martin, 1977). The correlation is of the form

$$\overline{Nu}_{D_n} = \frac{\frac{D_n}{D_h}\left[2 - 4.4\frac{D_n}{D_h}\right]}{\left[1 + 0.2\left(\frac{S}{D_n} - 6\right)\left(\frac{D_n}{D_h}\right)\right]} F(Re_{D_n}) Pr^{0.42} \quad (5.62\text{a})$$

where

$$F(Re_{D_n}) = 2Re_{D_n}^{1/2}\left[1 + \frac{Re_{D_n}^{0.55}}{200}\right]^{1/2} \quad (5.62\text{b})$$

To facilitate the comparison, Womac et al. (1993) defined the following equivalent diameter for their heat source:

$$D_{\text{equiv}} = 0.5(L_h + \sqrt{2}L_h) \quad (5.63)$$

Except for data corresponding to the largest nozzle ($D_n = 6.55$ mm), which correspond to $(D_{equiv}/D_n) = 2.34$ and, hence, are outside the range of applicability of Eq. 5.62, there is good agreement with the correlation. This result suggests that, with an appropriate Prandtl number exponent, correlations obtained for submerged gas jets may be used to a reasonable approximation for liquid jets.

Womac et al. (1993) compared heat transfer data for free-surface and submerged jets, and, for both FC-77 and water, Nusselt numbers for impingement within the potential core of the submerged jet exceeded those of the free-surface jet for $Re_{D_n} > 4000$. The larger Nusselt numbers obtained for the submerged jet were attributed to heat transfer enhancement by turbulence generated in the free shear layer of the jet.

Example 5.1

A circular, free-surface jet of FC-77 is used to cool an electronic chip of width $L_h = 15$ mm on a side. The jet is discharged from a tube whose length-to-diameter ratio is sufficient to maintain fully developed conditions at the tube exit. If the maximum allowable chip temperature is $T_s = 85°C$, what is the maximum allowable heat dissipation q for nominal operating conditions of $D_n = 3$ mm, $V_n = 3$ m/s, and $T_i = 15°C$? How is the heat dissipation affected by variations in D_n and V_n?

Solution

Known: Nominal operating conditions for a circular, free-surface jet of FC-77 used to cool a chip of prescribed width and maximum allowable temperature.

Find: Maximum allowable heat dissipation for the nominal operating conditions. Effect of variations in nozzle diameter and velocity on heat dissipation.

Assumptions:

1. Steady-state conditions.
2. Isothermal chip.
3. Heat transfer is exclusively by convection to the impinging jet.
4. Applicability of the Womac correlation, Eq. 5.33.
5. Negligible difference in nozzle discharge and jet impingement conditions ($D_n = D_i$, $V_n = V_i$).

Properties: FC-77 ($T_f = 50°C$): From Appendix B: $k_f = 0.061$ W/m · K, $\nu = 5.39 \times 10^{-7}$ m²/s, and $Pr = 16.5$.

Analysis: With the prescribed values of C_1, C_2, m, and n, Eq. 5.33 is of the form

$$\frac{\overline{Nu_L}}{Pr^{0.4}} = 0.516 Re_{D_i}^{0.5} \frac{L_h}{D_i} A_r + 0.491 Re_{L^*}^{0.532} \frac{L_h}{L^*}(1 - A_r)$$

where, for the nominal conditions,

$$A_r = \pi D_i^2/4L_h^2 = \pi(0.003 \text{ m})^2/4(0.015 \text{ m})^2 = 0.0314$$

and

$$Re_{D_i} = V_i D_i/\nu = 3 \text{ m/s}(0.003 \text{ m})/5.39 \times 10^{-7} \text{ m}^2/\text{s} = 16{,}700.$$

From Eq. 5.32, the average length of the wall jet region is

$$L^* = \frac{0.5(\sqrt{2}L_h - D_i) + 0.5(L_h - D_i)}{2} = 0.00755 \text{ m}$$

Hence, with $Re_{L^*} = V_i L^*/\nu = 3 \text{ m/s}(0.00755 \text{ m})/5.39 \times 10^{-7} \text{ m}^2/\text{s} = 42{,}000$,

$$\frac{\overline{Nu}_L}{Pr^{0.4}} = 0.516(16{,}700)^{0.5} \frac{0.015 \text{ m}}{0.003 \text{ m}}(0.0314)$$

$$+ 0.491(42{,}000)^{0.532} \frac{0.015 \text{ m}}{0.00755 \text{ m}}(0.969)$$

$$\overline{Nu}_L = 283 Pr^{0.4} = 283(16.5)^{0.4} = 868$$

$$\overline{h} = \overline{Nu}_L \frac{k_f}{L_h} = \frac{868(0.061 \text{ W/m} \cdot \text{K})}{0.015 \text{ m}} = 3530 \text{ W/m}^2 \cdot \text{K}$$

$$q = \overline{h}A_s(T_s - T_i) = 3530 \text{ W/m}^2 \cdot \text{K}(0.015 \text{ m})^2(85 - 15)°\text{C} = 55.6 \text{ W}$$

The corresponding value of the average heat flux for the surface is

$$q'' = \frac{q}{A_s} = \frac{55.6 \text{ W}}{(0.015 \text{ m})^2} = 2.47 \times 10^5 \text{ W/m}^2 = 24.7 \text{ W/cm}^2$$

If the jet diameter is fixed at $D_i = 3$ mm and the velocity is increased over the range $1 \leq V_i \leq 5$ m/s, $\overline{h}$, q, and q'' increase from 1970 to 4630 W/m²·K, 31.0 to 72.9 W, and 13.8 to 32.4 W/cm², respectively. The increase in heat transfer with increasing V_i is significant, but is limited by the small Reynolds-number exponents of $m, n \sim 0.5$. If the velocity is fixed at $V_i = 3$ m/s and the diameter is increased over the range $1 \leq D_i \leq 5$ mm, $\overline{h}$, q and q'' increase from 3330 to 3706 W/m²·K, 52.4 to 58.4 W, and 23.3 to 25.9 W/cm², respectively. This comparatively small effect is attributable to the fact that, even for $D_i = 5$ mm, the area ratio is small ($A_r = 0.087$) and heat transfer is dominated by conditions in the wall jet region.

Comments:

1. Conditions at the stagnation point may be estimated from Eq. 5.22. Having assumed that $D_n = D_i$ and $V_n = V_i$, $Re_{D_n} = Re_{D_i} = 16{,}700$ for the nominal conditions, in which case,

$$Nu_{o,D_n} = 0.278 Re_{D_n}^{0.633} Pr^{1/3} = 0.278(16,700)^{0.633}(16.5)^{1/3} = 333$$

$$h_o = Nu_{o,D_n} \frac{k_f}{D_n} = 333 \frac{0.061 \text{ W/m} \cdot \text{K}}{0.003 \text{ m}} = 6780 \text{ W/m}^2 \cdot \text{K}$$

The local heat flux at the stagnation point is then

$$q_o'' = h_o(T_s - T_i) = 6780 \text{ W/m}^2 \cdot \text{K}(70°\text{C})$$
$$= 4.74 \times 10^5 \text{ W/m}^2 = 47.4 \text{ W/cm}^2$$

Note that h_o and q_o'' are substantially larger than the average convection coefficient and heat flux, $\bar{h}$ and q'', for the entire surface

2. If cooling is effected within the potential core of a submerged jet, rather than by a free-surface jet, the average convection coefficient can be obtained from Eq. 5.60:

$$\frac{\overline{Nu_L}}{Pr^{0.4}} = 0.785 Re_{D_n}^{0.5} \frac{L_h}{D_n} A_r + 0.0257 Re_{L^*}^{0.8} \frac{L_h}{L^*}(1 - A_r)$$

For $D_n = 3$ mm, the average length of the wall jet region is

$$L^* = \frac{(0.5\sqrt{2}L_h - 1.9 D_n) + (0.5 L_h - 1.9 D_n)}{2}$$

$$= \frac{[0.5\sqrt{2}(0.015 \text{ m}) - 1.9(0.003 \text{ m})] + [0.5(0.015 \text{ m}) - 1.9(0.003 \text{ m})]}{2}$$

$$= 0.00335 \text{ m}$$

and the area ratio is $A_r = \pi(1.9 D_n)^2/L_h^2 = \pi(1.9 \times 0.003 \text{ m})^2/(0.015 \text{ m})^2 = 0.454$. For $V_n = 3$ m/s, $Re_{D_n} = 16,700$ and $Re_{L^*} = V_n L^*/\nu = 3$ m/s $\times 0.00335$ m/5.39×10^{-7} m^2/s $= 18,600$. Hence,

$$\overline{Nu_L} = (16.5)^{0.4} \left[0.785(16,700)^{0.5} \frac{0.015 \text{ m}}{0.003 \text{ m}}(0.454) \right.$$
$$\left. + 0.0257(18,600)^{0.8} \frac{0.015 \text{ m}}{0.0035 \text{ m}}(0.546) = 1190 \right]$$

$$\bar{h} = \overline{Nu_L} \frac{k_f}{L_h} = 1190 \frac{0.061 \text{ W/m} \cdot \text{K}}{0.015 \text{ m}} = 4830 \text{ W/m}^2 \cdot \text{K}$$

$$q = \bar{h} A_s (T_s - T_i) = 4830 \text{ W/m}^2 \cdot \text{K}(0.015 \text{ m})^2 (70 \text{ K}) = 76.1 \text{ W}$$

$$q'' = \frac{q}{A_s} = \frac{76.1 \text{ W}}{(0.015 \text{ m})^2} = 3.38 \times 10^5 \text{ W/m}^2 = 33.8 \text{ W/cm}^2$$

Contrasting the foregoing results with those for the free-surface jet, it is evident that a significantly larger convection coefficient and heat rate are associated with the submerged condition.

5.5 RECTANGULAR, UNCONFINED, SUBMERGED JETS

5.5.1 Stagnation Zone

Although heat transfer measurements for liquid planar jets are sparse, comprehensive measurements of local convection coefficients for planar gas jet impingement have been made by Gardon and Akfirat (1966) over the parameter ranges $1.59 < w_n < 6.35$ mm, $450 < Re_{w_n} < 50{,}000$, and $0.33 < S/w_n < 80$. For turbulent flow at the nozzle discharge ($Re_{w_n} > 1000$), a maximum in the stagnation convection coefficient existed for $S/w_n \approx 8$, which corresponds approximately to the length of the potential core. At these Reynolds numbers, turbulence along the centerline of the jet increases before the centerline velocity begins to decrease. Hence, even inside the potential core, where the centerline velocity is constant, an increase in separation distance for $S < S_{pc}$ increases jet turbulence and heat transfer at the impingement surface. For $S > S_{pc}$, the effect of the reduced jet arrival velocity exceeds the effect of enhancement resulting from increased turbulence, and the convection coefficient at the stagnation line decreases with increasing separation distance. A separation distance of eight jet widths represents an optimal combination of jet arrival velocity and turbulence intensity that maximizes heat transfer.

For $2000 < Re_{w_n} < 50{,}000$ and $8 < S/w_n < 60$, Gardon and Akfirat (1966) recommend a stagnation line correlation of the form

$$Nu_{o,w_n} = 1.2 Re_{w_n}^{0.58} \left(\frac{S}{w_n}\right)^{-0.62} \tag{5.64}$$

To a reasonable approximation, the result may be extended to liquid jets by including a Prandtl number effect and modifying the coefficient accordingly. Selecting $Pr^{0.4}$, the correlation becomes

$$Nu_{o,w_n} = 1.38 Re_{w_n}^{0.58} Pr^{0.4} \left(\frac{S}{w_n}\right)^{-0.62} \tag{5.65}$$

It is recommended that the value of Nu_{o,w_n} corresponding to $S/w_n = 8$ be used for $S/w_n < 8$.

At lower Reynolds numbers for which the nozzle discharge is laminar and turbulence generation in the shear layer is negligible, the stagnation convection coefficient was found to be approximately constant for impingement within the potential core and to decrease with increasing S/w_n for impingement outside the core (Gardon and Akfirat, 1966). Numerical simulations for laminar jet impingement also reveal that, for impingement within the potential core, the stagnation convection coefficient depends strongly on the velocity profile at the nozzle exit. For an equivalent Reynolds number, the convection coefficient associated with a parabolic profile is almost twice the value obtained for a uniform profile (Sparrow and Lee, 1975; Al-Sanea, 1992). However, although the higher midline momentum flux for the parabolic profile clearly enhances heat transfer in proximity to the stagnation line, results become virtually independent of the nozzle profile for $x/w_n \geq 1$.

5.5.2 Boundary Layer Development Regions

Results for local convection coefficients away from the stagnation line ($x > 0$) follow trends similar to those associated with submerged circular jets (Fig. 5.7). The representative results of Gardon and Akfirat (1966) reveal three distinct patterns in the streamwise variation of the coefficient for impinging *turbulent* jets. For $S/w_n \geq 14$, the distribution is bell shaped with a maximum at the stagnation line. For $8 \leq S/w_n \leq 14$, the coefficient still decreases monotonically with increasing x, but with an abrupt change in slope at $x/w_n \approx 4$, where the pressure gradient approaches zero. Finally, as the separation distance is reduced for $S/w_n \leq 8$, a secondary peak begins to form at $x/w_n \approx 7$ and becomes well defined for $S/w_n \leq 6$. The secondary peak is attributed to decay of the pressure gradient, which stabilizes the laminar boundary layer in the impingement zone, and to the attendant boundary layer transition from laminar to turbulent flow. Transition is manifested by an increase in the heat transfer coefficient, which is followed by a reduction as the boundary layer thickens. However, with increasing S, the attendant increase in the turbulence level of the jet resulting from mixing increases heat transfer coefficients in the stagnation region, diminishing the effect of boundary layer transition. For *laminar* flow, the convection coefficient decays monotonically with x (Gardon and Akfirat, 1966), although weak manifestations of a secondary peak exist for $S/w_n \leq 8$. Correlations for the longitudinal variation of the local convection coefficient have not been developed.

5.5.3 Average Heat Transfer Coefficients

Martin (1977) obtained the following correlation for the average convection coefficient associated with a planar gas jet:

$$\overline{Nu}_{w_n} = \frac{1.53}{0.5[(x/w_n) + (S/w_n) + 2.78]} Re_{D_h}^m Pr^{0.42} \qquad (5.66a)$$

where $D_h = 2w_n$ is the hydraulic diameter of the nozzle and

$$m = 0.695 - \left[\left(\frac{x}{2w_n}\right) + \left(\frac{S}{2w_n}\right)^{1.33} + 3.06\right]^{-1} \qquad (5.66b)$$

The range of conditions governed by the correlation are $4 \leq S/w_n \leq 20$, $1500 \leq Re_{w_n} \leq 45{,}000$, and $8 \leq x/w_n \leq 100$. As with Eq. 5.62a, inclusion of a Prandtl number term in the correlation permits its use as a first approximation for liquid jets.

Wadsworth and Mudawar (1990) measured average convection coefficients for a planar jet in the context of cooling a 3 × 3 array of 12.7 mm × 12.7 mm simulated chips. The cooling module is shown schematically in Figure 5.9. Coolant enters the module through a front cover plate, where it is deflected to establish a uniform pressure distribution in the chamber formed with the cooling block and, hence, uniform flow through each of the planar nozzles machined in the block. Equivalent jets impinge on each of the heaters in the component plate, and flow is confined to a

192 JET IMPINGEMENT COOLING

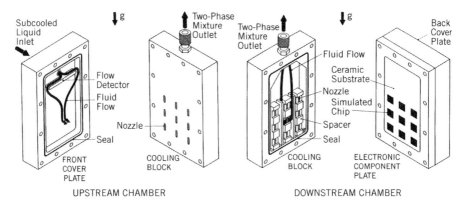

FIGURE 5.9 Schematic of module for planar jet impingement cooling of a multichip array (Wadsworth and Mudawar, 1990). Used with permission of ASME.

channel between the nozzle face and the chip (Figure 5.10). The V-shaped nozzle profile shown in the figure is one of three designs that were considered and for which equivalent heat transfer results were obtained. After exiting the channel, fluid (FC-72) is routed through the cooling block outlet. Experiments were performed for $0.127 \leq w_n \leq 0.508$ mm, $1000 \leq Re_{D_h} \leq 30{,}000$, and $0.127 \leq S \leq 5.08$ mm, with S/w_n in the range from 1 to 20.

Although the geometry of Figure 5.10 is more appropriately classified as semiconfined and the range of S/w_n exceeds that associated with the length of the potential core, the heat transfer data were virtually independent of S. Moreover, uniformity of cooling was demonstrated for each of the chips in the array. Accordingly, data for both a single chip and the entire array were correlated by the expression

$$\frac{\overline{Nu_L}}{Pr^{1/3}} = 3.06 Re_{D_h}^{0.50} + 0.099 Re_{D_h}^{0.664} \left[\frac{L - w_n}{w_n} \right]^{0.664} \quad (5.67)$$

The correlation was formulated using a superposition of contributions from the impingement and wall jet regions. The independence of $\overline{Nu_L}$ on S was attributed to the large values of L/w_n, which were in the range from 25 to 100. Gardon and Akfirat (1966) also found almost no dependence of average heat transfer on S for $L/w_n = 20$.

Example 5.2

A multichip module, similar to that illustrated in Figure 5.9, consists of chips that are of width $L_h = 15$ mm on a side and are each cooled by a single rectangular jet of width $w_n = 0.3$ mm, velocity $V_n = 1$ m/s, and temperature $T_i = 15°C$. The chip temperature T_h may not exceed 55°C, and either FC-72 or FC-77 may be used as a coolant. Which liquid is preferable? For this liquid, what are the effects of variations in w_n and V_n on the heat rate?

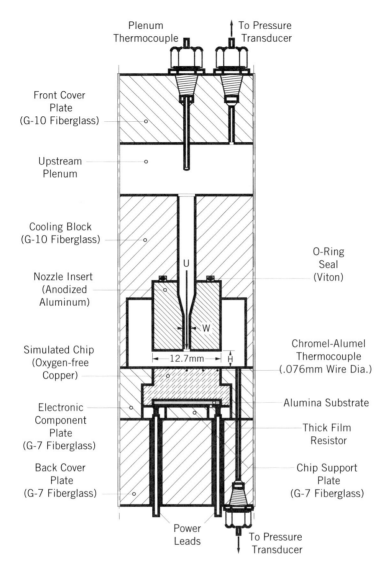

FIGURE 5.10 Sectional view of planar nozzle and heater geometry (Wadsworth and Mudawar, 1990). Used with permission of ASME.

Solution

Known: Operating conditions for a rectangular jet of dielectric liquid used to cool a chip. Chip dimensions and maximum allowable temperature.

Find: Preferred liquid, FC-72 or FC-77. Effect of jet width and velocity on heat transfer from a chip.

Assumptions:

1. Steady-state conditions.
2. Isothermal chips.
3. Heat transfer is exclusively by convection to the impinging jet.
4. Heat transfer is independent of the nozzle-to-chip spacing S and the nozzle type.

Properties: Appendix B ($T_f = 35°C$). FC-72: $k_f = 0.0561$ W/m·K, $\nu = 3.73 \times 10^{-7}$ m²/s, and $Pr = 11.7$. FC-77: $k_f = 0.0623$ W/m·K, $\nu = 6.85 \times 10^{-7}$ m²/s, and $Pr = 20.5$.

Analysis: The choice of a coolant may be made on the basis of which liquid provides the largest average convection coefficient and, hence, the largest heat rate. The convection coefficient may be determined from the correlation due to Wadsworth and Mudawar (1990), Eq. 5.67:

$$\frac{\overline{Nu}_L}{Pr^{1/3}} = 3.06 Re_{D_h}^{0.50} + 0.099 Re_{D_h}^{0.664} \left[\frac{L - w_n}{w_n}\right]^{0.664}$$

where $L = L_h = 0.015$ m, $w_n = 0.0003$, and $D_h = 2w_n = 0.0006$ m.

FC-72:

$$Re_{D_h} = \frac{V_n D_h}{\nu} = \frac{1 \text{ m/s}(0.0006 \text{ m})}{3.73 \times 10^{-7} \text{ m}^2/\text{s}} = 1610$$

$$\frac{\overline{Nu}_L}{Pr^{1/3}} = 3.06(1610)^{0.50} + 0.099(1610)^{0.664}\left[\frac{0.015 - 0.0003}{0.0003}\right]^{0.664} = 299$$

$$\overline{Nu}_L = 299(Pr^{1/3}) = 299(11.7)^{1/3} = 680$$

$$\bar{h} = \overline{Nu}_L \frac{k_f}{L} = 680\frac{0.0561 \text{ W/m·K}}{0.015 \text{ m}} = 2540 \text{ W/m}^2 \cdot \text{K}$$

$$q = \bar{h} A_s (T_h - T_i) = 2540 \text{ W/m}^2 \cdot \text{K}(0.015 \text{ m})^2(55 - 15)°C = 22.9 \text{ W}$$

FC-77:

$$Re_{D_h} = \frac{V_n D_h}{\nu} = \frac{1 \text{ m/s}(0.0006 \text{ m})}{6.85 \times 10^{-7} \text{ m}^2/\text{s}} = 876$$

$$\frac{\overline{Nu_L}}{Pr^{1/3}} = 3.06(876)^{0.50} + 0.099(876)^{0.664} \left[\frac{0.015 - 0.0003}{0.0003}\right]^{0.664} = 209$$

$$\overline{Nu_L} = 209(Pr^{1/3}) = 209(20.5)^{1/3} = 571$$

$$\bar{h} = \overline{Nu_L}\frac{k_f}{L} = 571\frac{0.0623 \text{ W/m} \cdot \text{K}}{0.015 \text{ m}} = 2370 \text{ W/m}^2 \cdot \text{K}$$

$$q = \bar{h} A_s (T_h - T_i) = 2370 \text{ W/m}^2 \cdot \text{K}(0.015 \text{ m})^2 (55 - 15)°\text{C} = 21.3 \text{ W}$$

Although it is characterized by superior thermal properties (larger values of k_f and P_r), the larger kinematic viscosity of FC-77 yields a much smaller Reynolds number and a slightly smaller convection coefficient and heat rate. Hence, FC-72 is the preferred fluid.

The effect of the nozzle width is manifested principally through the Reynolds number, Re_{D_h}, which increases with increasing w_n. If w_n is increased fivefold from 0.1 mm to 0.5 mm for a fixed value of $V_n = 1$ m/s, there is a 34% increase in $\overline{Nu_L}$, $\bar{h}$, and q, with values at 0.5 mm corresponding to 755, 2830 W/m² · K, and 25.4 W, respectively. The effect of the nozzle velocity is manifested exclusively through the Reynolds number. If V_n is increased fivefold from 1 m/s to 5 m/s, there is a 163% increase in $\overline{Nu_L}$, $\bar{h}$, and q, with values at 5 m/s corresponding to 1786, 6680 W/m² · K and 60.1 W, respectively.

Comments: Although FC-72 provides slightly better performance for equivalent values of T_h and V_n, its use is limited to $T_h < 56°$C, if boiling is to be avoided. FC-77 could be used to operate at chip temperatures approaching 100°C without boiling.

5.6 UNCONFINED MULTIPLE JETS

As discussed in Section 2.7.4, cooling with arrays of multiple jets in close proximity has the advantage of providing multiple stagnation zones, more uniform cooling, and a larger overall heat transfer coefficient. However, results depend on the nature of interactions between adjoining jets, as well as the manner in which spent fluid is discharged from the system. Discharge conditions for which jet impingement is unaffected by drainage of the spent fluid yield the largest convection coefficients.

5.6.1 Free-Surface Jets

Combining an approximate analysis of heat transfer from a symmetrical region of an isothermal surface cooled by a square array of free-surface jets with experimental results for a nine-jet array, Yonehara and Ito (1982) proposed a correlation for which dependence of the average Nusselt number on the Reynolds number and the dimensionless pitch (Fig. 2.22) was governed by a relation of the form

$$\overline{Nu}_{D_n} = 2.38 Re_{D_n}^{2/3} Pr^{1/3} \left(\frac{P_n}{D_n}\right)^{-4/3} \quad (5.68)$$

However, experimental conditions corresponded to large jet spacings, $13.8 < P_n/D_n < 330$, which would not be prototypic of electronic cooling requirements and for which uniformity of heat transfer under the array would not be achieved.

Working with a nine-jet square (in-line) array and a seven-jet triangular (staggered) array, and with water jet diameters of $D_n = 1, 2,$ and 3 mm, Pan and Webb (1994a, 1995) measured average heat transfer coefficients for $2 \leq P_n/D_n \leq 8$ and correlated the results by an expression of the form

$$\overline{Nu}_{D_n} = 0.225 Re_{D_n}^{2/3} Pr^{1/3} \exp\left[-0.095\left(\frac{P_n}{D_n}\right)\right] \quad (5.69)$$

The correlation applies for both the square and the triangular arrays, as well as for $2 \leq S/D_n \leq 5$ and $5000 \leq Re_{D_n} \leq 20{,}000$. In all cases, there was good drainage of the spent fluid, which flowed radially outward in lanes formed between the jets and originating at the center of the array. In the limit as $P_n/D_n \to 0$, the jets merge to form a single jet whose diameter is presumed to encompass the entire surface. The corresponding value of $\overline{Nu}_{D_n}$ may be estimated from Eq. 5.69. With increasing $P_n/D_n > 8$, predictions based on Eq. 5.69 merge with those based on Eq. 5.68, suggesting consistency in using the former for $P_n/D_n < 10$ and the latter for $P_n/D_n > 10$. Note, however, that both correlations pertain to conditions for which the area of the heater surface corresponds to that of the nozzle plate associated with the jet array and to conditions for which jet cross-flow effects are negligible.

Pan and Webb (1994a, b, 1995) also measured local convection coefficients on an isoflux surface for the square and triangular arrays of jets. The convection coefficient was measured as a function of the radial distance extending from the centerline of the central jet. The uniformity of the distribution improved with decreasing P_n/D_n and was maintained to within 5% for $P_n/D_n = 2$. Although the local Nusselt number was independent of P_n/D_n at the stagnation point, it decreased with increasing r and the decay became more pronounced with increasing P_n/D_n. The decay was not monotonic, however, with the Nusselt number subsequently increasing and reaching a secondary maximum, which occurred in the interaction zone between neighboring jets. The secondary maximum increased with decreasing P_n/D_n. Secondary maxima were also recorded in the interaction region between rows of closely spaced circular jets (Slayzak et al., 1994a).

Heat transfer measurements for arrays of circular jets have also been made by Jiji and Dagan (1988) and Womac et al. (1994). Both studies considered liquid jets of water and FC-77 emerging from tubular nozzles of 0.5 and 1 mm diameter and tube length-to-diameter ratios in the range ($4.7 \leq L_n/D_n \leq 12.8$). Hence, conditions corresponded to partially developed laminar or turbulent velocity profiles at the nozzle exit, depending on the Reynolds number. In both studies, spent fluid flowed radially from the array center.

Jiji and Dagan (1988) considered square arrays of four and nine jets and, for $2800 \leq Re_{D_n} \leq 20{,}000$, correlated their data by an expression of the form

$$\frac{\overline{Nu_L}}{Pr^{1/3}} = 3.84 \left[0.008 \frac{L_h}{D_n} N + 1\right] Re_{D_n}^{1/2} \quad (5.70)$$

where $L_h = 12.7$ mm is the length of a square heater and N is the number of jets. For the same total flow rate and nozzle diameter, surface temperature uniformity was observed to increase with increasing N. However, the effect of jet spacing (pitch) in the array was not considered, and, because of the vertical orientation of the impingement surface, some of the jets in the array were affected by the cross flow of fluid from other jets.

Womac et al. (1994) also considered heat transfer from a 12.7 mm × 12.7 mm source to 2 × 2 and 3 × 3 arrays of free-surface jets. As shown in Figure 5.11, three different configurations were considered, and, in each case, an *effective heater length*, $L_e = P_n$, was used to characterize a unit cell. In two cases, unit cells for outer jets extended beyond the edge of the actual heat source. The effective heater length was introduced to estimate the average distance L^* associated with radial flow in the wall jet region beyond the stagnation zone ($r > r_s$). Assuming $r_s = D_n/2$ for free-surface jets,

$$L^* = \frac{\left[\left(\sqrt{2} P_n/2\right) - (D_i/2)\right] + [(P_n/2) - (D_i/2)]}{2} \quad (5.71)$$

Adopting a superposition procedure to evaluate the Nusselt number as an area-weighted combination of correlations associated with the stagnation and wall jet regions, experimental results were correlated by the following expression (Womac et al., 1994):

$$\frac{\overline{Nu_L}}{Pr^{0.4}} = 0.516 Re_{D_i}^{0.5} \left(\frac{L_h}{D_i}\right) A_r + 0.344 Re_{L^*}^{0.579} \left(\frac{L_h}{L^*}\right)(1 - A_r) \quad (5.72)$$

where $A_r = N(\pi D_i^2)/4L_h^2$. For $500 \leq Re_{D_n} \leq 20{,}000$ and $5 \leq P_n/D_n \leq 20$, the data were correlated to within ±17%, with a standard deviation of 8%. The average Nusselt number was independent of the nozzle-to-heater spacing, S. However, because of a corresponding increase in the jet velocity for a fixed volumetric flow rate, $\overline{Nu_L}$ increased with reductions in both D_n and N. The average Nusselt number also increased with decreasing P_n or L^*, and the trend was attributed to increased jet interference and/or to a reduction in the extent of the wall jet region.

The only study of flow and heat transfer for an array of planar jets, Figure 2.22c, was performed by Slayzak et al. (1994b). The spent fluid was discharged laterally in

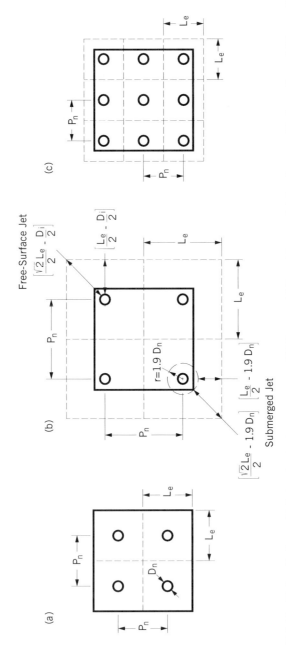

FIGURE 5.11 Unit cells for multiple jets impinging on a square (12.7 mm × 12.7 mm) heater: (a) 2 × 2 array with $L_e = P_n = 6.35$ mm; (b) 2 × 2 array with $L_e = P_n = 10.16$ mm; and (c) 3 × 3 array with $L_e = P_n = 5.08$ mm. Adapted from Womac et al. (1994).

the interference region between adjoining jets, which was also the location of a secondary maximum in the longitudinal distribution of the local convection coefficient.

5.6.2 Submerged Jets

For arrays of submerged jets, conditions are influenced by the same potential core and entrainment effects that characterize a single submerged jet, as well as by interactions between neighboring jets.

Using nozzle and array geometries equivalent to those considered for free-surface jets (Fig. 5.11), Womac et al. (1994) correlated their heat transfer measurements by an expression of the form

$$\overline{Nu}_L = 0.509 Re_{D_n}^{0.5} \left(\frac{L_h}{D_n}\right) A_r + 0.0363 Re_{L^*}^{0.8}(1 - A_r) \tag{5.73}$$

where

$$L^* = \frac{\left[\left(\sqrt{2}P_n/2\right) - 1.9D_n\right] + \left[(P_n/2) - 1.9D_n\right]}{2} \tag{5.74}$$

The stagnation zone was assumed to be bounded by $r_s = 1.9D_n$, in which case, $A_r = N\pi(1.9D_n)^2/L_h^2$. The correlation was obtained for $500 \leq Re_{D_n} \leq 20{,}000$, $5 \leq P_n/D_n \leq 10$, and $2 \leq S/D_n \leq 4$, and the data were correlated to within $\pm 30\%$ with a standard deviation of 12%. As with arrays of free-surface jets, for a fixed volumetric flow rate $\overline{Nu}_L$ increases with decreasing N, D_n, and P_n (or L^*).

For submerged jets, additional insights may be gained from results obtained for arrays of air jets. For example, Gardon and Cobonpue (1961) measured local heat transfer coefficients for 5×5 and 7×7 arrays of circular air jets and found local maxima to exist at locations midway between adjoining jets and minima to exist at the midpoints of diagonals between jets. Kercher and Tabakoff (1970) investigated the effects of nozzle diameter, D_n, pitch, P_n, separation distance, S, and cross flow for air jet impingement on a surface that was contained on three sides, thereby forcing spent fluid to exit in a single direction. Because of increasing cross-flow effects, local heat transfer coefficients were found to decrease in the direction of the exiting fluid. For fixed V_n, the average Nusselt number increased with decreasing P_n and D_n. From measurements performed with a 4×4 array of air jets, Hollworth and Berry (1978) also found the average Nusselt number to increase with decreasing P_n but to be independent of D_n. For the large nozzle pitches of the study ($P_n/D_n \geq 10$), jet interaction effects were negligible. Correlations for the average Nusselt number have been developed by Martin (1977) and Hollworth and Berry (1978).

Gardon and Akfirat (1966) and Korger and Krizek (1966) measured local convection coefficients for unconfined planar jet arrays. For $P_n/w_n = 16$ and 32 and $S/w_n < 6$ (impingement within the potential core), the identity of each jet was well preserved and interactions between the jets were confined to a wall jet interaction region (Gardon and Akfirat, 1966). For $S/w_n > 6$, however, jet interactions that occurred before impingement significantly reduced the maximum convection coefficient in the stagnation region. Flow separation in the wall jet interaction zone resulted in upflow from the wall and was manifested by a secondary maximum in h between the nozzles. Similar results were obtained by Korger and Krizek (1966), and, for $S/w_n = 6$ and $P_n/w_n = 15$, the secondary peak was equivalent to that measured at the midlines of the impinging jets. This result was attributed to upflow in the interaction zone, but may have also been influenced by transition to or intensification of turbulence.

Although average heat transfer correlations have yet to be developed for liquid planar jet arrays, estimates may be based on correlations obtained for air jets, if Prandtl number effects are included. Martin (1977) correlated average convection coefficients for an array of planar nozzles for which the associated outflow was along the impingement surface parallel to the nozzle length. The expression is of the form

$$\overline{Nu}_{w_n} = \frac{1}{3} f_o^{3/4} \left[\frac{4 Re_{w_n}}{\dfrac{w_n/P_n}{f_o} + \dfrac{f_o}{w_n/P_n}} \right]^{2/3} Pr^{0.42} \quad (5.75a)$$

where

$$f_o = \left\{ 60 + \left[\left(\frac{S}{w_n} \right) - 4 \right]^2 \right\}^{-1/2} \quad (5.75b)$$

The correlation applies for $750 \leq Re_{w_n} \leq 20{,}000$, $0.008 \leq w_n/P_n \leq 2.5 f_o$, and $2 \leq S/w_n \leq 80$. Saad et al. (1980) examined an array of confined air jets with outflow ports of width $4w_n$, which were parallel to the slot nozzles and located midway between each pair of jets. Local heat transfer coefficients were measured and integrated to obtain average values, and, presuming a Prandtl number dependence of the form $Pr^{0.42}$, the results are correlated by the expression

$$\overline{Nu}_{w_n} = 0.163 Re_{w_n}^{0.775} \left(\frac{S}{w_n} \right)^{-0.286} \left(\frac{P_n}{w_n} \right)^{-0.314} Pr^{0.42} \quad (5.76)$$

which applies for $3000 \leq Re_{w_n} \leq 30{,}000$, $8 \leq (S/w_n) \leq 24$, and $12.5 \leq (P_n/w_n) \leq 66.7$. Predictions based on this correlation (outflow between nozzles) exceed those based on the Martin (1977) correlation (outflow parallel to the nozzle length).

5.7 SEMICONFINED JETS

5.7.1 Rectangular Jets

Thermal conditions associated with jet impingement may also be influenced if the jet and ambient fluid are confined by a second surface that is in the plane of the nozzle

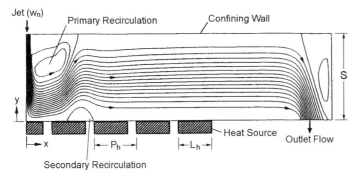

FIGURE 5.12 Streamlines associated with laminar, slot jet impingement under semiconfined conditions for a uniform jet velocity profile, $S/w_n = 1.5$ and $Re_{w_n} = 500$. Adapted from Schafer et al. (1992).

exit and parallel to the impingement surface. For such conditions, a two-dimensional conjugate numerical simulation has been performed to determine flow and heat transfer associated with using a single, laminar, slot jet to cool a linear array of nine discrete heat sources (Schafer et al., 1992). From symmetry about the midplane of the jet, modeling was restricted to a half section of the system, and a representative streamline plot is shown in Figure 5.12 for a jet that emerges from its nozzle with a uniform velocity profile. Subsequent to impingement on a heater that is centered with the jet, the redirected flow passes over four additional heaters before it is routed from the system. The outflow boundary was placed sufficiently far downstream of the fifth heater to render its effects on flow and heat transfer negligible.

As fluid enters the system, shear forces generated at the interface between the jet and the ambient induce a *primary* recirculation zone that inhibits spreading until the jet is beyond the recirculation cell. As the fluid is redirected by the impingement plate, it continues to be confined by the recirculation cell, and it is only at the downstream edge of the cell that the flow expands into the entire channel. An adverse streamwise pressure gradient accompanies the expansion and induces boundary layer separation that is manifested by a secondary recirculation at the surface.

The corresponding streamwise distribution of the local Nusselt number is shown in Figure 5.13 for heat sources mounted flush with a substrate and centered at distances of 0, 5, 10, 15, and 20 jet widths from the stagnation line. For the prescribed substrate-to-fluid conductivity ratio of $k_s/k_f = 0.34$, the substrate is a poor conductor, and more than 97% of the energy dissipated in each heater is transferred by convection from the surface of the heater.

The distribution exhibits a maximum at the stagnation line, a decay resulting from thermal boundary layer development on the surface of the first heater, and a precipitous decline on the substrate separating the trailing edge of the first heater from the leading edge of the second heater. Within this region, there is a significant reduction in boundary layer temperature gradients near the substrate surface such that, with restoration of significant heating at the leading edge of the second heater, there is

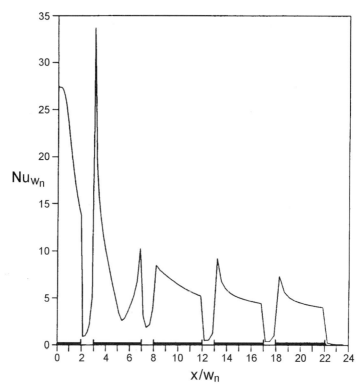

FIGURE 5.13 Streamwise distribution of the local Nusselt number for laminar, slot jet impingement under semiconfined conditions and a uniform jet velocity profile: $S/w_n = 1.5$, $Re_{w_n} = 500$, $L_h/w_n = 4$, $P_h/L_h = 1.25$, $Pr = 7.4$, and $k_s/k_f = 0.34$. Adapted from Schafer (1990).

a sharp increase in the local convection coefficient. This increase is enhanced by the secondary recirculation zone, which also enhances heat transfer at the trailing edge of the second source. For the other heaters, heat transfer is large at the leading edge, where the interrupted thermal boundary layer begins to grow again, and subsequently decays because of boundary layer growth. As the distance from the jet midline increases, heat transfer decreases as a result of the combined effects of upstream heating and diminished influence of the impinging jet.

The foregoing results apply for a uniform velocity profile at the nozzle discharge. The larger midline velocity associated with a parabolic profile enhances heat transfer at the stagnation line by approximately 67%. However, enhancement occurs only over approximately one half of the heater, and the average heat transfer for the first heater is only 20% larger for the parabolic profile. The foregoing results also apply for a comparatively small substrate-to-fluid thermal conductivity ratio ($k_s/k_f = 0.34$). The large resistance to heat conduction in the substrate causes most of the heat to be transferred by convection from the heater to the fluid. As shown in

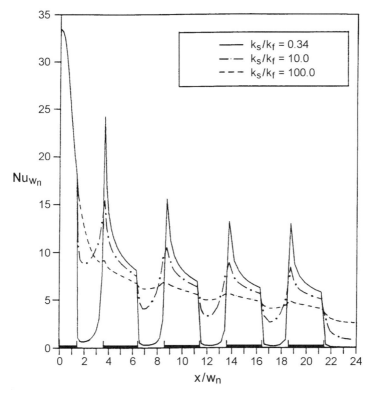

FIGURE 5.14 Effect of substrate-to-fluid thermal conductivity ratio on the local Nusselt number distribution for laminar, slot jet impingement under semiconfined conditions with a uniform jet velocity profile: $S/w_n = 1.0$, $Re_{w_n} = 500$, $L_h/w_n = 2.83$, $P_h/L_h = 1.76$, and $Pr = 7.4$. Adapted from Schafer (1990).

Figure 5.14, heat transfer by conduction through the substrate and, subsequently, by convection into the fluid becomes increasingly more important with increasing conductivity ratio. Although heat transfer from the first source is largely unaffected by the value of k_s/k_f, substrate conduction promotes thermal boundary layer development upstream of the second through the fifth heat sources, thereby attenuating heat transfer at the surface of each source. With increasing k_s/k_f, the temperature distribution along the entire impingement surface becomes more uniform. For $k_s/k_f = 100$, the entire impingement surface is nearly isothermal, and the heater/substrate interfaces are barely distinguishable.

Schafer et al. (1991) also measured average heat transfer coefficients for discrete 12.7 mm × 12.7 mm heaters cooled by a planar liquid jet. The purpose of the study was to assess the feasibility of using a single jet to effectively cool more than one row of chips in a multichip array. The merit of such an application depends on the ability to maintain nearly isothermal conditions on each of the chips cooled by the same jet, and prospects for success are enhanced by the existence of a secondary maximum in the distribution of the local convection coefficient. The secondary maximum could result from a transition to turbulence and/or to establishment of a secondary recirculation zone.

In one set of experiments, Schafer et al. (1991) measured average heat transfer coefficients for a single row of heaters when the row was centered at the nozzle midplane, as well as when the row was offset from the midplane. Results are shown in Figure 5.15. For $Re_L < 3000$, $\overline{Nu}_L$ is a maximum when the heater is centered at the jet stagnation line and decays monotonically with increasing x. At $Re_L = 7600$, a secondary peak is evident at $x/w_n \approx 3$, and, with increasing Re_L, the peak becomes more prominent as its location shifts to $x/w_n \approx 4$. At $Re_L = 37,200$, the secondary peak exceeds that associated with the stagnation line. In experiments performed for five rows of heaters centered at $x/w_n = 0, 5, 10, 15,$ and 20, a secondary peak was recorded on the second heater, and, for the largest Reynolds number ($Re_L \approx 125,000$), the Nusselt number exceeded that associated with the first heater (located under the impinging jet) by approximately 25%. Although in need of further study,

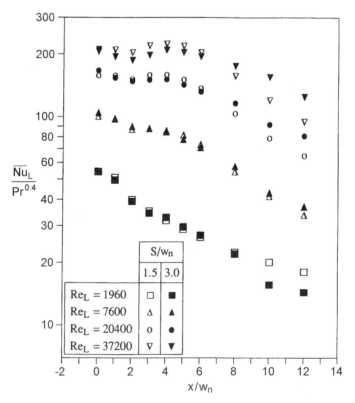

FIGURE 5.15 Variation of average Nusselt number with heater displacement for selected Reynolds numbers. Adapted from Schafer et al. (1991).

the data suggested that, at lower Reynolds numbers (Re_L < 17,000), the secondary peak was due to a secondary recirculation cell and that, at larger values of Re_L, it was due to a transition to turbulence or to a combination of both effects. For $1.0 \leq S/w_n \leq 3.0$, the effect of separation distance was found to be negligible.

The foregoing results suggest that, by designing a cooling system to place the secondary peak over a chip row that is downstream of jet impingement, several chip rows could be cooled by a single jet. For $P_h/L_h \approx 2$, $Re_L > 20,000$, and jet impingement occurring on the first row, the results of Figure 5.15 suggest the existence of nearly equivalent Nusselt numbers for the first three rows of chips and a reduction in $\overline{Nu}_L$ for subsequent rows.

For an array of jets, spent fluid is able to exit freely between and above the nozzles if the jets are unconfined. Using three slot jets with various pitches and fluid exiting between the jets, Korger and Krizek (1966) experimentally determined an increase in the convection coefficient at locations midway between two jets. The increase was attributed to upflow from the wall resulting from the interaction of the jets, and the magnitude was comparable to that at the stagnation line. However, space constraints in an electronic package do not allow for the use of unconfined jets. Thus, if no other location is provided for exiting fluid, the spent flow must travel either perpendicular or parallel to the jet midplanes. In the case of flow perpendicular to the midplanes, the spent fluid from one jet becomes a cross flow imposed on downstream jets. If the momentum of this cross flow is sufficiently large, the impingement of downstream jets is destroyed, because they are swept away by the cross flow before reaching the impingement surface (Fig. 2.21a). Florscheutz (1982) found that, by increasing the ratio of cross-flow momentum to jet momentum from 0.2 to 0.4, heat transfer was decreased by a factor of 2.5. Similarly, Sparrow et al. (1975) found that, for sufficiently large cross-flow rates, the peak in stagnation point heat transfer is eliminated. Al-Sanea (1992) performed a comparative study of flow and heat transfer for a laminar slot jet operating in the following modes: (i) unconfined, (ii) semiconfined without cross flow, and (iii) semiconfined with cross flow. For the range of conditions considered, results for cases (i) and (ii) were virtually identical, whereas cross flow degraded heat transfer by up to 60%.

The work of Schafer et al. (1991) was extended by Teuscher (1992) to multiple, semiconfined slot jets of FC-77 and included the effects of jet interaction and different outflow locations for the spent fluid. Two flow configurations were considered, each involving a linear array of five 12.7 mm × 12.7 mm heat sources (Fig. 5.16). The longitudinal pitch of the heat sources was $P_h = 15.88$ mm ($P_h/L_h = 1.25$). In one case, cooling was provided by five jets (one per source) and, in the other, by two jets. Operating conditions were restricted to an average maximum flow rate of 0.95 lpm (0.25 gpm) per source (to keep within acceptable levels of pump power and acoustical noise), a maximum jet velocity of 7 m/s (to avoid material erosion), and a limit of 340 kPa (50 psi) on the overall system pressure drop. For the multiple source/jet arrangement (Fig. 5.16b), each of the two jets impinges on an underlying chip and, following bifurcation, flows over adjoining chips. In this case, the maximum allowable flow rate per jet is 2.5×0.95 lpm $= 2.37$ lpm.

206 JET IMPINGEMENT COOLING

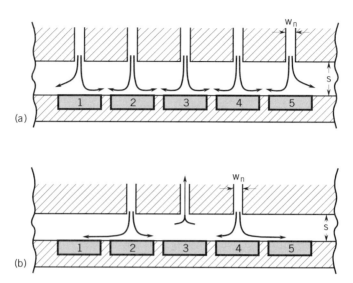

FIGURE 5.16 Schematic of flow configurations involving use of semiconfined, rectangular jets to cool a linear array of 12.7 mm × 12.7 mm heat sources: (a) one heat source per jet and (b) multiple heat sources per jet. Adapted from Teuscher (1992).

The actual nozzle design is shown in Figure 5.17. Each nozzle consisted of a rectangular slot of width $w_n = 0.25$ mm and depth $d_n = 12.7$ mm corresponding to the size of the heat source. The straight section of the nozzle was of length $L_n = 1.275$ mm and was separated from the inlet manifold by a 5.08-mm-long convergent ($\theta = 50°$) zone. The width of the channel into which the jets were discharged was twice the size of the heat source ($2L_h$), and the linear array of heat sources was centered in the channel. Hence, to a good approximation, the experimental results would be applicable to planar arrays for which the transverse (spanwise) pitch is $P_T = 2L_h = 25.4$ mm. The height of the channel was fixed at either $S = 1.27$ mm or $S = 2.54$ mm.

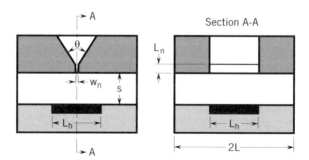

FIGURE 5.17 Schematic of nozzle-and-channel cross section used in experiments of Teuscher (1992).

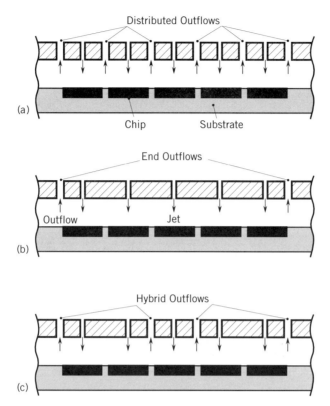

FIGURE 5.18 Outflow conditions for one heat source per jet: (*a*) distributed, (*b*) ends, and (*c*) hybrid (ends and interior). Adapted from Teuscher (1992).

The test section was fabricated with two nozzle plates to facilitate operation in either of the two flow configurations, as well as variation of the outflow locations. For the configuration corresponding to one heat source per jet (Fig. 5.16*a*), three outflow conditions were considered (Fig. 5.18): (*a*) on both sides of each source, (*b*) at the ends of the array, and (*c*) at the ends and on both sides of the center heater. For the configuration corresponding to multiple heat sources per jet (Fig. 5.16*b*), a single outflow condition was considered, with spent fluid discharged at both ends of the array and through a port above the center heater.

Heat transfer data for the configuration corresponding to one heat source per jet with a distributed outflow (Fig. 5.18*a*) are shown in Figure 5.19. The Nusselt number is defined as $\overline{Nu}_L = \overline{h} L_h / k_f$, where $\overline{h}$ is determined from Eq. 5.29, and the Reynolds number is defined as $Re_{D_h} = V_n D_h / \nu$, where $D_h = 2w_n$ is the hydraulic diameter of the nozzle exit. Because a nearly equivalent flow field is associated with each heat source, heater to heater variations in $\overline{Nu}_L$ are small. For each of the inner jets (2–4), bifurcated wall jets flow along the surface until they interact with neighboring wall jets and exit the system as upflows through discharge ports in the nozzle plate.

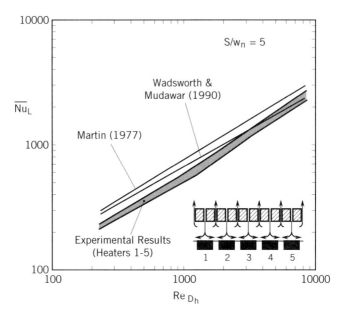

FIGURE 5.19 Variation of average Nusselt number with Reynolds number for heaters in the single source/jet configuration with distributed outflows and comparison with existing correlations. Adapted from Teuscher (1992).

Although one half of the bifurcated outer jets (1, 5) is routed from the system by the end walls, flow conditions are sufficiently similar to those of the inner jets to yield little difference in the Nusselt numbers. Additional data obtained for $S/w_n = 10$ are within 10% of the foregoing results, suggesting that, at least for $5 < S/w_n < 10$, the separation distance is not a critical parameter. The system pressure drop was proportional to the square of the nozzle velocity and achieved a peak value of $\Delta p \approx 200$ kPa for $Re_{D_h} \approx 8000$.

Because the foregoing configuration is similar to that considered by Wadsworth and Mudawar (1990), the data are contrasted with predictions based on Eq. 5.67. Although agreement is excellent for $Re_{D_h} > 2000$, the data are overpredicted by the correlation at lower Reynolds numbers. Teuscher (1992) attributed the difference to development of a slowly moving recirculation cell, which would reduce local heat transfer coefficients at the outer edges of the heaters. The data are also overpredicted by the Martin (1977) correlation, Eq. 5.66.

For outflow restricted to the ends of the array (Fig. 5.18b), jets 1, 2, 4, and 5 are affected by the cross flow resulting from upstream jets. Over the entire Reynolds number range ($200 < Re_{D_h} < 10{,}000$), Nusselt numbers for the end heaters (1, 5) were approximately 15% smaller than values of $\overline{Nu_L}$ associated with the interior heaters (Teuscher, 1992), and the cross flow clearly diminishes the effectiveness of the impinging jets at 1 and 5. For $Re_{D_h} < 3000$, the largest values of $\overline{Nu_L}$ were associated with heaters 2 and 4. The result was attributed to a maldistribution of flow to the nozzles, with the smaller flow attributed to the center nozzle, and not to the influence of spent flow from the center jet on jet impingement for heaters 2 and 4 (Teuscher, 1992). However, the cumulative effect of spent fluid from heaters 2, 3, and 4 was believed to retard jet impingement on heaters 1 and 5 and, hence, to reduce the corresponding Nusselt numbers. For $Re_{D_h} > 3000$, results for heaters 2, 3, and 4 were within a few percent of those obtained for all heaters and the distributed outflow condition (Fig. 5.19). The system pressure drop was slightly larger than that for the distributed outflow condition.

Relative to restriction of outflow at the ends of the array, the hybrid configuration of Figure 5.18c allows for additional outflow at both sides of the center heat source (3). This condition yielded nearly equivalent (within $\pm 5\%$) Nusselt numbers for the five heat sources, as well as for $S/w_n = 5$ and 10, and excellent agreement with results for the distributed outflow (Fig. 5.19). The system pressure drop was also equivalent to results obtained for the distributed outflow.

Because the hybrid outflow configuration yields results that are equivalent to those for the distributed condition, yet corresponds to a simpler manifold design, it would be the preferred configuration for the single source/jet mode of operation. The manner in which this configuration could be used for a large number of chips is illustrated in Figure 5.20. Outflow would be provided on each side of every *pair* of chips.

Although the multiple chip/jet configuration (Fig. 5.16b) is intended to reduce system flow requirements, it has the highly undesirable effect of yielding a nonuniform Nusselt number distribution (Fig. 5.21). Results for heaters 2 and 4, which are nearly equivalent to Nusselt numbers obtained for the distributed and hybrid outflows in the single source/jet system, exceed those for heaters 1, 3, and 5 by more than a factor of 2. Clearly, without an impinging jet, heat transfer is significantly diminished.

In Figure 5.22, the data are compared with correlations developed for jet impingement (Wadsworth and Mudawar, 1990) and channel flow (Incropera et al., 1986). Data for the fourth (and second) heater are in good agreement with the impingement correlation, Eq. 5.67, whereas data for the remaining sources are significantly underpredicted by the correlation for channel flow, Eq. 4.8. Because the leading edge of these sources is 37 jet widths from the midline of the nearest jet, it is plausible to assume that the corresponding wall jet fills the channel, providing fully developed flow upstream of the heater. Use of Eq. 4.8 was based on this premise,

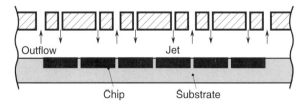

FIGURE 5.20 Preferred outflow arrangement for a linear chip array. Adapted from Teuscher (1992).

210 JET IMPINGEMENT COOLING

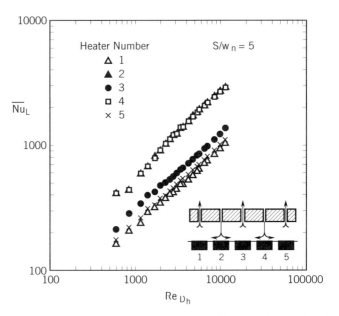

FIGURE 5.21 Variation of average Nusselt number with Reynolds number for heaters in the multiple source/jet configuration. Adapted from Teuscher (1992).

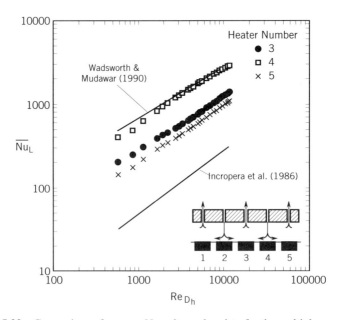

FIGURE 5.22 Comparison of average Nusselt number data for the multiple source/jet configuraton with correlations for jet impingement and channel flow. Adapted from Teuscher (1992).

with the average channel velocity determined from mass conservation requirements ($u_{ch} = V_n w_n/2S$). Underprediction of the data is attributed to the actual existence of a developing flow, as well as to heat transfer enhancement associated with outflow above the heater. For the middle heater (3), enhancement is larger because of the mixing that results from the interaction of adjoining jets.

If the multiple source/jet configuration is compared with the single source/jet configuration of Figure 5.18a or c on the basis of equivalent nozzle flow rates (velocities), the multiple source/jet configuration has a *smaller* system flow rate but heat transfer is significantly reduced for those sources on which jet impingement does not occur. In contrast, if the comparison is made on the basis of equivalent system flow rates (Teuscher, 1992), relative to Nusselt numbers corresponding to the configurations of Figure 5.18a and c, the larger nozzle flow rates (by a factor of 2.5) for the multiple source/jet configuration yield larger values of $\overline{Nu_L}$ at those sources experiencing impingement (2, 4) and only slightly smaller values at those sources for which there is no impingement (1, 3, 5). Although this result would seem to favor the use of the multiple/jet arrangement, the configuration fails to meet the important requirement of cooling uniformity among the heat sources.

Of the configurations considered by Teuscher (1992), the hybrid outflow arrangement (Fig. 5.18c) may represent the best choice from the standpoint of providing uniform cooling equivalent to that of the distributed outflow (Fig. 5.18a), while requiring a simpler manifold design.

5.7.2 Circular Jets

Garimella and Rice (1995) performed local heat transfer measurements for impingement of a submerged, semiconfined circular jet of FC-77 on a square, 10 mm × 10 mm heat source. The jet was discharged from a cylindrical hole ($0.79 \leq D_n \leq 6.35$ mm, $4000 \leq Re_{D_n} \leq 23{,}000$) drilled in a sharp-edged nozzle plate of fixed thickness ($L_n = 6.35$ mm), and the effects of confinement of the radial flow by the plate were considered for plate-to-heater spacings in the range $1 \leq S/D_n \leq 14$. A secondary maximum in the local convection coefficient was observed at $r/D_n \approx 2$ and became more pronounced with increasing D_n (for fixed Re_{D_n}), increasing Re_{D_n}, and decreasing S/D_n. The radial location of the maximum increased with increasing S/D_n. Correlations of the stagnation and average Nusselt numbers for $1.59 \leq D_n \leq 6.35$ mm revealed a weak dependence on S/D_n for $1 \leq S/D_n \leq 5$ and a significant reduction with increasing S/D_n, $Nu \sim (S/D_n)^{-1/2}$, for $6 \leq S/D_n \leq 14$. For $D_n = 3.18$ and 6.35 mm, $L_n/D_n = 1$, and $2 < S/D_n < 4$, turbulence measurements based on laser–Doppler velocimetry revealed the location of maximum turbulence in the wall jet to be in the range $2 < (r/D_n) < 2.5$ and to increase with increasing S/D_n (Fitzgerald and Garimella, 1996). This location coincides with that of the secondary peak in the local heat transfer distribution.

The foregoing study was extended by Garimella and Nenaydykh (1995) to include the effect of the nozzle plate thickness, L_n, for nozzle thickness-to-diameter ratios in the range $0.25 \leq (L_n/D_n) \leq 12$. Heat transfer was maximized for $L_n/D_n < 1$, decreased sharply with increasing L_n/D_n from 1 to 4, and increased slightly with increasing L_n/D_n from 4 to 12. Variations were most pronounced for impingement within the potential core ($S/D_n < 5$). For $1.59 \leq D_n \leq 6.35$ mm, $4000 \leq Re_{D_n} \leq 23{,}000$, and $0.25 \leq L_n/D_n \leq 12$, the following correlations were proposed for the average Nusselt number:

$1 < (S/D_n) < 5$:

$$\overline{Nu}_{D_n} = 0.169 Re_{D_n}^{0.689} Pr^{0.4} \left(\frac{S}{D_n}\right)^{-0.07} \left(\frac{L_n}{D_n}\right)^{-0.12} \quad (5.77a)$$

$5 < (S/D_n) < 14$:

$$\overline{Nu}_{D_n} = 0.183 Re_{D_n}^{0.778} Pr^{0.4} \left(\frac{S}{D_n}\right)^{-0.566} \left(\frac{L_n}{D_n}\right)^{-0.067} \quad (5.77b)$$

All properties were evaluated at the film temperature.

For a circular jet, chip cooling under semiconfined conditions is conveniently maintained by routing spent fluid from the impingement region through a concentric annulus. Such a scheme was used in cooling the Fujitsu FACOM M-780 series of computers (Yamamoto et al., 1988a, b).

Assuming discharge of a laminar jet from a tubular nozzle of diameter D_n, Besserman et al. (1991a) computed flow and thermal conditions for impingement on a circular heat source that is of diameter D_h and concentric with a confining wall of diameter D_c (Fig. 5.23a). For a parabolic profile at the nozzle discharge, representative streamlines are shown in Figure 5.23b. As the jet approaches the impingement surface, the axial velocity component is transformed to a radial component with acceleration in the radial direction. The confining wall acts as a second impingement surface, turning the fluid another 90°. Because of shear forces generated between the jet and return flows, a large toroidal recirculation cell develops near the impingement surface. The cell strengthens with increasing Reynolds number, and, with increasing S/D_n, it enlarges in the radial direction, providing a "pseudo" nozzle wall, which inhibits significant spreading of the jet, thereby causing impingement flow conditions to be approximately independent of S/D_n.

Although barely visible in Figure 5.23b, a small recirculation (clockwise) cell was predicted to exist at the corner formed by the confining wall and the impingement surface. The cell strengthened with increasing Re_{D_n}, and velocities approached those of the impinging jet.

With a uniform profile at the nozzle exit, a large discontinuity in velocity exists between the jet and the return flows, and the resulting shear stress enhances fluid entrainment from the recirculation cell into the impinging jet. In addition, the spacing between streamlines along the impingement surface is slightly larger for the uniform profile, implying that the wall jet is concentrated in a region closer to the impingement surface for the parabolic profile.

Radial distributions of the local Nusselt number are shown in Figure 5.24 for a circular heater with $D_h = D_c$. For a parabolic profile, Figure 5.24a, Nu_{D_n} is

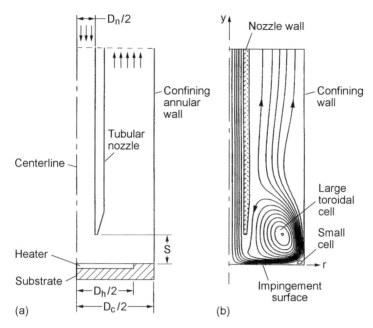

FIGURE 5.23 Circular jet impingement cooling of a circular heat source with annular collection of the spent fluid: (*a*) system geometry and (*b*) streamline pattern for a parabolic jet velocity profile ($Re_{D_n} = 500$, $S/D_n = 1$, and $D_c/D_n = 5$). Adapted from Besserman et al. (1991a).

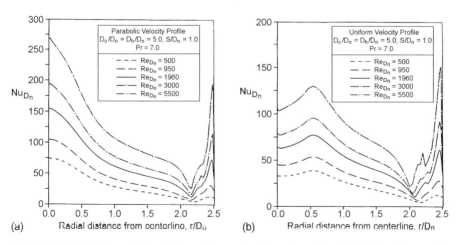

FIGURE 5.24 Effect of Reynolds number on the Nusselt number distribution for a semi-confined circular jet: (*a*) parabolic velocity profile and (*b*) uniform velocity profile. Adapted from Besserman et al. (1991a).

predicted to decrease with increasing radius up to $r/D_n \approx 2$, beyond which a sharp decline is followed by a pronounced increase. The local minimum and maximum in the region $2 < r/D_n < 2.5$ is due to the small, but intense cell in the corner. The maximum at $r/D_n \approx 2.4$ results from fluid impingement on the surface; the minimum at $r/D_n \approx 2.2$ is due to the interaction and subsequent upwelling of warm fluid moving radially outward and inward from the stagnation point and confining wall, respectively. With increasing Reynolds number, local Nusselt numbers increase across the entire heater surface because of thinning of the boundary layers.

For a uniform velocity profile, Figure 5.24b, the centerline velocity is one half that for the parabolic profile and heat transfer at the stagnation point is significantly smaller. The off-axis maximum at $r/D_n \approx 0.5$ is attributed to the sharp discontinuity in velocity at the edge of the jet, which induces large shear forces between the jet and the recirculating flow. The corresponding increase in the momentum flux of fluid impinging on the surface enhances heat transfer. For $r/D_n > 1.0$, the effect of the jet velocity profile becomes less pronounced, and the heat transfer distributions for the uniform and parabolic velocity profiles are of similar shape. When local results are integrated over the entire surface, the average Nusselt number is approximately 20% smaller for the uniform velocity profile than for the parabolic profile.

For the test cell represented schematically by Figure 5.23a, experiments were performed to determine the average Nusselt number for a circular heater of diameter $D_h = 22.2$ mm (Besserman et al., 1991b) and a square heater of length $L_h = 12.7$ mm (Besserman et al., 1992). For the square heater with $D_n = 4.42$ mm and confinement diameters of 22.2 and 38.1 mm, the effects of nozzle spacing ($1.0 \leq S/D_n \leq 5.0$) and confinement were negligible. For $D_n = 9.27$ mm, however, the jet is more constrained by the annular wall and the Nusselt number is influenced by S/D_n, with a maximum existing for $S/D_n \approx 3$. Tighter confinement brings the impinging jet and counterflow in closer proximity, enhancing turbulence production in the adjoining mixing zone and reducing the length of the potential core. For $S/D_n = 3$, impingement may still occur in the potential core, thereby preserving the centerline velocity, while enhancing heat transfer resulting from increased turbulence. For $S/D_n = 5$, however, impingement may occur beyond the potential core, and the attendant decay in the centerline velocity may reduce heat transfer by more than the increase associated with enhanced mixing. For the square heater, average Nusselt number data were in excellent agreement ($\pm 5\%$) with Eq. 5.60 for the smaller nozzle ($D_n = 4.42$ mm) and annular confinements corresponding to $D_c/D_n = 5.03$ and 8.62. For $D_n = 9.27$ mm and $D_c/D_n = 2.40$, however, heat transfer was influenced by the confining wall and agreement between the data and the correlation was less satisfactory ($\pm 20\%$).

5.8 SPECIAL CONSIDERATIONS

The results of this chapter are restricted to conditions for which the impinging liquid jet is steady, oriented normal to the heated surface, and devoid of an entrained gas phase. The influence of unsteady behavior in the form of periodic jet pulsations has been considered by Zumbrunnen and Aziz (1993) and Sheriff and Zumbrunnen (1994). For a planar free-surface water jet, heat transfer was found to be enhanced by up to 33% if the flow was intermittent (periodically terminated) and the dimensionless frequency, $w_n f / V_n$, exceeded 0.26. Enhancement was attributed to periodic renewal of the surface boundary layer. However, jet pulsation was found to reduce heat transfer if the flow rate was periodically (sinusoidally) varied but never terminated. Physical mechanisms that affect convection heat transfer in pulsatile liquid jet flows and their effect on heat transfer enhancement have been reviewed by Zumbrunnen (1996) for both on/off and sinusoidally varying flow velocities.

The effects of jet inclination have been reviewed by Webb and Ma (1995). The effects include displacement of the location of maximum heat transfer, a slight reduction in the value of the maximum, and asymmetry in the radial distribution of the convection coefficient.

A two-phase (liquid/gas) jet may be created by injecting gas into a liquid jet. This option was considered by Zumbrunnen and Balasubramanium (1995), who inserted capillary tubes into a planar nozzle and injected air into a water jet. Heat transfer coefficients were enhanced by factors of 2.2 and 1.6 at the stagnation line and $x/w_n = 1.6$, respectively. Two-phase flow may also be established by shrouding a circular liquid jet with an annular air jet, which atomizes the liquid and creates a liquid/air *mist jet*. Such a condition was considered by Graham and Ramadhyani (1996), who measured surface-averaged heat fluxes as large as 60 W/cm^2 with methanol/air and water/air mists.

5.9 SUMMARY

Liquid jets provide an extremely effective means of dissipating large heat fluxes from electronic chips, while maintaining the chips at an acceptably low operating temperature. One mode of operation involves discharge of the jet into a gaseous ambient, thereby providing a *free surface* at the liquid/gas interface. The largest local convection coefficient exists within the *stagnation zone* of the target surface, and its value depends strongly on the streamwise velocity gradient at the surface, as well as on levels of turbulence within the jet. In turn, these conditions depend on the nature of the nozzle used to produce the jet. Although the stagnation zone is characterized by the largest local convection coefficient, the extent of the zone is small and, even for characteristically small chip dimensions, it may not contribute significantly to total heat transfer from the chip. The local convection coefficient decreases precipitously in a *wall jet region* downstream of the stagnation zone, although the decline may be reversed by transition to turbulence.

An alternative mode of operation involves discharge of the jet into an ambient consisting of the same liquid (a *submerged jet*). As for the free-surface condition, heat transfer in the stagnation zone depends on turbulence levels within the jet and on the streamwise velocity gradient. Maximum convection coefficients are achieved by maintaining a nozzle-to-target separation S that is within the potential core of the jet ($S < L_{pc}$), where $L_{pc}/D_n \sim 6$ and $L_{pc}/w_n \sim 8$ for circular and rectangular jets, respectively. Operation within the potential core also yields secondary maxima in the convection coefficient at locations removed from the stagnation zone. For both free-surface and submerged jets, correlations for the average convection coefficient are obtained by superposing contributions associated with the stagnation and wall jet regions. However, larger convection coefficients are associated with the use of a submerged jet.

In lieu of using a single jet to cool a chip, multiple jets may be used to achieve more uniform cooling and a larger overall average heat transfer coefficient. However, thermal performance depends on the manner in which spent fluid is discharged from the surface of the chip, as well as the nature of interactions between adjoining jets on the surface. Discharge conditions for which there is no interaction between adjoining, preimpingement jets and jet impingement is unaffected by drainage of the spent fluid yield the largest convection coefficients. In particular, impingement of one jet should not be obstructed by the cross flow imposed by a neighboring jet.

Alternatively, a single jet may be used effectively to cool more than one chip. For example, by exploiting the secondary maximum associated with a submerged rectangular jet characterized by $S < L_{pc}$, nearly uniform cooling could be maintained for the impingement chip and adjacent chips in a linear chip array. Such options may be used to simplify the design of coolant distribution units and to reduce overall flow requirements for an array of chips.

CHAPTER 6

HEAT TRANSFER ENHANCEMENT

6.1 INTRODUCTION

A common method for heat transfer enhancement involves using extended surfaces to increase the surface area available for convection, and applications to air cooling are common. In the IBM 4381 multichip module, for example, cooling is effected by air jet impingement on a heat sink consisting of 256 hollow, integrally connected aluminum pin fins (Biskeborn et al., 1984). Application has also been made to indirect, liquid cooling schemes, as in the FACOM VP 2000 multichip module (Suzuki et al., 1989). Each chip is cooled by a single water jet impinging on a cup-shaped spreader plate that is attached to the chip and acts as an extended surface.

The greatest potential for using extended surfaces to enhance heat transfer by liquid immersion is in a passive cooling system, such as a liquid-filled rectangular cavity with heat sources mounted to one wall and the opposite wall consisting of a cold plate (Chapter 3). Because convection coefficients associated with buoyancy-driven flows are inherently small, there is considerable advantage in using extended surfaces such as plate and pin fins. Although the effectiveness of extended surfaces is reduced in forced-convection applications characterized by large convection coefficients, as in channel flows (Chapter 4) or impinging jets (Chapter 5), significant benefits may still be derived by increasing the surface area for convection. Heat transfer is also enhanced when extended, interrupted, or roughened surfaces disrupt the thermal boundary layer, thereby increasing the convection coefficient.

In this chapter, we consider the means by which extended surfaces (heat sinks) may be used to enhance single-phase free-, forced-, or mixed-convection heat transfer from discrete sources. However, successful implementation depends as much on manufacturability, reliability, and cost issues as it does on increases in thermal performance. Reliability depends on the method of attaching the heat sink to the source.

Whether attachment is in the form of a metallurgical (soldered or brazed) joint or a bonding agent, the attachment interface should be characterized by a small thermal contact resistance and the ability to withstand mechanical loads imposed by the coolant flow.

In the following sections, results are provided for the thermal resistance and effectiveness of different flow conditions and heat sink geometries. In each case, the presumption is that the thermal contact resistance between the heat source and sink is negligible.

6.2 NATURAL CONVECTION

Liquid cooling by natural convection provides a means of thermal control that does not require a pump and, hence, provides a vibration- and noise-free environment. However, relative to forced convection, natural convection is characterized by larger thermal resistances and, hence, lower heat fluxes for a prescribed surface temperature. Even if a chip were immersed in a dielectric liquid, the maximum allowable heat flux corresponding to a representative upper temperature limit of 85°C would only be on the order of 1 W/cm^2 (Kraus and Bar-Cohen, 1983). To exceed this limit in a passive cooling system, some form of heat transfer enhancement must be implemented. In this section, the use of extended surfaces is considered, with emphasis placed on parallel-plate fin arrays.

6.2.1 Parallel-Plate Fins

A common heat sink geometry involves an array of parallel plates that extend from a common base, as shown in Figure 6.1. The dimensions of the base are assumed to correspond to those of a discrete heat source to which it is attached, and the geometry of the array is determined by the fin thickness t_f, spacing S_f (or pitch P_f), length L_f, and height H_f. The fin tips are separated by the distance S_w from a confining shroud, which could be the opposing wall of a rectangular cavity.

For the vertical orientation shown in the figure, adjoining fins form a parallel-plate channel, as illustrated in Figure 2.11. Because the fin temperature exceeds that of the fluid ($T > T_\infty$), positive buoyancy forces induced by the temperature difference cause fluid to be entrained from the ambient and to ascend within the channels. If there is a fin tip clearance ($S_w > 0$), fluid may be entrained at the edges ($y = L_f$, $0 < z < H_f$) of the channels, whereas entrainment is restricted to the bottom ($0 < y < L_f$, $z = 0$) of the channel for zero clearance ($S_w = 0$). However, the effect of conditions at the tip, as well as the base, of the fins on flow and heat transfer within the channels decreases with increasing L_f/S_f and may be neglected as $L_f/S_f \to \infty$.

Heindel et al. (1996b) extended their study of natural-convection heat transfer from a 3 × 3 array of discrete heat sources mounted flush to one wall of a rectangular cavity (Fig. 3.1) to include the effect of attaching an array of parallel-plate fins to each source, as well to the opposing cold plate. The cavity was again filled with FC-77, and heater and cavity dimensions were fixed at $L_{h,x} = L_{h,z} = 12.7$ mm,

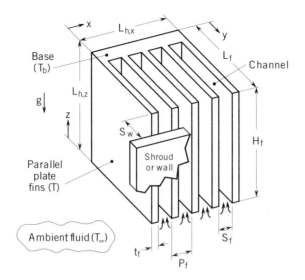

FIGURE 6.1 Heat sink consisting of a parallel-plate fin array.

$P_x = P_z = 15.9$ mm, $H = 95.3$ mm, $H_1 = H_3 = S = 25.4$ mm, $W = 57.2$ mm, and $W_1 = W_3 = 6.35$ mm. Except for the cavity spacing S, these dimensions are equivalent to those used in the prior study of unaugmented heat sources (Heindel et al., 1995b). The new value of $S = 25.4$ mm was chosen to provide some clearance (1.4 mm) between the tips of fin arrays associated with the heat sources and the cold plate. Although the corresponding value of the major aspect ratio ($A_z = 3.75$) differed from that of the prior study ($A_z = 7.50$), the variation is within a range for which there is no influence of A_z on heat transfer (Polentini et al., 1993). Hence, a direct comparison of results from the two studies may be made to assess the effect of the fin array on heat transfer enhancement.

Copper fins were used for the heat sources and the cold plate, and, in each case, the fin length was fixed at $L_f = 12.0$ mm. However, values of the fin thickness and spacing were determined from an approximate analysis that used the Elenbaas equation, Eq. 2.66, to evaluate the average convection coefficient for the fin surfaces. The temperature difference in this equation was approximated as $(T_s - T_\infty) = (T_b - T_\infty)$, which presumes ideal fin behavior ($\eta_f = 1$). However, although isothermal fins were assumed to evaluate $\overline{h}$, nonideal fin behavior was considered by expressing the fin heat transfer rate as

$$q_f = \eta_f \overline{h} A_{s,f}(T_b - T_\infty) \tag{6.1}$$

where $A_{s,f}$ is the total surface area (both sides) of a fin and T_b is the temperature at its base. The fin efficiency was evaluated from Eq. 2.21, which assumes an adiabatic tip. Neglecting transfer from the exposed portion of the base, the total rate of heat transfer from the fin array is

$$q_t = N_f q_f = N_f \eta_f \overline{h} A_{s,f}(T_b - T_\infty) \qquad (6.2)$$

where N_f, the number of fins in an array, depends on S_f and t_f.

Results of parametric calculations based on the foregoing model are shown in Figure 6.2 for representative ranges of S_f and t_f. Although the fin efficiency decreases with decreasing t_f, a corresponding increase in N_f, and, hence, the total surface area of the array, yields an increase in q_t with decreasing t_f. The number of fins also increases with decreasing S_f. However, because the average convection coefficient decreases with decreasing S_f, there is an optimum value of S_f that maximizes the product $N_f \eta_f \overline{h} A_{s,f}$ and, hence, the total heat rate q_t. For the conditions associated with Figure 6.2, this optimum spacing ranges from approximately 0.35 to 0.45 mm, depending on the value of t_f. Performing similar calculations for parallel-plate fins attached to the cold plate, Heindel et al. (1996b) determined that all of the heat dissipated by the 3×3 array of heat sources could be absorbed by the cold plate with fin spacings in the range $0.55 \leq S_f \leq 0.75$ mm for $0.2 \leq t_f \leq 0.5$ mm.

To standardize fin fabrication requirements for the heat sources and the cold plate, Heindel et al. (1996b) selected common values of $S_f = 0.51$ mm and $t_f = 0.2$ mm, thereby providing values of $N_f = 18$ and $N_f = 72$ for the heaters and cold plate, respectively, and increasing the corresponding surface areas more than 30-fold. The fin arrays were manufactured using a copper bonding process involving stamped

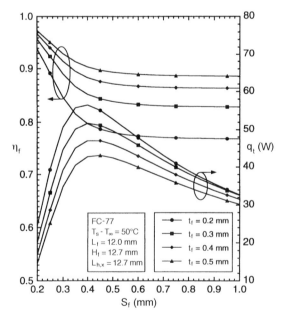

FIGURE 6.2 Effect of fin spacing and thickness on fin efficiency and total heat rate for a parallel-plate array. Adapted from Heindel et al. (1996b).

copper shin stock. A resistive film heater was soldered to the back of each 12.7 mm × 12.7 mm fin array, which was mounted to the wall of the cavity such that the base of the fins was flush with the substrate surface.

In the experiments of Heindel et al. (1996b), the performance of each finned heater was assessed by evaluating its unit thermal resistance from the expression

$$R_{th}'' = \frac{A_{s,h}(\overline{T}_h - T_c)}{q_h} \quad (6.3)$$

where $\overline{T}_h$ is the average temperature of the heater at the base of the fins and T_c is the temperature of the cold plate. The base area $A_{s,h}$ of the heater was chosen to facilitate comparison with experimental results for the corresponding unfinned heater array (Heindel et al., 1995b).

Row-averaged values of R_{th}'' are presented in Figure 6.3 for each row of heaters mounted to the vertical wall of a rectangular cavity with an opposing cold wall. Cavity, heater, and fin dimensions were previously enumerated, and results encompass the Rayleigh number range, $5 \times 10^8 < Ra_{L_{h,z}}^* < 8 \times 10^{11}$, which corresponds to $0.5 < q_h < 38$ W. The maximum heater power of 38 W is equivalent to a heat flux of 23.6 W/cm².

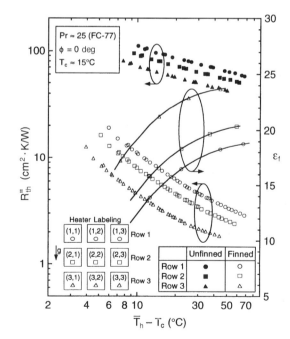

FIGURE 6.3 Comparison of row-averaged thermal resistances for a 3 × 3 array of 12.7 mm-square heaters mounted flush to the vertical wall of a rectangular cavity *with* (Heindel et al., 1996b) and *without* (Heindel et al., 1995b) parallel-plate fins. Adapted from Heindel et al. (1996b).

Variations in R''_{th} among heaters of a row are negligible, and, for each row R''_{th} decreases with increasing temperature difference. This behavior is due to an attendant increase in buoyancy forces and, hence, the velocity of flow through the channels formed by the fins. The smallest value of $R''_{th} \approx 2.0$ cm²·K/W corresponds to the lowermost row and the largest temperature difference. The increase in R''_{th} with decreasing row number is due to an increase in the temperature of the fluid as it ascends along the heated wall, which increases $\overline{T}_h$ for fixed q_h. For $q_h = 38$ W, maximum surface temperatures of $\overline{T}_h = 84.0, 73.9$, and $52.8°C$ were obtained for rows 1 to 3, respectively.

Thermal resistances associated with the unfinned heater array (Heindel et al., 1995b) are also shown in Figure 6.3. Although the fins increase the heat transfer surface area 35-fold, the reduction in R''_{th} is smaller, ranging from approximately 12 to 24 at $(\overline{T}_h - T_c) = 45°C$ for rows 1 and 3, respectively. The ratio of thermal resistances for the unfinned and finned heaters is approximately equal to the fin effectiveness, $\varepsilon_f \approx R''_{th}$ (unfinned)/R''_{th} (finned), which increases with increasing row number and temperature difference. The reduction in ε_f with decreasing row number is due to heating of the fluid as it ascends the wall; the increase in ε_f with increasing $(\overline{T}_h - T_c)$ is due to an increase in the velocity of fluid flow in the channels formed by the fins.

Flow and heat transfer in a rectangular cavity with the 3×3 array of finned heaters mounted to one vertical wall and the opposing wall consisting of a finned cold plate have also been considered theoretically, with the finned regions modeled as porous media (Heindel et al., 1996b). For $Ra^*_{L_{h,z}} = 10^6$, fluid flow was largely restricted to recirculation within the unfinned region above the top row of heaters, whereas flow was weak within the fin arrays. However, with increasing $Ra^*_{L_{h,z}}$, the fin arrays experienced increased fluid penetration and recirculation extended well into the arrays. The corresponding increase in convection coefficients for the finned surfaces enhanced the fin effectiveness, which is consistent with the experimental results of Figure 6.3.

As noted in Chapter 3 and affirmed by the results of Figure 6.3, a serious deficiency of cooling schemes for which an array of heaters is mounted to the vertical wall of a cavity is the significant nonuniformity of heat transfer from row to row of the array. However, as shown in Figure 6.4, this nonuniformity may be essentially eliminated by inclining the cavity 90° from the vertical, thereby orienting the finned heaters and cold plate on the bottom and top walls, respectively. Fluid motion is symmetric about row 2, ascending from the row after descending to the substrate outside rows 1 and 3 and passing through these rows. Because the fluid is heated as it flows through rows 1 and 3, the thermal resistance of row 2 is slightly larger than the equivalent resistances for the two outside rows. For $q_h = 45$ W ($q''_h = 28$ W/cm²) and $T_c = 15°C$, minimum thermal resistances of 2.2, 2.3, and 2.2 cm²·K/W were measured for rows 1 to 3, respectively.

As shown in Figure 6.4, heat transfer enhancement associated with the use of fins, and, hence, the fin effectiveness, $\varepsilon_f \approx R''_{th}$ (unfinned)/R''_{th} (finned), increases with increasing temperature difference because of the attendant increase in buoy-

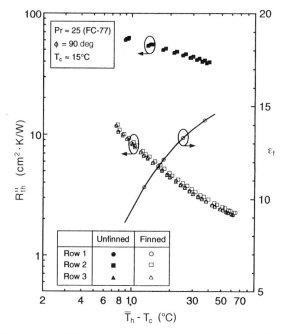

FIGURE 6.4 Comparison of thermal resistances for a 3 × 3 array of 12.7-mm-square heaters mounted flush to the bottom wall of a rectangular cavity *with* (Heindel et al., 1996b) and *without* (Polentini et al., 1993) parallel-plate fins. Adapted from Heindel et al. (1996b).

ancy forces and fluid penetration within the fin arrays. Thermal resistances for the horizontal arrays are within approximately 10% of those for row 2 of the vertical array.

It is noteworthy that results achieved by Heindel et al. (1996b) for heat transfer by single-phase, natural convection from the finned 3 × 3 heat source array are comparable to those associated with pool boiling from unaugmented sources. For systems involving boiling of FC-72 on the chip surfaces and heat removal by a submerged condenser, Geisler et al. (1996) report volumetric heat rates ranging from 2 to 4 W/cm^3. In contrast, Heindel et al. (1996b) achieve a heat dissipation of 2.6 W/cm^3 within their rectangular cavity. Moreover, further enhancement could be readily achieved by adding more heater rows or by reducing the values of H_1 and H_3. A good deal of research has been performed on optimizing the geometric arrangement of parallel plates for natural convection, including the effect of staggering plates in the vertical direction (Bar-Cohen and Rohsenow, 1984; Anand et al., 1992; Ledezma and Bejan, 1997).

6.2.2 Pin Fins

Pin fin arrays, as illustrated in Figure 6.5, may also be used to enhance liquid immersion cooling by free convection. The heat sink is attached to a heat source at its

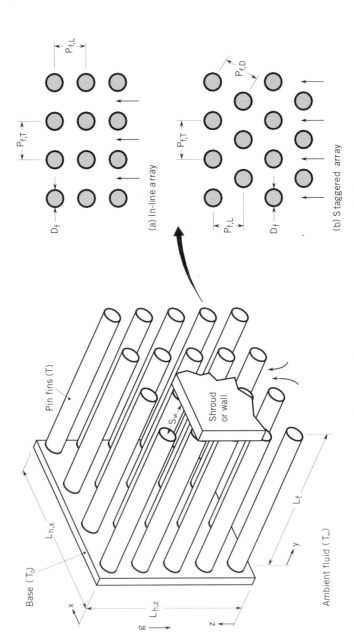

FIGURE 6.5 Heat sink consisting of a pin fin array: (*a*) in-line array and (*b*) staggered array.

base, and the pin tips may or may not be in close proximity to a shroud. If the heat source is attached to a vertical surface, positive buoyancy forces associated with the temperature difference $(T - T_\infty) > 0$ induce an ascending flow over the pins. If an *in-line* array is used, its geometry is determined by the transverse and longitudinal fin pitches, $P_{f,T}$ and $P_{f,L}$, as well as by the diameter, D_f. If the pins are *staggered*, a diagonal pitch, $P_{f,D}$, is also specified.

Heindel et al. (1996c) performed an approximate analysis to determine the extent to which free convection from a 12.7 mm × 12.7 mm heat source mounted flush to a vertical wall may be enhanced by using an in-line array of copper pin fins. The fin length and dimensionless longitudinal pitch were fixed ($L_f = 12.0$ mm, $P_{f,L}/D_f = 2.0$), and the following correlation developed by Aihara et al. (1990) for a densely populated, in-line array was used to estimate the average convection coefficient for the array

$$\overline{Nu}_{P_{f,T}} = \frac{(P_{f,L}/D_f)(\eta_f Ra_{P_{f,T}})^{3/4}}{\pi \eta_f} \left[1 - \exp\left(-\frac{120}{\eta_f Ra_{P_{f,T}}}\right)\right]^{1/2}$$

$$+ \frac{(P_{f,L}/D_f)}{100\pi}(\eta_f Ra_{P_{f,T}})^{1/4} \qquad (6.4)$$

The average Nusselt number and Rayleigh number are defined as

$$\overline{Nu}_{P_{f,T}} = \frac{\overline{h} P_{f,T}}{k_f} \qquad (6.5a)$$

and

$$Ra_{P_{f,T}} = \left[\frac{g\beta(T_b - T_\infty) P_{f,T}^3}{\alpha_f \nu}\right] \times \left[\frac{P_{f,T}}{(N_L - 1)P_{f,L} + D_f}\right] \qquad (6.5b)$$

where T_b is the base temperature of the pins and N_L is the number of pin rows in the longitudinal direction. The length scale $P_{f,T}$ used in the correlation is the transverse pitch of the array. The correlation was developed for conditions corresponding to $1.7 \leq (P_{f,L}/D_f) \leq 3.5$, $0.3 \leq \eta_f Ra_{P_{f,T}} \leq 600$, and $0.16 \leq \{L_f/[(N_L - 1)P_{f,L} + D_f]\} \leq 1.37$. Fluid properties were evaluated at the base temperature T_b.

Assuming isothermal pins ($\eta_f = 1$) for the purpose of evaluating $\overline{h}$ from Eq. 6.4, Heindel et al. (1996c) used Eqs. 6.2 and 2.21 to determine the total heat rate for the array, q_t, and the actual pin fin efficiency, η_f, as a function of D_f and $P_{f,T}$ for FC-77, a temperature difference of $(T_b - T_\infty) = 50°C$, and base dimensions of $L_{h,z} = L_{h,x} = 12.7$ mm. Although an optimum value of $(P_{f,T} - D_f)$ was predicted for $D_f \leq 0.3$ mm, the heat rate increased monotonically with decreasing $(P_{f,T} - D_f)$ for $D_f \geq 0.4$ mm. For $(P_{f,T} - D_f) > 0.6$ mm, q_t was independent of D_f. The calculations revealed that, for $(P_{f,T} - D_f) \leq 0.25$ mm, the total heat rate could be maintained in excess of approximately 50 W for $0.2 \leq D_f \leq 0.4$ mm. Fin efficiencies, which increased with decreasing $(P_{f,T} - D_f)$ and increasing D_f, ranged from approximately 70% to 80%. However, the manufacturing costs associated with fabricating an array of copper pin fins with $L_f = 12$ mm and $D_f \leq 0.4$ mm may be excessive, whereas the fins themselves may be unduly fragile and prone to damage. In contrast, dense arrays of parallel-plate fins are easier and less costly to fabricate.

6.3 CHANNEL FLOW

Several studies have considered the means by which heat transfer may be enhanced for liquids flowing through channels used to facilitate electronic cooling. Enhancement may be achieved by increasing the surface area for convection and/or by disrupting thermal boundary layer development, thereby increasing the convection coefficient. Experimental and theoretical studies have been performed for extended surfaces attached to simulated chips mounted flush with one wall of a rectangular channel, as well as for flow through microchannels with pin fin arrays.

6.3.1 Discrete Heat Sources in Forced Convection

In Section 4.2, we considered an array of chiplike heat sources mounted flush to one wall of a rectangular channel (Fig. 4.1), with cooling provided by passage of a liquid through the channel. As for any convection cooling scheme, conditions are complicated by the fact that, from row to row of the array, thermal boundary layer development is intermittent, with resumption occurring at the leading edge of each successive row. Because the substrate is conducting, thermal boundary layer development is also influenced by convection heat transfer from the substrate, which is sustained by conduction from the heat sources to the substrate (*conjugate heat transfer*). In addition, for small heat sources, boundary layer development is influenced by three-dimensional edge effects, which, in concert with conjugate effects, may cause *apparent* convection coefficients to exceed classical results based on two-dimensional boundary layer theory for flow over a heated surface.

If extended surfaces are attached to the heat sources, an additional complication relates to bypass effects, where fluid, following the path of least resistance, favors motion along lanes between heat sources. The corresponding reduction of flow through the extended surfaces reduces their effectiveness in enhancing heat transfer. Complications are also associated with differences between local convection coefficients associated with the extended and heated (base) surfaces.

One means of enhancing heat transfer from the array of Figure 4.1 involves attaching pin fins to the heat sources. In addition to increasing the surface area available for convection heat transfer, the pins disrupt thermal boundary layer development on the heater surface, thereby increasing local convection coefficients. As shown in Figure 6.6, an in-line array of pin fins may be attached to each heat source by means of a permanent metallurgical (soldered or brazed) joint or by a detachable (nonpermanent) bonding agent. In either case, the interface should be characterized by a low thermal resistance and the ability to withstand mechanical loads imposed by the coolant flow.

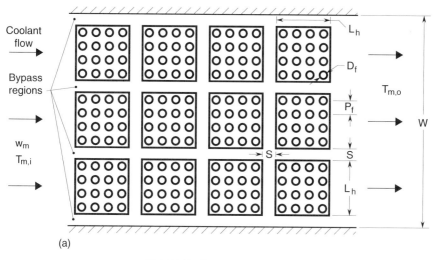

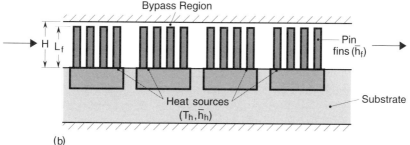

FIGURE 6.6 Schematic of heat source array with pin fins in channel flow: (*a*) top view and (*b*) side view.

Allowing for convection heat transfer from the tip, Eq. 2.17b may be used to estimate the rate of heat transfer from a single fin (Kelecy et al., 1987):

$$q_\mathrm{f} = M \frac{\sinh m L_\mathrm{f} + (\overline{h}_\mathrm{f}/m k_\mathrm{f}) \cosh m L_\mathrm{f}}{\cosh m L_\mathrm{f} + (\overline{h}_\mathrm{f}/m k_\mathrm{f}) \sinh m L_\mathrm{f}} \qquad (6.6)$$

where

$$M = (\overline{h}_\mathrm{f} P k_\mathrm{f} A_{\mathrm{c},\mathrm{f}})^{1/2} \theta_\mathrm{h} = \left(\frac{\pi}{2}\right) (\overline{h}_\mathrm{f} k_\mathrm{f} D_\mathrm{f}^3)^{1/2} \theta_\mathrm{h} \qquad (6.7\mathrm{a})$$

$$m = \left(\frac{\overline{h}_\mathrm{f} P}{k_\mathrm{f} A_{\mathrm{c},\mathrm{f}}}\right)^{1/2} = 2 \left(\frac{\overline{h}_\mathrm{f}}{k_\mathrm{f} D_\mathrm{f}}\right)^{1/2} \qquad (6.7\mathrm{b})$$

The subscript f refers to the pin fin material (k), geometry (A_c, D), and surface condition ($\overline{h}$), and θ_h is the difference between the temperature of the heat source and a reference fluid temperature. Although Eq. 2.17b is based on assuming a uniform convection coefficient over the fin surface, some variation in the local convection coefficient is likely and an average value, $\overline{h}_f$, is used to represent the entire surface.

If an array of N_f pins is attached to a single source and an equivalent convection coefficient is assumed for each pin, the rate of heat transfer from the source to the coolant may be expressed as

$$q_h = N_f q_f + \overline{h}_h \left(A_{s,h} - N_f A_{c,f} \right) \theta_h = N_f q_f + \overline{h}_h \left(L_h^2 - \frac{N_f \pi D_f^2}{4} \right) \theta_h \qquad (6.8)$$

where the second term on the right-hand side accounts for heat transfer from the exposed portion of the heater surface and the corresponding convection coefficient $\overline{h}_h$ is presumed to differ from $\overline{h}_f$. Assuming equivalent conditions for each of the N_h heat sources in an array of sources, the total heat rate for the array may be expressed as

$$q_t = N_h q_h = N_h \left[N_f q_f + \overline{h}_h \left(L_h^2 - \frac{N_f \pi D_f^2}{4} \right) \theta_h \right] \qquad (6.9)$$

Because flow over an array of pin fins resembles a tube bank in cross flow, $\overline{h}_f$ may be estimated by using the Grimison (1937) or Zhukauskas (1972) correlation, Eq. 2.98 or 2.99, respectively. For an in-line array of pins, $Re_{D,\max} = w_{\max} D_f / \nu$ and the maximum velocity of flow through the array may be expressed as $w_{\max} = P_f w_m / (P_f - D_f)$. This evaluation presumes that the velocity of fluid entering the array is equivalent to that associated with flow through the bypass regions along the sides and above the array.

Performing experiments for a channel of width $W = 19.1$ mm and height $H = 11.9$ mm with water flow over a single, 12.7 mm × 12.7 mm, flush-mounted heat source to which a 4 × 4 in-line array of 2.03-mm-diameter, 11.2-mm-long copper pins was attached, Kelecy et al. (1987) used Eqs. 6.6 to 6.9 to obtain values of $\overline{h}_f$ that were approximately 25% smaller than predictions based on Eq. 2.98. The difference was attributed to bypass effects, which reduced flow through the array, and to the formation of hydrodynamic and thermal boundary layers along the heated wall, which reduced local convection coefficients near the base of the pins.

Kelecy et al. (1987) also found that $\overline{h}_h \approx \overline{h}_f$ for the water flow, but that $\overline{h}_f > \overline{h}_h$ for comparable flow conditions involving FC-77. This disparity was attributed to a combination of (i) significant differences in the active surface area of the pins for water and FC-77 and (ii) the manner in which the local convection coefficient varies along the pin surface. In water, $\overline{h} \approx 10^4$ W/m² · K and $\eta_f \approx 0.4$. Hence, most of the fin heat transfer occurs near the base, within a short distance of which the pin temperature approaches that of the coolant. Conversely, for FC-77, $\overline{h} \approx 2000$ W/m² · K, $\eta_f \approx 0.75$, and the pin temperature exceeds the coolant temperature along its entire length. Hence, although the pin is only active near its base for water, it is active over the entire length for FC-77. Because the local convection coefficient of a fin with tip clearance increases from the base to the tip, the value of $\overline{h}_f$ obtained for FC-77 should exceed the corresponding value of $\overline{h}_h$, whereas $\overline{h}_f \approx \overline{h}_h$ for water.

Relative to using Eqs. 6.6 to 6.9 for design calculations, an important implication of the foregoing results is that, if $\overline{h}_f$ is estimated from a standard tube bank correlation and equivalence between $\overline{h}_f$ and $\overline{h}_h$ is assumed, the total heat rate will be overpredicted by as much as approximately 25%. Overprediction would be larger for water than for a dielectric liquid.

In using Eqs. 6.6 to 6.9, the temperature difference may be approximated as $\theta_h = T_h - T_{m,i}$, if the fluid temperature rise across the array of heat sources is small ($T_{m,o} \approx T_{m,i}$). If the temperature rise is significant, however, a log-mean temperature difference should be used, where

$$\theta_h = \theta_{lm} \equiv \frac{(T_h - T_{m,i}) - (T_h - T_{m,o})}{\ln\left[\frac{(T_h - T_{m,i})}{(T_h - T_{m,o})}\right]} \quad (6.10)$$

and all of the heat sources are assumed to have the same temperature T_h.

With the thermal resistance defined as

$$R_{th} = \frac{\overline{T}_h - T_{m,i}}{q_h} \quad (6.11)$$

experiments have been performed to determine R_{th} for each row of the heat source array shown in Figure 6.6, as well as for each row of an array with square fins (Fig. 6.7) applied to the pins (Kelecy et al., 1987; Ramadhyani and Incropera, 1987). A

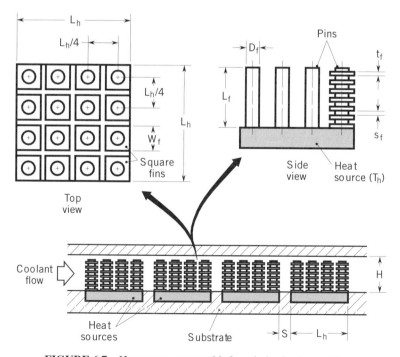

FIGURE 6.7 Heat source array with finned pins in channel flow.

4 × 4 in-line array of copper pins, 2.03 mm in diameter and 11.2 mm long, was attached to each of the 12.7 mm × 12.7 mm heat sources in a 3 × 4 array mounted flush to one wall of a rectangular channel for which $W = 50.8$ mm, $H = 11.9$ mm, $D_h = 19.3$ mm, and $S = 3.18$ mm. The addition of square fins to the circular pins was motivated by the desire to increase the active surface area and to turbulate the flow further. Fin dimensions of $W_f = 2.31$ mm and $t_f = S_f = 0.7$ mm were selected from an approximate conduction analysis, which indicated improved performance if (i) t_f and S_f are small, (ii) a fin tip clearance is maintained, and (iii) the first fin is in contact with the base. The surface area of an unaugmented heat source is increased eightfold by the addition of the pins, and an additional 60% increase in surface area is associated with use of the square fins.

Thermal resistance data for the unpinned, pinned, and finned-pin arrays are plotted as a function of Reynolds number in Figure 6.8. To reduce overlap with data

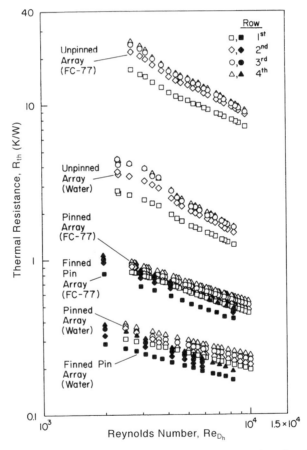

FIGURE 6.8 Row-by-row variations in the thermal resistance of unpinned, pinned, and finned-pin 12.7 mm × 12.7 mm heat source arrays for forced convection in channel flow. Adapted from Ramadhyani and Incropera (1987).

for the pinned array, only selected results are plotted for the finned pins. For water, use of the pins reduces R_{th} by factors of approximately 10 and 6, respectively, at the lower and higher extremities of the Reynolds number range. For FC-77, the reduction is larger, ranging from factors of approximately 20 to 15 for the lower and higher Reynolds numbers. The reductions are due to both the increased surface area and the increased convection coefficients associated with use of the pins. The larger reduction for FC-77 is attributed to the larger fin efficiency, for which more of the pin surface is active. As for the unaugmented array, thermal resistances for the pinned array increase with increasing row number, although row-to-row variations are more uniform.

The additional reduction in R_{th} associated with using the finned pins is much smaller, ranging from approximately 18% to 25% for water and from 17% to 22% for FC-77 at $Re_{D_h} \approx 8000$. The fact that the improvement is substantially less than the increase in surface area is attributed primarily to bypass effects and a corresponding reduction in the average convection coefficient associated with use of the fins (Ramadhyani and Incropera, 1987). A secondary effect is related to the fact that any increase in the area of an extended surface near its base decreases the activity of surfaces further removed from the base, thereby partially negating the effect of increasing surface area. Although use of the pin and finned-pin arrays increased the pressure drop across the channel, it remained small ($\Delta p < 1500$ N/m^2) for the Reynolds numbers ($Re_{D_h} < 10{,}000$) associated with the experiments (Ramadhyani and Incropera, 1987). Hence, it may be possible to reduce the deleterious influence of flow bypass by placing obstructions in the bypass lanes, without incurring an unacceptable penalty in pressure losses.

The effects of attaching strip fins (Brinkman et al., 1988) and low-profile microstuds (Maddox and Mudawar, 1988) to a single, 12.7-mm-square heat source have also been considered. In the strip fin study, the smallest thermal resistance was provided by an offset arrangement, and, for an equivalent duct Reynolds number in water or FC-77, the resistance was approximately 20% less than that for the heat source with finned pins (Ramadhyani and Incropera, 1987). In FC-72, 1.02-mm-long microstuds reduced the thermal resistance by up to a factor of 6, but, at comparable Reynolds numbers, resistances exceeded those associated with the finned pin or offset strip fin arrangements. Thermal resistances associated with the different enhancement schemes are compared in Figure 6.9. The thermal resistance of 0.09 K/W achieved by Tuckerman and Pease (1981) for water flow through microchannels in a 20-mm-square silicon substrate is also included.

Example 6.1

A 3×4 array of chips is mounted to a substrate that forms one wall of a rectangular channel, as shown in Figure 6.6. To enhance heat transfer from the chips, a 4 × 4 in-line array of copper pin fins ($k_f = 390$ W/m·K) is attached to each chip, also as shown in Figure 6.6. The chips are of length $L_h = 10$ mm on a side, and the diameter, pitch, and length of the pins are $D_f = 2$ mm, $P_f = 2.50$ mm, and $L_f = 10$ mm, respectively. The channel width and height are $W = 48$ mm and

232 HEAT TRANSFER ENHANCEMENT

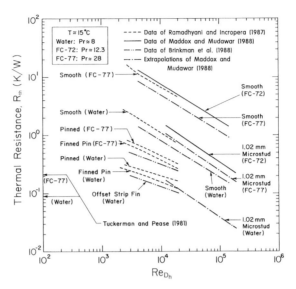

FIGURE 6.9 Thermal resistances of a 12.7 mm × 12.7 mm heat source with pins, finned pins, offset strip fins, and microstuds for forced convection in channel flow.

$H = 11$ mm, respectively, and the chips are cooled by FC-77, which has an upstream temperature of $T_{m,i} = 15°C$ and a mean velocity of $w_m = 1.0$ m/s. If the maximum allowable chip temperature is $T_h = 85°C$, what is the maximum allowable chip heat rate q_h?

Solution

Known: Dimensions of pin fin and chip arrays mounted to one wall of rectangular channel. Channel dimensions and coolant inlet conditions. Maximum allowable chip temperature.

Find: Maximum allowable chip heat rate.

Assumptions:

1. Steady-state conditions.
2. Negligible effect of chip protrusion.
3. Negligible bypass of flow around pin fin arrays.
4. Average convection coefficient $\bar{h}_f$ on pin surfaces may be obtained by approximating the pins as circular cylinders in cross flow (Section 2.6.2), with the upstream velocity V approximated as the mean velocity w_m of the channel flow.
5. Equivalent convection coefficients for the pin and exposed base surfaces ($\bar{h}_f = \bar{h}_h$).

6. Negligible variation of conditions from chip to chip in a row or from row to row.
7. Negligible contact resistance between the base of a pin and the chip to which it is attached.
8. Properties may be evaluated by assuming a negligible increase in the coolant temperature ($T_{m,o} \approx T_{m,i}$) and a uniform fin temperature equivalent to the chip temperature ($\eta_f \approx 1$).
9. Negligible flow work effects in channel

Properties: From Appendix B: FC-77 ($\bar{T}_f \approx (T_h + T_{m,i})/2 = 50°C$): $\rho = 1716$ kg/m^3, $c_p = 1087$ J/kg · K, $k = 0.061$ W/m · K, $\alpha = k/\rho c_p = 3.27 \times 10^{-8}$ m^2/s, $\mu = 9.25 \times 10^{-4}$ kg · s/m, $\nu = \mu/\rho = 5.39 \times 10^{-7}$ m^2/s, and $Pr = \nu/\alpha = 16.5$.

Analysis: With $\bar{h}_h = \bar{h}_f$, the rate of heat transfer from a single chip, Eq. 6.8, can be expressed as

$$q_h = N_f q_f + \bar{h}_f \left(L_h^2 - \frac{N_f \pi D_f^2}{4} \right) \theta_h$$

where $N_f = 16$ and, from Eqs. 6.6 and 6.10, respectively,

$$q_f = M \frac{\sinh mL_f + (\bar{h}_f/mk_f) \cosh mL_f}{\cosh mL_f + (\bar{h}_f/mk_f) \sinh mL_f}$$

$$\theta_h = \frac{(T_h - T_{m,i}) - (T_h - T_{m,o})}{\ln \left[\frac{(T_h - T_{m,i})}{(T_h - T_{m,o})} \right]}$$

The outlet temperature of the coolant may be obtained from an overall energy balance, which equates the total heat rate for the chip array, $q_t = N_h q_h$, to the increase in enthalpy for flow over the chips. Hence, with negligible flow work and $c_v = c_p$ for an incompressible liquid, Eq. 2.132 yields

$$q_t = \dot{m} c_p (T_{m,o} - T_{m,i})$$

where $\dot{m} = \rho w_m A_c = \rho w_m (WH) = 1716$ kg/m^3(1 m/s)(0.048 m × 0.011 m) = 0.906 kg/s.

The convection coefficient $\bar{h}_f$ may be evaluated from Eq. 2.98:

$$\overline{Nu}_{D_f} = 1.13 C_1 C_2 Re_{D,\max}^m Pr^{1/3}$$

where $C_1 = 0.348$, $C_2 = 0.90$, and $m = 0.592$ for $P_f/D_f = 1.25$ (Incropera and DeWitt, 1996). The value of $C_2 = 0.90$, which presumes four rows of pins, would be larger if multiple chip rows were considered. Hence, selection of $C_2 = 0.90$ provides a slightly conservative (low) estimate of heat transfer.

From Eqs. 2.101 and 2.100,

$$V_{max} = \frac{P_f}{P_f - D_f} w_m = \frac{2.5 \text{ mm}}{(2.5 - 2.0) \text{ mm}} 1 \text{ m/s} = 5 \text{ m/s}$$

$$Re_{D,max} = \frac{V_{max} D_f}{\nu} = \frac{5 \text{ m/s}(0.002 \text{ m})}{5.39 \times 10^{-7} \text{ m}^2/\text{s}} = 18{,}550$$

Hence,

$$\overline{Nu}_{D_f} = 1.13(0.348)(0.90)(18{,}550)^{0.592}(16.5)^{1/3} = 303$$

and

$$\bar{h}_f = \overline{Nu}_{D_f} \frac{k}{D_f} = 303 \frac{0.061 \text{ W/m} \cdot \text{K}}{0.002 \text{ m}} = 9240 \text{ W/m}^2 \cdot \text{K}$$

From Eqs. 6.7b and 6.7a, respectively, it follows that

$$m = 2\left(\frac{\bar{h}_f}{k_f D_f}\right)^{1/2} = 2\left(\frac{9240 \text{ W/m}^2 \cdot \text{K}}{390 \text{ W/m} \cdot \text{K} \times 0.002 \text{ m}}\right)^{1/2} = 217.7 \text{ m}^{-1}$$

$$M = \frac{\pi}{2}\left(\bar{h}_f k_f D_f^3\right)^{1/2} \theta_h = \frac{\pi}{2}\left(9240 \text{ W/m}^2 \cdot \text{K} \times 390 \text{ W/m} \cdot \text{K}(0.002 \text{ m})^3\right)^{1/2} \theta_h$$

$$= 0.267\theta_h$$

With the foregoing results, the expressions for q_f, q_h, q_t, and θ_h may be solved iteratively to obtain

$$T_{m,o} = 17.9°C$$
$$\theta_h = 68.5°C$$
$$q_f = 17.9 \text{ W}$$
$$q_h = 318.1 \text{ W}$$
$$q_t = 2863 \text{ W}$$

where $q_h = 318.1$ W is the maximum allowable chip heat rate.

Comments:

1. The temperature rise of the coolant, $T_{m,o} - T_{m,i} = 2.9°C$, is small, justifying its neglect in evaluating fluid properties.
2. Note that the subscript f has been used exclusively to designate characteristics of the fins, including the thermal conductivity. To avoid confusion, the subscript was not used with the fluid properties.

3. For the prescribed conditions, the fin efficiency is $\eta_f = 0.428$. This small value suggests that there is room for optimizing D_f and P_f to increase q_h.
4. With increasing w_m from 0.5 to 2.0 m/s, $\bar{h}_f$ and q_f increase from 6130 to 13,940 W/m²·K and from 14.0 to 22.5 W, respectively. The fact that the increase in q_f is not commensurate with the increase in $\bar{h}_f$ is due to a reduction in η_f from 0.512 to 0.354.

6.3.2 Discrete Heat Sources in Mixed Convection

The influence of longitudinal (parallel plate) fins on heat transfer from a 4 × 3 array of flush-mounted, 12.7-mm-square heaters has been considered experimentally and theoretically for mixed-convection conditions (Mahaney et al., 1990b, 1991). Each heat source contained five equally spaced fins of 10.6 mm height and 1.08 mm thickness. Fin performance was evaluated in terms of a unit thermal resistance

$$R''_{th} = \frac{\overline{T}_h - T_{m,i}}{q_h/L_h^2} \qquad (6.12)$$

and experimental results are compared with those for an unaugmented (bare) heat source array in Figure 6.10. The 10-fold increase in surface area associated with the longitudinal fins results in the thermal resistance of the four rows being reduced by factors of 4.6 to 5.7 at the lowest flow rate and by factors of 13.3 to 14.2 at the highest flow rate. The fact that heat transfer enchancement at the lowest flow rate is less than the surface area augmentation suggests that local heat transfer coefficients for the finned surfaces are less than that associated with convection from a heated horizontal surface facing upward. The fact that heat transfer enhancement at the highest flow rate exceeds the increase in surface area suggests that acceleration of the fluid resulting from the flow restriction provided by the fins and possible turbulation of the flow at the downstream rows causes local convection coefficients on the finned surfaces to exceed those associated with convection from the unfinned surface. The thermal resistance increases with increasing row number and decreasing Reynolds number. However, more than a 50-fold change in Re_{L_h} results in less than a four-fold change in R''_{th}. A numerical simulation of mixed convection heat transfer from an array of discrete heat sources with longitudinal fins (Mahaney et al., 1991) predicts buoyancy-induced flows in the form of pairs of counterrotating circulation cells between neighboring fins, with upflow adjacent to the fins and downflow between fins.

6.3.3 Microchannels

Heat transfer in microchannels may be enhanced by using interrupted, instead of continuous, walls. Such an arrangement prevents continuous boundary layer development, thereby maintaining larger convection coefficients. However, the attendant reduction in the surface area has the opposing effect of reducing heat transfer.

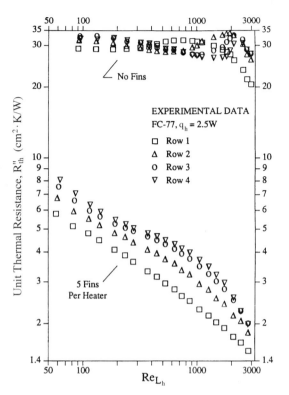

FIGURE 6.10 Effect of longitudinal fins on the thermal resistance of a heat source array for mixed convection in channel flow. Adapted from Mahaney et al. (1990b).

Tuckerman (1984) used precision mechanical sawing to create a microstructure of square pin fins (studs) in silicon, with a fin pitch of $P_f = 80$ μm. A maximum heat flux of $q_h'' = 1300$ W/cm² was achieved in water, substantially exceeding results for uninterrupted channel walls. Kishimoto and Sasaki (1987) also used mechanical sawing to create a staggered array of diamond-shaped pin fins. The diamond fin profile was intended to minimize flow separation and, thereby, to derive maximum benefit from disruption of boundary layer development.

Using a three-dimensional numerical simulation of laminar flow in an array of square pin fins, Yin and Bau (1995) compared predictions with results for continuous channel walls. The comparative performance depended on the channel/pin height ($H = L_f$), with the lower thermal resistance corresponding to the continuous walls and the pin fins for deep ($H > 1000$ μm) and shallow ($H < 100$ μm) channels, respectively.

6.4 IMPINGING JETS

Despite the fact that large heat transfer coefficients are associated with impinging liquid jets, surface modifications may still be used to achieve heat transfer enhancement. As for channel flow, the modifications may increase heat transfer by (i) disrupting boundary layer development, thereby advancing the transition to turbulence and/or increasing the turbulence intensity, and (ii) increasing the total surface area for convection. With surface modification, the effect of nozzle geometry is diminished.

6.4.1 Circular Jets

Although, by itself, liquid jet impingement provides an effective means of transferring heat from electronic devices, extended surfaces may still be used to good advantage. One approach involves attaching a *spreader plate* to an electronic chip and using a submerged circular jet to cool the plate. Such a scheme has been considered by Sullivan et al. (1992a), who performed experiments with FC-77 for a range of nozzle diameters and flow rates (Fig. 6.11). The configuration would be particularly

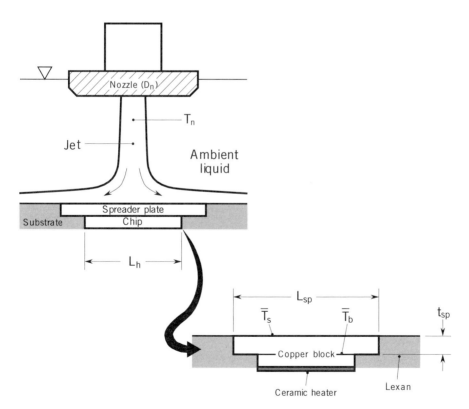

FIGURE 6.11 Impingement jet cooling of a simulated electronic chip and spreader plate assembly. Adapted from Sullivan et al. (1992a).

effective if the spreader plate encompassed a secondary peak in the local heat transfer coefficient, which would otherwise not occur on the smaller chip. The spreader/chip assembly was machined from a single block of copper, with a ceramic heater joined to the back surface of a 12.7 mm × 12.7 mm simulated chip. Two spreader plate configurations, with planform areas of 15 mm × 15 mm and 20 mm × 20 mm, were considered, each with a nearly optimum thickness of $t_{sp} = 2$ mm. Experiments were performed for wetted surfaces that were *smooth* and *roughened* (Fig. 6.12).

Roughness elements can enhance heat transfer by increasing the surface area and by either turbulating a laminar boundary layer or disrupting the viscous sublayer of an already turbulent flow. From Schlichting (1960), the minimum roughness height required for protrusion through, and, hence, disruption of, a viscous sublayer is $e_{min} \approx 100\nu/u_\infty$. One roughness condition (Fig. 6.12a) consisted of a saw-cut pattern of *millistuds*, which doubled the surface area of the spreader plate. The other roughness condition was formed by *peening* the surface with a 3-mm-diameter rod to create dimples ranging from approximately 0.10 to 0.30 mm deep.

Defining a unit thermal resistance as

$$R''_{th} = \frac{A_{s,h}(\overline{T}_b - T_n)}{q_h} \qquad (6.13)$$

where $A_{s,h} = L_h^2$ and $\overline{T}_b$ is the average temperature at the spreader/chip interface, experiments were performed to determine the effect on R''_{th} of the nozzle diameter D_n, the jet volumetric flow rate $\dot{V}$, the area $A_{s,sp}$ of a smooth spreader, and the

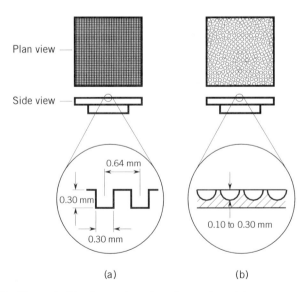

FIGURE 6.12 Plan and side views of roughened spreaders: (*a*) saw cut and (*b*) dimpled. Adapted from Sullivan et al. (1992a).

TABLE 6.1 Effect of Nozzle Diameter and Heater Surface Condition on Unit Thermal Resistance, R''_{th} (cm² · K/W), for a Submerged Jet of FC-77 ($S/D_n = 3$, $\dot{V} = 1$ lpm).

	D_n (mm)		
	1.65	3.11	6.55
Unaugmented heat source (12.7 mm × 12.7 mm)	1.2	2.3	5.5
Smooth spreader (15 mm × 15 mm)	1.0	1.8	4.2
Smooth spreader (20 mm × 20 mm)	0.7	1.3	2.5
Dimpled spreader (20 mm × 20 mm)	0.6	1.3	2.8
Saw-cut spreader (20 mm × 20 mm)	0.5	0.6	1.2

roughened surface conditions. In all cases, the nozzle-to-spreader spacing was fixed at $S/D_n = 3$.

Representative results for $\dot{V} = 1$ lpm are presented in Table 6.1, which also includes data obtained by Womac et al. (1990) for an unaugmented heat source. Use of the spreader plate reduces the thermal resistance, and, although there is little effect of a dimpled surface, further reduction is effected by the saw-cut condition. For fixed $\dot{V}$, R''_{th} decreases with decreasing D_n because of the attendant increase in V_n. The smallest value of $R''_{th} = 0.5$ is achieved for the largest spreader, the saw-cut surface condition, and the smallest nozzle diameter.

The effect of the volumetric flow rate on the thermal resistance is shown in Figure 6.13 for smooth and roughened spreaders with 20 mm × 20 mm planform dimensions. For the smallest nozzle ($D_n = 0.98$ mm), there is little effect of surface modification on R''_{th} over the entire range of flow rates. This behavior is attributed to the fact that, with transition to turbulence occurring at $r \approx 1.9 D_n$ for a smooth surface, nearly the entire surface experiences turbulent flow for $D_n = 0.98$ mm. Hence, any effect that roughness may have on enhancing heat transfer by accelerating the transition to turbulence is small. However, assuming $u_\infty = V_n$, the smallest nozzle operating at $\dot{V}_{min} = 0.1$ lpm yields a maximum value of $e_{min} \approx 0.04$ mm, which is less than the height of the millistuds. Hence, the laminar sublayer of the turbulent flow is disrupted and, in combination with the increased surface area afforded by the studs, this effect acts to enhance heat transfer slightly.

With increasing D_n, the portion of the smooth surface that experiences turbulent flow decreases, and the effect that the studs have on accelerating the transition to turbulence becomes more prominent, thereby decreasing the value of R''_{th} associated with use of the saw-cut pattern. Over the entire range of conditions, dimpling has a negligible effect on the thermal resistance.

The influence of surface roughness on stagnation point heat transfer for a *free-surface* water jet has been studied by Gabour and Lienhard (1994). The root mean square (rms) roughness was varied from 4.7 to 28.2 μm, and, for nozzle diameters of 4.4, 6.0, and 9.0 mm, results were compared to those for a smooth surface over the Reynolds number range $20{,}000 \leq Re_{D_n} < 84{,}000$. The data indicated an increasing heat transfer enhancement with increasing roughness and a maximum enhancement of approximately 50% for the smallest nozzle diameter and largest Reynolds number.

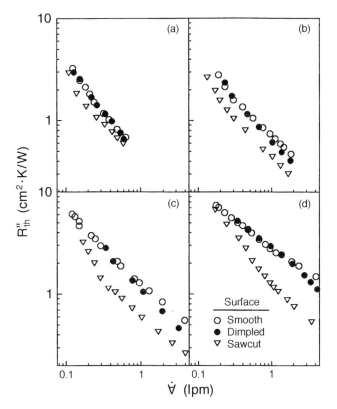

FIGURE 6.13 Effect of volumetric flow rate on thermal resistance for smooth and roughened 20 mm × 20 mm spreader plates (FC-77, $S/D_n = 3$): (a) $D_n = 0.98$ mm, (b) $D_n = 1.65$ mm, (c) $D_n = 3.11$ mm, and (d) $D_n = 6.55$ mm. Adapted from Sullivan et al. (1992a).

From boundary layer considerations, Gabour and Lienhard (1994) also concluded that heat transfer enhancement resulting from surface roughness becomes more pronounced with increasing Prandtl number. Based on data obtained for water and the use of boundary layer arguments to infer a Prandtl number dependence, they obtained the following correlation for a dimensionless critical roughness, $(e/D_n)_c$, at which heat transfer is increased by 10%:

$$\left(\frac{e}{D_n}\right)_c = 12.1 Re_{D_n}^{-0.713} Pr^{-1/3} \tag{6.14}$$

For values of $(e/D_n)_c$ below this threshold, the surface may be approximated as smooth. In using Eq. 6.14, Lienhard (1995) cautions that roughness effects can depend on wall material and roughness geometry, as well as on roughness height. The study by Gabour and Lienhard (1994) is marked by its consideration of surface roughness effects to the exclusion of enhancement resulting from extended surfaces (increased surface area).

In an extension of their investigation of surface roughness effects (Sullivan et al., 1992a), Sullivan et al. (1992b) considered heat transfer enhancement associated with using square pin fins (*millistuds*) and radial grooves. The pin fin geometry is shown in Figure 6.14. The large heat transfer coefficient associated with the stagnation zone may be maintained, while circumventing attenuation associated with the wall jet region, by leaving the impingement zone unaltered and adding pin fins to the wall jet region. The end face of the nozzle makes contact with the fin tips, providing semiconfined submerged jet conditions for which the fins turbulate the flow, as well as increase the heat transfer surface area. The lengths of the square chip and spreader plate were fixed at $L_h = 12.7$ mm and $L_{sp} = 15$ mm, respectively, while three different pin dimensions were considered (Table 6.2).

The 0.3-mm-square-by-0.3-mm-tall pins provided an extremely compact arrangement with nearly isothermal surfaces ($\eta_f \approx 1$), whereas the 0.3-mm-square-by-1.0-mm-tall pins provided more than a threefold increase in surface area with only a small reduction in fin efficiency ($\eta_f \approx 0.9$ for a nominal flow rate of 0.95 lpm), as well as a significant reduction in pressure drop for a prescribed flow rate. The 0.9-mm-square-by-3.0-mm-tall array provided a slightly larger total surface area, $A_{s,t}$, and a negligible pressure drop. For the two taller pin fin arrays, a 4.5-mm-diameter clearing was created in the center of the impingement surface to allow for full penetration of the jet. For the 0.3-mm-long pins, the array covered the entire surface.

The planform of the radial grooves is shown in Figure 6.15. The grooves were machined in the wall jet region and were aligned with the radial flow. The surface was fabricated by taking eight 3-mm-deep passes with a 0.9-mm slitting saw and leaving a 4.1-mm-diameter clearance in the center region. The configuration was intended to provide moderate heat transfer enhancement with negligible pressure drop.

Experiments were performed for FC-77 with four nozzles having diameters of $D_n = 0.98, 1.65, 3.11$, and 6.55 mm. Unit thermal resistances were evaluated from Eq. 6.13, and, for $\dot{V} = 0.95$ lpm, results are compared with those for an unaugmented heat source (Womac et al., 1990) in Figure 6.16. Both augmented surfaces provide significant heat transfer enhancement, with the 0.3-mm-square-by-1.0-mm-long pins and the 0.98-mm-diameter jet providing the smallest thermal resistance. For a fixed flow rate, the thermal resistance decreased with decreasing D_n, and, for a fixed diameter, it decreased with increasing $\dot{V}$. Similar trends characterized the radial grooves. Conversely, the pressure drop increased with decreasing D_n and increasing $\dot{V}_n$. The effect of volumetric flow rate on the thermal resistance of each configuration is shown in Figure 6.17 for $D_n = 1.65$ mm.

Priedeman et al. (1994) also studied the effect of using square pin fins to enhance heat transfer for *free-surface* jets of FC-77 and water. Fin thicknesses, t_f, of 0.8 and 1.59 mm, with corresponding pitches, P_f, of 1.6 and 3.2 mm, respectively, were considered, along with a pyramidal surface having a 3.18-mm-square base. Heat transfer enhancement afforded by the extended surfaces was as high as a factor of 2.8 for water and 4.5 for FC-77. Relative to an unaugmented surface, enhancement was attributed to the increased surface area, as well as to an increase in the average convection coefficient. Square pin fins in an impinging jet have also been considered by Ledezma et al. (1996), who developed relations for the optimal pin spacing and

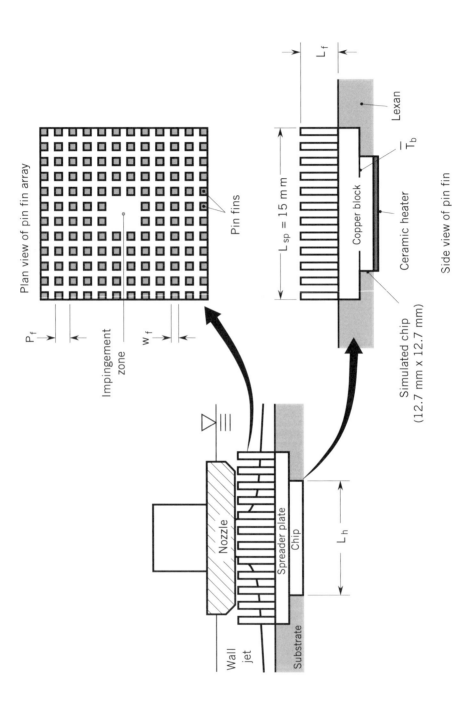

FIGURE 6.14 Impingement jet cooling of a simulated electronic chip and spreader plate assembly with an array of square pin fins (studs) machined in the spreader plate. Adapted from Sullivan et al. (1992b).

TABLE 6.2 Pin Fin Array Geometries for the Heat Transfer Enhancement Scheme of Figure 6.14

L_f (mm)	P_f (mm)	w_f (mm)	$A_{s,t}$ (cm^2)
0.3	0.6	0.3	4.4
1.0	0.6	0.3	9.4
3.0	1.8	0.9	9.7

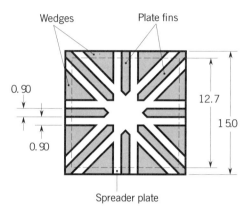

FIGURE 6.15 Planform of surface enhancement with radial grooves (dimensions in mm). Adapted from Sullivan et al. (1992b).

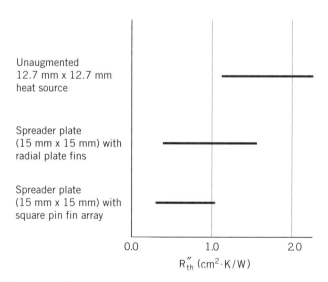

FIGURE 6.16 Range of unit thermal resistances for impingement of a semiconfined submerged jet of FC-77 on unaugmented and augmented surfaces ($\dot{V} = 0.95$ lpm; $D_n = 0.98$, 1.65, 3.11, and 6.55 mm). Adapted from Sullivan et al. (1992b).

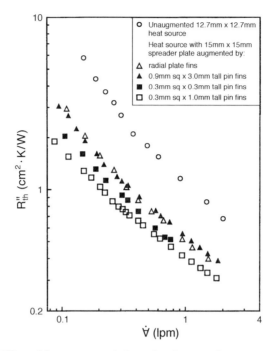

FIGURE 6.17 Effect of flow rate on unit thermal resistance of unaugmented and augmented 12.7-mm-square heat sources for impingement of a semi-confined submerged FC-77 jet ($D_n = 1.65$ mm). Adapted from Sullivan et al. (1992b).

minimum thermal resistance. Although their experimental and theoretical work was restricted to air, the relations include a Prandtl number term to account for variations in fluid type.

Radial fins have also been considered as a means of enhancing heat transfer under conditions for which there is annular collection of the spent fluid (Heindel et al., 1992b). The test cell was equivalent to that used by Besserman et al. (1991b) and is shown schematically in Figure 6.18. However, in this case, a circular copper heater of diameter $D_h = 22.2$ mm was machined with 36 integral radial fins, the leading edges of which were located 5.1 mm from the stagnation point, thereby maintaining a smooth heater surface in the jet impingement zone. Each fin was of width $W_f = 6.0$ mm, length $L_f = 3.6$ mm, and thickness $t_f = 0.5$ mm, and, collectively, the fins provided more than a fivefold increase in the surface area. Experiments were performed for FC-77 and parameter variations corresponding to $4.42 < D_n < 9.27$ mm, $22.23 < D_c < 30.16$ mm, $0.5 < S/D_n < 5$, $0.2 < \dot{V} < 11$ lpm, and $700 < Re_{D_n} < 44,000$.

For prescribed values of D_c and S/D_n, the thermal resistance decreased with increasing jet volumetric flow rate, $\dot{V}$, or mean velocity, V_n, and either increased or decreased with increasing nozzle diameter, depending on whether $\dot{V}$ or V_n was fixed. For $D_c = 22.23$ mm and laminar jets ($Re_{D_n} < 2000$), the effect of separation

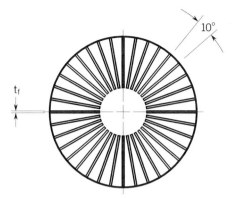

Plan view of finned heater

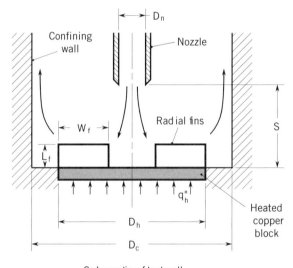

Schematic of test cell

FIGURE 6.18 Schematic of test cell for jet impingement cooling a circular heat source with radial plate fins and annular collection of the spent fluid. Adapted from Heindel et al. (1992b).

distance was negligible. However, for turbulent flow, R''_{th} decreased with decreasing S/D_n, and the effect increased with increasing D_n. Over the range of confinement diameters, D_c, the thermal resistance was found to decrease by approximately 20% with decreasing D_c for laminar flow and by 10% to 20% for turbulent flow, depending on the value of S/D_n. This trend differs from that reported for an unaugmented surface (Besserman et al., 1991a), which revealed a slight increase in thermal resistance with decreasing D_c.

In Figure 6.19, the results of Heindel et al. (1992b) are compared with those obtained by Sullivan et al. (1992b) for an augmented surface with 16 radial grooves,

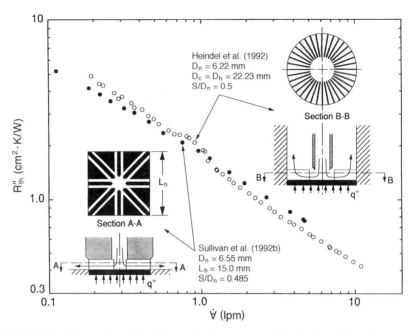

FIGURE 6.19 Comparison of unit thermal resistances associated with surface enhancement by means of radial grooves (Sullivan et al., 1992b) and radial fins (Heindel et al., 1992b). Adapted from Heindel et al. (1992b).

each 3.18 mm deep and 0.9 mm wide. The data of Sullivan et al. (1992b) were renormalized with respect to the spreader, rather than the heater, surface area to be consistent with the manner in which Eq. 6.13 was used by Heindel et al. (1992b). There is little difference between the results, which is perhaps surprising in view of the fact that the radial fins provide 2.3 times more surface area. However, the thick fins formed by the radial grooves are more efficient than the thin fins of Heindel et al. (1992b), and the flow conditions differ significantly. Nevertheless, more than a threefold enhancement in heat transfer may be effected by using the radial fins, with a minimum thermal resistance of 0.4 cm^2·K/W having been obtained.

The foregoing radial fin configuration is one of several extended surfaces discussed in a patent (Yamamoto et al., 1988b) concerning heat transfer enhancement for electronic chips cooled by liquid jet impingement with annular collection of the spent fluid (Yamamoto et al., 1988a). In the FACOM M-780 multichip module (Yamamoto et al., 1988a), each 9.3-mm-square chip is cooled by a single water jet, 3 mm in diameter (Fig. 1.2). The flow rate through the nozzle is 0.34 lpm, and each chip dissipates up to 9.5 W. A flat spreader plate is attached to the chip with an elastic material, and spent water from the impinging jet exits through the annulus formed by the nozzle and the bellows. The chip-to-fluid thermal resistance is 2.1 cm^2·K/W. The VP 2000 (Suzuki et al., 1989) is an improved version of the M-780 design and dissipates up to 30 W/chip with a nozzle diameter and flow rate of 4.5 mm and 0.86 lpm, respectively. The corresponding value of the chip-to-fluid thermal resistance is 1.0 cm$^2 \cdot$ K/W.

6.4.2 Rectangular Jets

Wadsworth and Mudawar (1992) extended their study of rectangular submerged jets (Wadsworth and Mudawar, 1990) to consider the effect of using low-profile extended surfaces in the form of microgrooves and microstuds (Fig. 6.20) to enhance heat transfer from a simulated 12.7 mm × 12.7 mm chip. Experiments were performed

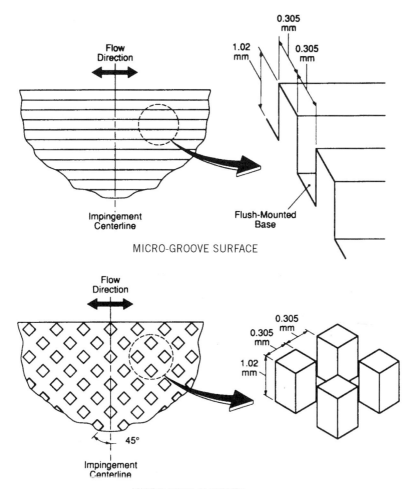

FIGURE 6.20 Microgroove and microstud enhancement geometries (Wadsworth and Mudawar, 1992). Used with permission of ASME.

for FC-72, a nozzle width of $w_n = 0.254$ mm, a nozzle-to-chip spacing of $S = 2.54$ mm, and Reynolds numbers in the range $2000 \leq Re_{D_h} \leq 30{,}000$. Relative to an unaugmented (smooth) surface, heat transfer enhancement increased with increasing Re_{D_h}, ranging from 240% to 406% and from 230% to 316% for the microgroove and microstud surfaces, respectively. Heat transfer correlations for each of the augmented chips are of the form

$$\overline{Nu}_L = 2.95 Re_{D_h}^{0.756} Pr^{1/3} \quad \text{(microgroove)} \tag{6.15}$$

$$\overline{Nu}_L = 4.12 Re_{D_h}^{0.698} Pr^{1/3} \quad \text{(microstud)} \tag{6.16}$$

which may be contrasted with Eq. 5.67 for the smooth surface. In addition to providing slightly superior heat transfer enhancement relative to the microstuds, the microgrooves yielded a smaller increase in the pressure drop (3450 N/m² relative to 6900 N/m² at the largest Reynolds number) and a smaller variation in temperature across the chip surface (0.8°C for a heat flux of 61.8 W/cm² relative to 3.2°C for a heat flux of 52.6 W/cm²). In electronic cooling, surface temperature uniformity is a desired feature.

Heat transfer enhancement for planar liquid (FC-77) jet impingement was also considered by Teuscher et al. (1993) for an in-line array of five simulated chips, each of length $L_h = 12.7$ mm on a side. As shown in Figure 6.21, spent liquid was discharged between nozzle exits in the jet orifice plate, and augmentation was

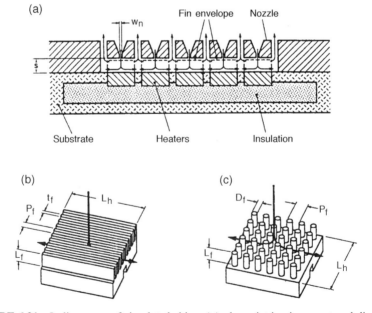

FIGURE 6.21 In-line array of simulated chips: (*a*) planar jet impingement and discharge, (*b*) plate fin array, and (*c*) pin fin array. Adapted from Teuscher et al. (1993).

achieved by using a plate or pin fin array. Two plate fin thicknesses were considered ($t_f = 0.203$ mm, 0.406 mm), each with a dimensionless fin pitch of $P_f/t_f = 2$, along with a single pin diameter ($D_f = 0.20$ mm) and a dimensionless pitch of $P_f/D_f = 2$. The fin length was fixed at $L_f = 2.54$ mm, and two nozzle widths ($w_n = 0.254$ mm, 0.508 mm) were considered.

As shown in Figure 6.22, superior performance is associated with plate fins, which, like the microgrooves of Wadsworth and Mudawar (1992), are aligned in the direction of the wall jet. The thermal resistance is reduced by decreasing the fin thickness, for which the number of fins increases from 16 to 32 and the fin surface area increases from 10.3 to 20.6 cm^2. For a fixed flow rate, the thermal resistance also decreases with decreasing nozzle width because of the corresponding increase in jet velocity.

Measurements of the pressure drop between upstream and downstream manifolds revealed nearly equivalent results for the three extended surfaces and the smaller nozzle ($w_n = 0.254$ mm), but a significant ($\sim 60\%$) reduction for the larger nozzle ($w_n = 0.508$). Hence, the dominant contribution to the pressure drop is made by the nozzle and, for the smaller slot, the effect of the extended surfaces is negligible. At the largest and nominal flow rates of 3 lpm/jet and 0.95 lpm/jet, the pressure drop,

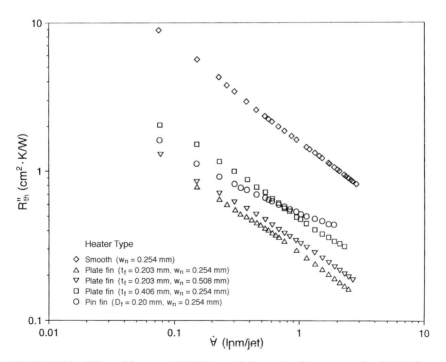

FIGURE 6.22 Effect of flow rate (FC-77) on unit thermal resistance associated with planar jet impingement on surfaces enhanced by means of plate and pin fins. Adapted from Teuscher et al. (1993).

TABLE 6.3 Comparison of Unit Thermal Resistances Associated with Liquid Jet Impingement and Various Fin Arrays

Extended Surface	R''_{th} [cm^2·K/W]
Teuscher et al. (1993), $\dot{V} = 0.95$ *lpm/jet (FC-77)*	
0.203-mm plate fins	0.284
0.406-mm plate fins	0.458
0.2-mm pin fins	0.345
Sullivan et al. (1992b), $\dot{V} = 0.95$ *lpm/jet (FC-77)*	
0.3-mm-square × 1.0-mm-tall pin fins	0.32–0.65
0.3-mm-square × 0.3-mm-tall pin fins	0.4–1.2
0.9-mm-square × 3.0-mm-tall pin fins	0.4–0.75
Radial fins	0.4–1.65
Commercial Packages	
Fujitsu VP 2000 (0.86 lpm)	1.0
Fujitsu FACOM M-780 (0.34 lpm)	2.1

Δp, for the smallest slot was approximately 200 kPa (2 atm) and 20 kPa (0.2 atm), respectively.

In Table 6.3, the results of Teuscher et al. (1993) are compared to those of Sullivan et al. (1992b) for a nominal flow rate of $\dot{V} = 0.95$ lpm/jet, as well as to those reported for commercial packages developed by Fujitsu. Values reported for the pin and radial fin arrays of Sullivan et al. (1992b) correspond to results for nozzle diameters in the range from 0.98 to 6.55 mm. The results of Teuscher et al. (1993) and Sullivan et al. (1992b) are both well below those of the commercial systems, suggesting the potential for improvements in cooling system designs. The unit thermal resistances of Teuscher et al. (1993) are smaller than those of Sullivan et al. (1992b), even though the cross-sectional area of the slot nozzle ($A_{c,n} = 3.23$ mm^2) exceeds that of the smallest circular nozzle ($A_{c,n} = 0.75$ mm^2). The improvement is, at least in part, attributable to the larger wetted areas associated with the extended surfaces of Teuscher et al. (1993), but it may also be due to the strictly two-dimensional flow lanes afforded by the slot jets.

For $t_f = 0.203$ mm, $w_n = 0.254$ mm, and $\dot{V} = 0.95$ lpm/jet, the plate fins of Teuscher et al. (1993) yield a unit thermal resistance of $R''_{th} = 0.284$ cm^2 · K/W and, hence, an allowable chip heat dissipation of 176 W/cm^2 for an assumed 50°C temperature difference between the fluid and the chip surface. At 3 lpm/jet, $R''_{th} \approx 0.15$ cm^2 · K/W and the allowable heat dissipation would be 333 W/cm^2. Even at the largest flow rates, the fin efficiencies estimated by Teuscher et al. (1993) for their extended surfaces exceed 80%, suggesting that significant additional reductions in R''_{th} could be achieved by increasing the fin length beyond the fixed value of $L_f = 2.54$ mm. However, manufacturing difficulties and packaging constraints could limit this option in an actual electronic application.

When considering extended surfaces for enhancing jet impingement heat transfer, attention should be placed on maximizing the surface area, while ensuring good *wetting* of the entire surface and maintaining acceptable pressure losses. Effective wetting implies maximum exposure of the surfaces to the fluid and would be achieved when flow lanes provided by the extended surfaces are consistent with flow patterns normally resulting from jet impingement and when the flow is *confined* to movement over the fins. Flow confinement is ensured by maintaining the fin tips in close proximity to the nozzle plate ($L_f \approx S$).

6.5 SUMMARY

Heat sinks with extended surfaces may be used to significantly enhance heat transfer by single-phase convection in liquids used for electronic cooling. Many configurations are possible, although work to date has focused on the use of plate and pin fin geometries.

Cooling by free convection may be effected in a closed system, as, for example, a liquid-filled rectangular cavity with the heat-dissipating elements mounted to one wall and a cold plate used for the opposing wall. Parallel-plate fins have been found to enhance heat transfer as much as 24-fold when attached to an array of discrete sources on one of the vertical walls, and, for FC-77, thermal resistances as low as $R''_{th} \approx 2$ cm$^2 \cdot$ K/W have been obtained for the lowermost row of heat sources in the array. However, progressively larger values of R''_{th} are associated with each of the upper rows, rendering the option unattractive for electronic cooling. In contrast, row-to-row variations are negligible for a horizontal orientation in which the heat sources are mounted to the bottom wall and the opposing (top) wall is composed of a cold plate. Although parallel-plate fins enhance heat transfer by a smaller amount (approximately 15-fold for this orientation), heat fluxes of approximately 28 W/cm^2 may be achieved for each heat source in the array, with FC-77 and heater and cold plate temperatures maintained at 80 and 15°C, respectively. This result is comparable to heat fluxes achieved for systems involving boiling and condensation of a dielectric liquid.

For forced-convection flow of FC-77 in a rectangular channel with a heat source array mounted to one wall, pin fins may be used to enhance heat transfer by factors ranging from 15 to 20, depending on the Reynolds number. Enhancement is due to increased mixing (turbulation) of the flow, and, hence, to increasing values of $\overline{h}$, as well as to the increased surface area afforded by the fins. However, enhancement factors are smaller for water, ranging from 6 to 10, because of the smaller fin efficiencies η_f associated with the larger values of $\overline{h}$. As for the heat source array without extended surfaces (Chapter 4), thermal resistances for FC-77 increase with increasing row number in the flow direction. The reduction in R''_{th} associated with intermittently including square fins along the axis of each pin is not commensurate with the increase in surface area because of a reduction in η_f, as well as a reduction in fluid flow (bypass effects) through the fin array. Bypass effects may be circumvented by placing barriers in flow lanes adjoining the heat sources, but not without an additional increase in pressure losses.

For mixed convection, parallel-plate fins have been found to enhance heat transfer by factors ranging from approximately 5 to 14, with values of R''_{th} as low as 1.4 cm^2·K/W having been obtained for FC-77.

For circular jets, significant heat transfer enhancement has been demonstrated through the use of spreader plates, which are smooth, roughened, or machined with radial grooves, as well as through the use of radial and pin fin arrays. For FC-77, the lowest thermal resistances are associated with using a roughened spreader plate consisting of square millistuds extending into the flow by up to approximately a millimeter. Thermal resistances decrease with decreasing nozzle diameter D_n (for fixed $\dot{V}$) and with increasing volumetric flow rate $\dot{V}$ (for fixed D_n), and values as low as 0.2 cm^2·K/W have been reported for FC-77.

For rectangular jets, surface roughness generated in the form of microgrooves or microstuds has been found to enhance heat transfer by up to 400%, and thermal resistances as low as 0.2 cm^2·K/W have been achieved with FC-77. For a fixed volumetric flow rate, R''_{th} decreases with decreasing nozzle width.

There are many other options for enhancing heat transfer from electronic components and systems. Various surface roughness elements, such as ribbed or V-shaped turbulators, can, for example, advance transition to turbulence and, when used for natural convection from a vertical surface, can enhance heat transfer by as much as 40% (Bhavani and Bergles, 1990; Wirtz and Zhong, 1991; Misumi and Kitamura, 1993). Heat sink geometries may also vary greatly and, in the most general case, could be viewed as porous solids. Specific forms could, for example, include a layered wire mesh, a sintered metal, or a honeycomb structure, as well as densely populated pin or plate fin arrays. Performance depends on the porosity and thermal conductivity of the solid structure, as well as on convection coefficients and pressure losses associated with fluid flow through the structure. In all cases, selection of an enhancement option is strongly influenced by manufacturing requirements, and costs. A comprehensive treatment of the analysis and design of heat sinks is provided by Kraus and Bar-Cohen (1995).

APPENDIX A

NOMENCLATURE

ROMAN LETTER SYMBOLS

A aspect ratio
A_c cross-sectional area, m^2
A_{ex} exposed surface area of protruding heat source ($L_h^2 + 4B_h L_h$), m^2
A_f fin surface area, m^2
A_h heater aspect ratio, $L_{h,x}/L_{h,z}$
A_{min} minimum area, m^2
A_n nozzle cross-sectional area, m^2
A_p plenum cross-sectional area, m^2
A_r ratio of nozzle or jet impingement cross-sectional area to that of the impingement surface or a unit cell for an array of nozzles; ratio of heater-to-subtrate surface areas
A_s surface area, m^2
$A_{s,b}$ surface area of exposed base with a fin array, m^2
$A_{s,h}$ heater surface area, m^2
$A_{s,sub}$ substrate surface area, m^2
$A_{s,t}$ total surface area of a fin array, m^2
B_h height of protruding heat source, m
B_{sub} embedded depth of heat source in substrate, m
C_1, C_2 coefficients in mixed-convection correlation
C_f friction coefficient
$C_{f,L}$ average friction coefficient for plate of length L

NOMENCLATURE

C_r radial free-stream velocity gradient in the stagnation zone of a circular jet, du_∞/dr, s^{-1}

C_x lateral free-stream velocity gradient in the stagnation zone of a planar jet, du_∞/dx, s^{-1}

c_p specific heat at constant pressure, J/kg · K

c_v specific heat at constant volume, J/kg · K

D_c diameter of confining tube, m

D_e laminar equivalent diameter, m

D_{equiv} equivalent diameter of a square heater, m

D_f pin fin diameter, m

D_h hydraulic diameter, $4A_c/P$, m; heater diameter, m

D_i jet impingement diameter, m

D_n nozzle exit diameter, m

d_n depth of slot nozzle, m

$\dot{E}_g$ rate of energy generation, W

e surface roughness, m

F_F figure of merit for forced convection, $k\, Pr^n/\nu^m$

F_N figure of merit for natural convection, $k(\beta/\alpha\nu)^m$

Fr_S Froude number based on nozzle-to-plate spacing, $V_n/(gS)^{1/2}$

f friction factor; similarity variable; jet pulsation frequency, s^{-1}; heat transfer parameter, Nu/Pr^n

f_D Darcy friction factor

f_F Fanning friction factor

G_r dimensionless radial free-stream velocity gradient in the stagnation zone of a circular jet, $(D_i/V_i)Cr$

G_x dimensionless lateral free-stream velocity gradient in the stagnation zone of a planar jet, $(w_i/V_i)C_x$

Gr_{L_o} Grashof number based on characteristic length, $g\beta\,\Delta T L_o^3/\nu^2$

$Gr_{L_o}^*$ modified Grashof number based on characteristic length, $g\beta q_s'' L_o^4/k_f \nu^2$

g acceleration due to gravity, 9.8 m/s^2

H height of rectangular channel, parallel plate channel, or rectangular cavity, m

h local convection coefficient, W/m^2·K

$\overline{h}$ average convection coefficient, W/m^2·K

$\overline{h}_{ad}$ average adiabatic heat transfer coefficient, W/m^2·K

j row of heater array

j_H Colburn j factor, $St\, Pr^{2/3}$

$\overline{j}_H$ average Colburn j factor, $\overline{St}\, Pr^{2/3}$

K loss coefficient

k thermal conductivity, W/m·K

NOMENCLATURE

L thickness of plane wall, m; length of plate or channel, m
L^* average length of wall jet region for a square heat source, m; dimensionless channel length, $L/(D_h Re_{D_h} Pr)$
L_c length of chip carrier, m
L_{cf} length of confining wall for jet impingement, m
L_e length of unit cell for multiple jets, m
L_f fin length, m
$L_{fd,h}$ channel length required to achieve fully developed hydrodynamic conditions, m
L_h length of a heated surface, m; width of square chip, m
L_i length of unheated inlet region for channel flow, m
L_n length of tubular or slot nozzle, m
L_o characteristic length, m
L_{pc} length of potential core for a submerged jet, m
L_{sp} width of spreader plate, m
L_x, L_z base dimensions of plate fin array, m
M fin parameter, $(hPkA_c)^{1/2}\theta_b$, W
m Reynolds or Rayleigh number exponent; fin parameter, $(hP/kA_c)^{1/2}$, m^{-1}
$\dot{m}$ mass flow rate, kg/s
N number of jets, fins, or heat sources in an array
N_L number of longitudinal rows in an array
N_T number of transverse rows in an array
n Prandtl number exponent; exponent in mixed convection correlation
Nu Nusselt number
Nu_D local Nusselt number based on channel hydraulic diameter, hD_h/k_f
Nu_{D_i} local Nusselt number based on diameter of impinging circular jet, hD_i/k_f
Nu_{D_n} local Nusselt number based on nozzle diameter, hD_n/k_f
Nu_{L_o} local Nusselt number based on characteristic length, hL_o/k_f
Nu_r local Nusselt number based on radial coordinate, hr/k_f
Nu_{w_i} local Nusselt number based on width of impinging rectangular jet, hw_i/k_f
Nu_{w_n} local Nusselt number based on width of planar nozzle, hw_n/k_f
Nu_x local Nusselt number based on longitudinal coordinate, hx/k_f
$\overline{Nu}_{D_f}$ average Nusselt number based on pin fin diameter, $\bar{h}D_f/k_f$
$\overline{Nu}_{D_n}$ average Nusselt number based on nozzle diameter, $\bar{h}D_n/k_f$
$\overline{Nu}_H$ average Nusselt number based on channel height, $\bar{h}H/k_f$
$\overline{Nu}_L$ average Nusselt number based on length of surface, $\bar{h}L/k_f$
$\overline{Nu}_{L_o}$ average Nusselt number based on characteristic length, $\bar{h}L_o/k_f$

$\overline{Nu}_S$ average Nusselt number based on spacing between parallel plates, $\overline{h}S/k_f$

$\overline{Nu}_{w_n}$ average Nusselt number based on width of planar nozzle, $\overline{h}w_n/k_f$

P perimeter, m; pitch, m

P_f fin pitch, m

P_n nozzle pitch, m

p pressure, N/m^2

Pe Péclet number, $Re \cdot Pr$

Pr Prandtl number, ν/α_f

Q heater power, W

q heat rate, W

q_t total heat rate of fin array, W

q' heat rate per unit length, W/m

q'' local heat flux, W/m^2

$\bar{q}''$ average heat flux around perimeter of a duct, W/m^2

$\dot{q}$ rate at which thermal energy is generated per unit volume, W/m^3

R conductivity ratio

R_{th} thermal resistance, K/W

R''_{th} unit or areal thermal resistance, $m^2 \cdot K/W$

r radial coordinate, m

r_c critical radius for onset of turbulence in wall jet region for a circular impinging jet, m

r_s radial extent of stagnation zone for a circular jet, m

r_t radial location for onset of fully turbulent flow in wall jet region for a circular impinging jet, m

r_{th} radial location at which thermal boundary layer thickness reaches film thickness in wall jet region for an impinging circular jet, m

r_v radial location at which hydrodynamic boundary layer thickness reaches film thickness in wall jet region for an impinging circular jet, m

Ra^*_H modified Rayleigh number based on cavity height, $g\beta q''_h H^4/k_f\alpha_f\nu$

Ra_{L_h} Rayleigh number based on heater length and temperature difference between heated and chilled surfaces, $g\beta(T_h - T_c)L_h^3/\alpha_f\nu$

$Ra^*_{L_h}$ modified Rayleigh number based on heater length, $g\beta q''_h L_h^4/k_f\alpha_f\nu$

Ra_{L_o} Rayleigh number based on characteristic length and temperature difference, $g\beta\Delta T L_o^3/\alpha_f\nu$

$Ra^*_{L_o}$ modified Rayleigh number based on characteristic length, $g\beta q''_s L_o^4/k_f\alpha_f\nu$

Ra_S Rayleigh number based on spacing and temperature difference between vertical surfaces of a cavity, $g\beta(T_h-T_c)S^3/\alpha_f\nu$; Rayleigh number based on spacing between parallel plates and difference between plate and ambient temperatures, $g\beta(T_s - T_\infty)S^3/\alpha_f\nu$

Ra_S^* modified Rayleigh number based on plate spacing, $g\beta q_s'' S^4 / k_f \alpha_f \nu$
Re_{D_e} Reynolds number based on laminar equivalent diameter, $w_m D_e/\nu$
Re_{D_f} Reynolds number based on pin fin diameter, $V D_f/\nu$
Re_{D_h} Reynolds number based on channel hydraulic diameter, $w_m D_h/\nu$, or nozzle hydraulic diameter, $V_n D_h/\nu$
$Re_{D_h,c}$ critical Reynolds number for transition to turbulence
Re_{D_i} Reynolds number based on velocity and diameter of an impinging circular jet, $V_i D_i/\nu$
$Re_{D,\max}$ Reynolds number based on maximum velocity in an array of pin fins, $V_{\max} D_f/\nu$
Re_{D_n} Reynolds number based on nozzle diameter, $V_n D_n/\nu$
Re_H Reynolds number based on channel height, $w_m H/\nu$
Re_{L^*} Reynolds number based on jet velocity and average length of wall jet region, $V_i L^*/\nu$
Re_{L_h} Reynolds number based on length of heater, $w_m L_h/\nu$
Re_{L_o} Reynolds number based on characteristic length and velocity, $U_o L_o/\nu$
Re_r Reynolds number based on free stream velocity and radial coordinate, $u_\infty r/\nu$
Re_{w_i} Reynolds number based on velocity and width of an impinging planar jet, $V_i w_i/\nu$
Re_{w_n} Reynolds number based on nozzle width and exit velocity, $V_n w_n/\nu$
Re_x Reynolds number based on free stream velocity and longitudinal coordinate, $u_\infty x/\nu$
$Re_{x,c}$ critical Reynolds number for transition to turbulence
S nozzle-to-plate spacing, m; spacing between heaters in an array, m; spacing between parallel plates, m; spacing between walls of a rectangular cavity, m
St local Stanton number, $h/\rho U_o c_{p,f}$
$\overline{St}$ average Stanton number, $\bar{h}/\rho U_o c_{p,f}$
S_f spacing between parallel-plate fins in an array, m
S_o critical nozzle-to-plate spacing for splattering of a free-surface, circular jet, m
S_w spacing between fin tip and adjoining wall, m
T temperature, K
$\overline{T}$ average temperature, K
T_{ad} temperature assumed by an unheated element in an array of heated elements, K
T_f fluid or film temperature, K
T_m mixed mean fluid temperature, K
ΔT_{lm} log-mean temperature difference, K

t thickness, m; time, s
t_f fin thickness, m
U_o characteristic velocity, m/s
u velocity component in the x direction, m/s
u_∞ free-stream velocity at outer edge of hydrodynamic boundary layer, m/s
V velocity upstream of an array of circular rods, m/s
V_i jet impingement velocity, m/s
V_{max} maximum velocity within an array of circular rods, m/s
V_n mean velocity at nozzle exit, m/s
V_s free-surface velocity of a wall jet, m/s
$\overline{V}$ time mean velocity component, m/s
$\forall$ volume, m³
$\dot{\forall}$ volumetric flow rate, m³/s
v velocity component in the y direction, m/s
v' rms fluctuation in velocity component normal to surface, m/s
$\tilde{v}$ turbulence intensity, $v'/\overline{V}$
W width of rectangular plate, channel or cavity, m
$\dot{W}$ pumping power, W
We_{D_n} Weber number of free-surface, circular jet, $\rho V_n^2 D_n / \sigma$
W_f width of square or rectangular fin, m
w velocity component in the z direction, m/s
w_f width of square pin fins (studs), m
w_i width of impinging rectangular (planar or slot) jet, m
w_m mean streamwise (z direction) velocity in a channel, m/s
w_{max} maximum velocity through a pin fin array, m/s
w_n width of rectangular (planar or slot) nozzle, m
x Cartesian coordinate; flow streamwise or spanwise coordinate, m
y Cartesian coordinate; coordinate normal to surface, m
z Cartesian coordinate; flow streamwise or longitudinal coordinate, m
z^+ dimensionless coordinate, $z/D_h Re_{D_h}$
z^* dimensionless longitudinal coordinate for channel flow, $z/D_h Re_{D_h} Pr$

GREEK LETTER SYMBOLS

α thermal diffusivity, m²/s
β thermal expansion coefficient, K⁻¹
δ hydrodynamic boundary layer thickness, m
δ_{th} thermal boundary layer thickness, m
ε effectiveness

η dimensionless similarity parameter; efficiency
ξ fraction of flow that is splattered
ζ distance along a solid/fluid interface, m
μ dynamic viscosity, N·s/m^2
ν kinematic viscosity, m^2/s
ω splattering scaling parameter
ρ density, kg/m^3
ψ stream function, m^3/s
ψ^* dinensionless stream function
ϕ angle of inclination relative to vertical orientation, °
σ surface tension, N/m; area ratio
θ temperature difference, K
θ^* dimensionless temperature difference
τ shear stress, N/m^2
$\bar{\tau}$ average shear stress over plate of length L or perimeter of channel, N/m^2
χ correction factor for pressure drop across an array of pins

SUBSCRIPTS

b base of fin, spreader, or channel
c contact; cross-sectional; critical value for transition to turbulence; coolant; contraction; chilled
cnd conduction heat transfer
cns constriction of heat flow
cnv convection heat transfer
D diagonal
d downstream; electronic device
e expansion
F pure forced convection
f fin; fluid; film
fd fully developed conditions
h heater; hydrodynamic conditions
i impinging jet; column of heater array; inlet condition; interface condition; initial condition
j row of heater array
L longitudinal or streamwise direction
lm log mean
m mean value over channel cross section
N pure natural convection

n   nozzle exit condition
o   stagnation point or line; outlet condition; overall performance of a fin array
p   pump
ref   reference condition
s   surface condition; stagnation region; substrate condition
sp   spreader plate or spreading effect
sub   substrate
T   transverse or spanwise direction
t   total; turbulent
th   thermal
u   upstream
vis   viscous
x, y, z Cartesian coordinate directions
∞ free-stream or ambient condition

APPENDIX B

THERMOPHYSICAL PROPERTY FUNCTIONS

When performing calculations for liquid-cooled electronic packages, it is often convenient to have access to functional relations for the temperature dependence of pertinent thermophysical properties. Such relations are provided for the mass density ρ(kg/m^3), specific heat c_p (J/kg·K), thermal conductivity k (W/m·K), dynamic viscosity μ(N·s/m^2), and thermal expansion coefficient β(1/K) of two of the Fluorinert class of electronic liquids (FC-72 and FC-77), as well as for water. Except for the density and thermal expansion coefficient of water, Eqs. B.11 and B.15, where the temperature is expressed in degrees Celsius (°C), each expression is given as a function of absolute temperature in kelvins (K). The relations may also be used to evaluate the kinematic viscosity ($\nu = \mu/\rho$), the thermal diffusivity ($\alpha = k/\rho c_p$), and the Prandtl number ($Pr = c_p \mu/k$) from their definitions.

FLUORINERT FC-72

Density (kg/m^3)

$$\rho(T) = 2453 - 2.61T \quad \text{(B.1)}$$

Specific Heat (J/kg·K)

$$c_p(T) = 585 + 1.550T \quad \text{(B.2)}$$

Thermal Conductivity (W/m·K)

$$k(T) = 0.090 - 1.10 \times 10^{-4}T \quad \text{(B.3)}$$

Dynamic Viscosity (N·s/m²)

$$\mu(T) = 1.0017 \times 10^{-2} - 5.12375 \times 10^{-5}T + 6.7252 \times 10^{-8}T^2 \tag{B.4}$$

Thermal Expansion Coefficient (1/K)

$$\beta(T) = \frac{0.00261}{2.453 - 0.00261T} \tag{B.5}$$

FLUORINERT FC-77

Density (kg/m³)

$$\rho(T) = 2507 - 2.45T \tag{B.6}$$

Specific Heat (J/kg·K)

$$c_p(T) = 579 + 1.572T \tag{B.7}$$

Thermal Conductivity (W/m·K)

$$k(T) = 0.0842 - 6.302 \times 10^{-5}T - 2.600 \times 10^{-8}T^2 \tag{B.8}$$

Dynamic Viscosity (N·s/m²)

$$\mu(T) = \rho(T)(1.4347 \times 10^{-5} - 7.7391 \times 10^{-8}T + 1.0725 \times 10^{-10}T^2) \tag{B.9}$$

Thermal Expansion Coefficient (1/K)

$$\beta(T) = \frac{0.00245}{2.507 - 0.00245T} \tag{B.10}$$

WATER (SATURATED)

Density (kg/m³)

$$\rho(T) = \frac{a_0 + a_1T + a_2T^2 + a_3T^3 + a_4T^4 + a_5T^5}{1 + bT} \tag{B.11}$$

$a_0 = 999.8396$ $\qquad a_3 = -5.54485 \times 10^{-5}$
$a_1 = 18.22494$ $\qquad a_4 = 1.49756 \times 10^{-7}$
$a_2 = -7.92221 \times 10^{-3}$ $\qquad a_5 = -3.93295 \times 10^{-10}$
$b = 1.81597 \times 10^{-2}$ $\qquad T$ has units of °C

Specific Heat (J/kg·K)

$$c_p(T) = 8958.9 - 40.535T + 0.11243T^2 - 1.0138 \times 10^{-4}T^3 \quad (B.12)$$

Thermal Conductivity (W/m·K)

$$k(T) = -0.58166 + 6.3555 \times 10^{-3}T - 7.9643 \times 10^{-6}T^2 \quad (B.13)$$

Dynamic Viscosity (N·s/m²)

$$\mu(T) = 2.414 \times 10^{-5} \times 10^{(247.8/(T-140))} \quad (B.14)$$

Thermal Expansion Coefficient (1/K)

$$\beta(T) = \frac{b}{1+bT} - \frac{a_1 + 2a_2T + 3a_3T^2 + 4a_4T^3 + 5a_5T^4}{a_0 + a_1T + a_2T^2 + a_3T^3 + a_4T^4 + a_5T^5} \quad (B.15)$$

where the coefficients a_0 to a_5 and b are equivalent to those associated with Eq. B.11 and T again has units of degrees Celsius (°C).

REFERENCES

Aihara, T., S. Maruyama, and S. Kobayakawa (1990). Free Convective/Radiative Heat Transfer from Pin-Fin Arrays with a Vertical Base Plate (General Representation of Heat Transfer Performance). *Int. J. Heat Mass Transfer*, 33, 1223–1232.

Al-Sanea, S. (1992). A Numerical Study of the Flow and Heat Transfer Characteristics of an Impinging Laminar Slot-Jet Including Crossflow Effects. *Int. J. Heat Mass Transfer*, 35, 2501–2513.

Anand, N. K., S. H. Kim, and L. S. Fletcher (1992). The Effect of Plate Spacing on Free Convection Between Heated Parallel Plates. *J. Heat Transfer*, 114, 515–518.

Aung, W. (1987). Mixed Convection in Internal Flow. *Handbook of Single-Phase Convective Heat Transfer*. S. Kakac, R. K. Shah, and W. Aung, Eds., Wiley–Interscience, New York, Chapter 15.

Baker, E. (1972). Liquid Cooling of Microelectronic Devices by Free and Forced Convection. *Microelectronics and Reliability*, 11, 213–222.

Baker, E. (1973). Liquid Immersion Cooling of Small Electronic Devices. *Microelectronics and Reliability*, 12, 163–173.

Bar-Cohen, A. (1991). Thermal Management of Electronic Components with Dielectric Liquids. *Proc. ASME/JSME Thermal Engineering Joint Conference*. J. R. Lloyd and Y. Kurosaki, Eds., Vol. 2, xv–xxxix.

Bar-Cohen, A., and W. M. Rohsenow (1984). Thermally Optimum Spacing of Vertical, Natural Convection Cooled, Parallel Plates. *J. Heat Transfer*, 106, 116–123.

Bergles, A. E., and A. Bar-Cohen (1990). Direct Liquid Cooling of Microelectronic Components. *Advances in Thermal Modeling of Electronic Components and Systems*, Vol. 2, A. Bar-Cohen and A. D. Kraus, Eds., ASME Press, New York, 233–342.

Berkovsky, B. M., and V. K. Polevikov (1977). Numerical Study of Problems in High Intensive Free Convection. *Heat Transfer and Turbulent Buoyant Convection*, Vol. 2, D. B. Spalding and N. Afgan, Eds., Hemisphere, Washington, DC, 443–455.

Besserman, D. L., S. Ramadhyani, and F. P. Incropera (1991a). Numerical Simulation of Laminar Flow and Heat Transfer for Liquid Jet Impingement Cooling of a Circular Heat Source with Annular Collection of the Spent Fluid. *Numerical Heat Transfer*, Part A, 20, 263–278.

Besserman, D. L., F. P. Incropera, and S. Ramadhyani (1991b). Experimental Study of Heat Transfer from a Discrete Source to a Circular Liquid Jet with Annular Collection of the Spent Fluid. *Experimental Heat Transfer*, 4, 41–51.

Besserman, D. L., F. P. Incropera, and S. Ramadhyani (1992). Heat Transfer from a Square Source to an Impinging Liquid Jet Confined by an Annular Wall. *J. Heat Transfer*, 114, 284–287.

Bhavani, S. H., and A. E. Bergles (1990). Effect of Surface Geometry and Orientation on Laminar Natural Convection Heat Transfer from a Vertical Flat Plate with Transverse Roughness Elements. *Int. J. Heat Mass Transfer*, 33, 965–981.

Bhunia, S. K., and J. H. Lienhard V (1993). Surface Disturbance Evolution and the Splattering of Turbulent Liquid Jets. *J. Fluids Eng.*, 116, 721–727.

Bier, W., W. Keller, G. Linder, D. Siedel, and K. Schubert (1990). Manufacturing and Testing of Compact Micro Heat Exchangers with High Volumetric Heat Transfer Coefficients. *Microstructures, Sensors and Actuators*, ASME DSC-19, 189–197.

Biskeborn, R. G., J. L. Horvath, and E. B. Hultmark (1984). Integral Cap Heat Sink Assembly for the IBM 4381 Processor. *Proc. Fourth Annual Int. Elec. Packaging Conf.*, 468–474.

Brinkman, R., S. Ramadhyani, and F. P. Incropera (1988). Enhancement of Convective Heat Transfer from Small Heat Sources to Liquid Coolants Using Strip Fins. *Experimental Heat Transfer*, 1, 315–330.

Chen, L., M. Keyhani and D. R. Pitts (1988). An Experimental Study of Natural Convection Heat Transfer in a Rectangular Enclosure with Protruding Heaters. ASME HTD-Vol. 96–2, H.R. Jacobs, Ed., 125–133.

Chen, T. S., and B. F. Armaly (1987). Mixed Convection in External Flow. *Handbook of Single-Phase Convective Heat Transfer*. S. Kakac, R. K. Shah, and W. Aung, Eds., Wiley–Interscience, New York, Chapter 14.

Choi, S. B., R. R. Barron, and R. O. Warrington (1991). Liquid Flow and Heat Transfer in Microtubes. *Micromechanical Sensors, Actuators and Systems*, ASME DSC-32, 123–134.

Chu, H. H. S., S. W. Churchill, and C. V. S. Patterson (1976). The Effect of Heater Size, Location, Aspect Ratio, and Boundary Conditions on Two-Dimensional, Laminar, Natural Convection in Rectangular Channels. *J. Heat Transfer*, 98, 194–201.

Chu, R. C., U. P. Hwang, and R. E. Simons (1982). Conduction Cooling for an LSI Package: A One-Dimensional Approach. *IBM Journal of Research and Development*, 26, 45–54.

Churchill, S. W., and H. H. S. Chu (1975). Correlating Equations for Laminar and Turbulent Free Convection from a Vertical Plate. *Int. J. Heat Mass Transfer*, 18, 1323–1330.

Copeland, D. (1995). Manifold Microchannel Heat Sinks: Analysis and Optimization. *ASME/JSME Thermal Engineering Conference*, Vol. 4, 169–174.

Copeland, D., H. Takahira, W. Nakayama, and B.-C. Pak (1995). Manifold Microchannel Heat Sinks: Theory and Experiment. *Advances in Electronic Packaging*, ASME EEP–Vol. 10–2, 829–835.

Cowen, G. H., P. C. Lovegrove, and G. L. Quarini (1982). Turbulent Natural Convection Heat Transfer in Vertical Single Water-Filled Cavities. *Heat Transfer*—1982, Vol. 2, U. Grigull et al., Ed., Hemisphere, Washington, DC, 195–203.

Danielson, R. D., N. Krajewski, and J. Brost (1986). Cooling of a Superfast Computer. *Elec. Packaging and Production*, July, 44–45.

Danielson, R. D., L. Tousignant, and A. Bar-Cohen (1987). Saturated Pool Boiling Characteristics of Commercially Available Perfluorinated Liquids. *Proc. Second ASME/JSME Thermal Engineering Joint Conference*, P. J. Marto and I. Tanasawa, Eds., 419–430.

DiMarco, P., W. Grassi, and A. Magrini (1994). Unsubmerged Jet Impingement Heat Transfer at Low Liquid Speed. *Proc. 10th Int. Heat Transfer Conf.*, Vol. 3, 59–64.

Elenbaas, W. (1942). Heat Dissipation of Parallel Plates by Free Convection. *Physica*, 9, 1–28.

Evans, H. L. (1962). Mass Transfer through Laminar Boundary Layers—Further Similar Solutions to the B-Equation, for the Case B=O. *Int. J. Heat Mass Transfer*, 5, 35–37.

Faggiani, S., and W. Grassi (1990). Round Liquid Jet Impingement Heat Transfer: Local Nusselt Number in the Region with Non-Zero Pressure Gradient. *Proc. Ninth Int. Heat Transfer Conf.*, Vol. 4, 197–202.

Falkner, V. M., and S. W. Skan (1931). Some Approximate Solutions of the Boundary Layer Equations. *Philos. Mag.*, 12, 865–896.

Fitzgerald, J. A., and S. V. Garimella (1996). Flow Field Measurements in Confined and Submerged Jet Impingement. ASME HTD-333, Vol. 2, 121–129.

Florschuetz, L. W. (1982). Jet Array Impingement Flow Distributions and Heat Transfer Characteristics. NASA CR-3630.

Fox, R. W., and A. T. McDonald (1985). *Introduction to Fluid Mechanics*. 3rd ed., Wiley, New York.

Gabour, L. A., and J. H. Lienhard V (1994). Wall Roughness Effects on Stagnation-Point Heat Transfer Beneath an Impinging Jet. *J. Heat Transfer*, 116, 81–87.

Gardon, R., and J. Cobonpue (1961). Heat Transfer Between a Flat Plate and Jets of Air Impinging on It. *Proc. Second Int. Heat Transfer Conf.*, 454–460.

Gardon, R., and J. C. Akfirat (1965). The Role of Turbulence in Determining the Heat Transfer Characteristics of Impinging Jets. *Int. J. Heat Mass Transfer*, 8, 1261–1272.

Gardon, R., and J. C. Akfirat (1966). Heat Transfer Characteristics of Impinging Two-Dimensional Air Jets. *J. Heat Transfer*, 88, 101–107.

Garimella, S. V., and P. A. Eibeck (1990). Heat Transfer Characteristics of an Array of Protruding Elements in Single Phase Forced Convection. *Int. J. Heat Mass Transfer*, 33, 2659–2669.

Garimella, S. V., and P. A. Eibeck (1991). Effect of Spanwise Spacing on the Heat Transfer from an Array of Protruding Elements in Forced Convection. *Int. J. Heat Mass Transfer*, 34, 2427–2430.

Garimella, S. V., and D. J. Schlitz (1995). Heat Transfer Enhancement in Narrow Channels Using Two and Three-dimensional Mixing Devices. *J. Heat Transfer*, 117, 590–596.

Garimella, S. V., and R. A. Rice (1995). Confined and Submerged Liquid Jet Impingement Heat Transfer. *J. Heat Transfer*, 117, 871–877.

Garimella, S. V., and B. Nenaydykh (1995). Influence of Nozzle Geometry on Heat Transfer in Submerged and Confined Liquid Jet Impingement. *Cooling and Thermal Design of Electronic Systems*, ASME HTD-319/EEP-15, 49–57.

Gauntner, J. W., J. N. B. Livengood, and P. Hrycak (1969). Survey of Literature of Flow Characteristics of a Single Turbulent Jet Impinging on a Flat Plate. National Aeronautics and Space Administration, NASA TN D-5657.

Gebhart, B., Y. Jaluria, R. L. Mahajan, and B. Sammakia (1988). *Buoyancy-Induced Flows and Transport*, Hemisphere, New York.

Geisler, K. J., D. Kitching, and A. Bar-Cohen (1996). A Passive Immersion Cooling Module with a Finned Submerged Condenser. *Process, Enhanced and Multiphase Heat Transfer*, R. M. Manglik and A. D. Kraus, Eds., Begell House, New York, 193–206.

Gersey, C. O., and I. Mudawar (1992). Effects of Orientation on Critical Heat Flux from Chip Arrays During Flow Boiling. *J. Electronic Packaging*, 114, 290–299.

Gersey, C. O., and I. Mudawar (1993). Nucleate Boiling and Critical Heat Flux from Protruded Chip Arrays During Flow Boiling. *J. Electronic Packaging*, 115, 78–88.

Glauert, M.G. (1956). The Wall Jet. *J. Fluid Mech.*, 1, 625–643.

Globe, S., and D. Dropkin (1959). Natural Convection Heat Transfer in Liquids Confined by Two Horizontal Plates and Heated from Below. *J. Heat Transfer*, 81, 24–28.

Goodson, K. E., K. Kurabayaski, and R. F. W. Pease (1997). Improved Heat Sinking for Laser-Diode Arrays Using Microchannels in CVD Diamond. *IEEE Trans.*, CPMT-B, 20, 104–109.

Graham, K. M., and S. Ramadhyani (1996). Experimental and Theoretical Studies of Mist Jet Impingement Cooling. *J. Heat Transfer*, 118, 343–349.

Grassi, W., and A. Magrini (1991). Effect of the Free Jet Fluid Dynamics on Liquid Jet Impingement Heat Transfer. *Proc. Ninth Natl. Heat Transfer Conf., Italy*, 410–433.

Grimison, E. D. (1937). Correlation and Utilization of New Data on Flow Resistance and Heat Transfer for Cross Flow of Gases over Tube Banks. *Trans. ASME*, 59, 583–594.

Gudapati, S. (1993). Influence of Geometry on Convective Heat Transfer from Discrete Sources. MS Thesis, University of Wisconsin, Milwaukee.

Harley, J. and H. H. Bau (1989). Fluid Flow in Micron and Submicron Size Channels. *IEEE Trans.*, THO 249–3, 25–28.

Harms, T. M., M. Kazmierczak, F. M. Gerner, A. Hölke, H. T. Henderson, J. Pilchowski, and K. Baker (1997). Experimental Investigation of Heat Transfer and Pressure Drop Through Deep Microchannels in a (110) Silicon Substrate. ASME HTD-351, 347–357.

Harpole, G. M., and J. E. Eninger (1991). Micro-Channel Heat Exchanger Optimization. *Proc. Seventh IEEE Semi-Therm Symposium*, 59–63.

Heindel, T. J., F. P. Incropera, and S. R. Ramadhyani (1992a). Liquid Immersion Cooling of a Longitudinal Array of Discrete Heat Sources in Protruding Substrates: 1—Single-Phase Forced Convection. *J. Electronic Packaging*, 114, 55–62.

Heindel, T. J., S. Ramadhyani, F. P. Incropera, and A. Campo (1992b). Surface Enhancement of a Heat Source Exposed to a Circular Liquid Jet with Annular Collection of the Spent Fluid. *Topics in Heat Transfer*, ASME HTD-206-2, 111–118.

Heindel, T. J. (1994). A Numerical and Experimental Study of Three-Dimensional Natural Convection in a Discretely Heated Cavity. Ph.D. Thesis, Purdue University, West Lafayette, IN.

Heindel, T. J., S. Ramadhyani, and F. P. Incropera (1995a). Laminar Natural Convection in a Discretely Heated Cavity: I—Assessment of Three-Dimensional Effects. *J. Heat Transfer*, 117, 902–909.

Heindel, T. J., F. P. Incropera, and S. Ramadhyani (1995b). Laminar Natural Convection in a Discretely Heated Cavity: II—Comparisons of Experimental and Theoretical Results. *J. Heat Transfer*, 117, 910–917.

Heindel, T. J., S. Ramadhyani, and F. P. Incropera (1995c). Conjugate Natural Convection from an Array of Discrete Heat Sources: Part 1—Two and Three-Dimensional Model Validation. *Int. J. Heat and Fluid Flow*, 16, 501–510.

Heindel, T. J., F. P. Incropera, and S. Ramadhyani (1995d). Conjugate Natural Convection from an Array of Discrete Heat Sources: Part 2—A Numerical Parametric Study. *Int. J. Heat Fluid Flow*, 16, 511–518.

Heindel, T. J., S. Ramadhyani, and F. P. Incropera (1996a). Conjugate Natural Convection from an Array of Protruding Heat Sources. *Numerical Heat Transfer*, 29A, 1–18.

Heindel, T. J., F. P. Incropera, and S. Ramadhyani (1996b). Enhancement of Natural Convection Heat Transfer from an Array of Discrete Heat Sources. *Int. J. Heat Mass Transfer*, 39, 479–490.

Heindel, T. J., F. P. Incropera, and S. Ramadhyani (1996c). Heat Transfer Enhancement from Arrays of Discrete Heat Sources. *Process, Enhanced, and Multiphase Heat Transfer*, R. M. Manglik and A. D. Kraus, Eds., Begell House, New York, 207–216.

Hollworth, B. R., and R. D. Berry (1978). Heat Transfer from Arrays of Impinging Jets with Large Jet-to-Jet Spacing. *J. Heat Transfer*, 100, 352–357.

Hoopman, T. L. (1990). Microchanneled Structures. *Microstructures, Sensors and Actuators*, ASME DSC-19, 117–174.

Hrycak, P., D. T. Lee, J. W. Gauntner, and J. N. B. Livengood (1970). Experimental Characteristics of a Single Turbulent Jet Impinging on a Flat Plate. National Aeronautics and Space Administration, NASA TN D-5690.

Hwang, L. T., I. Turlik, and A. Reisman (1987). A Thermal Module Design for Advanced Packaging. *J. Electronic Materials*, 16, 347–355.

Inada, S., Y. Miyasaka, and R. Izumi (1981). A Study on the Laminar-Flow Heat Transfer Between a Two-Dimensional Water Jet and a Flat Surface with a Constant Heat Flux. *Bulletin of JSME*, 24, 1803–1810.

Incropera, F. P. (1986). Research Needs in Electronic Cooling. *Proc. National Science Foundation Workshop*, Purdue University, West Lafayette, IN.

Incropera, F. P. (1988). Convection Heat Transfer in Electronic Equipment Cooling. *J. Heat Transfer*, 110, 1097–1111.

Incropera, F. P., and D. P. DeWitt (1996). *Fundamentals of Heat and Mass Transfer*, 4th ed., Wiley, New York.

Incropera, F. P., J. Kerby, D. F. Moffatt, and S. Ramadhyani (1986). Convection Heat Transfer from Discrete Sources in a Rectangular Channel. *Int. J. Heat Mass Transfer*, 29, 1051–1058.

Incropera, F. P., A. L. Knox, and J. R. Maughan (1987). Mixed-Convection Flow and Heat Transfer in the Entry Region of a Horizontal Rectangular Duct. *J. Heat Transfer*, 109, 434–439.

Janna, W.S. (1993). *Introduction to Fluid Mechanics*. PWS Publishing, Boston.

Jiji, L. M., and Z. Dagan (1988). Experimental Investigation of Single Phase Multi-Jet Impingement Cooling of an Array of Microelectronic Heat Sources. *Cooling Technology for Electronic Equipment*, W. Aung, Ed., Hemisphere, 333–351.

Jones, Jr., O. C. (1976). An Improvement in the Calculation of Turbulent Friction in Rectangular Ducts. *J. Fluids Eng.*, 98, 173–181.

Joshi, Y., M. D. Kelleher, and T. J. Benedict (1990). Natural Convection Immersion Cooling of an Array of Simulated Electronic Components in an Enclosure Filled with a Dielectric

Liquid. *Heat Transfer in Electronic and Microelectronic Equipment*, A. E. Bergles, Ed., Hemisphere, New York, 445–468.

Joy, R. C., and E. S. Schlig (1970). Thermal Properties of Very Fast Transistors. *IEEE Trans. Electron. Devices*, ED-17, 586–594.

Kays, W. M., and A. L. London (1984). *Compact Heat Exchangers*. McGraw–Hill, New York.

Kelecy, F. J., S. Ramadhyani, and F. P. Incropera (1987). Effect of Shrouded Pin Fins on Forced Convection Cooling of Discrete Heat Sources by Direct Liquid Immersion. *Proc. Second ASME/JSME Thermal Engineering Joint Conference*, P. J. Marto and I. Tanasawa, Eds., Vol. 3, 387–394.

Kendall, S. R., and H. V. Rao (1997). Experimental Study of Flow Through Micro-Passages. *Proc. Fourth International Conference on Experimental Heat Transfer, Fluid Mechanics and Thermodynamics*, M. Giot, F. Mayinger, and G. P. Celata, Eds., Vol. 4, 2423–2430.

Kercher, D. M., and W. Tabakoff (1970). Heat Transfer by a Square Array of Round Air Jets Impinging Perpendicular to a Flat Surface Including the Effect of Spent Air. *ASME J. Engineering for Power*, 92, 73–82.

Kestin, J. (1966). The Effect of Free-Stream Turbulence on Heat Transfer Rates. *Advances in Heat Transfer*, J. P. Hartnett and T. F. Irvine, Eds., Vol. 3, 1–32, Academic Press, New York.

Keyhani, M., V. Prasad, and R. Cox (1988a). An Experimental Study of Natural Convection in a Vertical Cavity with Discrete Heat Sources. *J. Heat Transfer*, 110, 616–624.

Keyhani, M., V. Prasad, R. Shen, and T. T. Wong (1988b). Free Convection Heat Transfer from Discrete Heat Sources in a Vertical Cavity. *Natural and Mixed Convection in Electronic Equipment Cooling*, ASME HTD-Vol. 100, R. A. Wirtz, Ed., 13–24.

Keyhani, M., L. Chen, and D. R. Pitts (1991). The Aspect Ratio Effect on Natural Convection in an Enclosure with Protruding Heat Sources. *J. Heat Transfer*, 113, 883–891.

Kishimoto, T., and T. Ohsaki (1986). VLSI Packaging Technique Using Liquid-Cooled Channels. *IEEE Trans. Components, Hybrids and Manufacturing*, CHMT-9, 328–335.

Kishimoto, T., and S. Sasaki (1987). Cooling Characteristics of Diamond-Shaped Interrupted Cooling Fin for High-Power LSI Devices. *Electron. Lett.*, 23, 456–457.

Korger, M., and F. Krizek (1966). Mass Transfer Coefficient in Impingement Flow from Slotted Nozzles. *Int. J. Heat Mass Transfer*, 9, 337–344.

Kraus, A. D., and A. Bar-Cohen (1983). *Thermal Analysis and Control of Electronic Equipment*. McGraw–Hill, New York.

Kraus, A. D., and A. Bar-Cohen (1995). *Design and Analysis of Heat Sinks*. Wiley–Interscience, New York.

Kuhn, D., and P. H. Oosthuizen (1986) Three-Dimensional Natural Convective Flow in a Rectangular Enclosure with Localized Heating. *Natural Convection in Enclosures*, ASME HTD–Vol. 63, R. S. Figliola and I. Catton, Eds., 55–67.

Ledezma, G., A. M. Morega, and A. Bejan (1996). Optimal Spacing Between Pin Fins with Impinging Flow. *J. Heat Transfer*, 118, 570–577.

Ledezma, G., and A. Bejan (1997). Optimal Geometric Arrangement of Staggered Vertical Plates in Natural Convection. *J. Heat Transfer*, 119, 700–707.

Lee, J. J., K. V. Liu, K. T. Yang, and M. D. Kelleher (1987). Laminar Natural Convection in a Rectangular Enclosure due to a Heated Protrusion on One Vertical Wall–Part II: Numerical Simulations. *Proc. Second ASME/JSME Thermal Engineering Joint Conference*, Vol. 2, 179–185.

Levy, S. (1952). Heat Transfer to Constant Property Laminar Boundary Layer Flows with Power Function Free-Stream Velocity and Wall Temperature Variation. *J. Aero. Sci.*, 19, 341–348.

Lienhard, J. H., V, X. Liu, and L. A. Gabour (1992). Splattering and Heat Transfer During Impingement of a Turbulent Liquid Jet. *J. Heat Transfer*, 114, 362–372.

Lienhard, J. H., V (1995). Liquid Jet Impingement. *Ann. Rev. Heat Transfer*, C. L. Tien, Ed., Vol. 6, Begell House, New York.

Liu, K. V., K. T. Yang, and M. D. Kelleher (1987). Three-Dimensional Natural Convection Cooling of an Array of Heated Protrusions in an Enclosure Filled with a Dielectric Liquid. *Proc. Int. Symp. on Cooling Technology for Electronic Equipment*, 486–497.

Liu, X., and J. H. Lienhard V (1989). Liquid Jet Impingement Heat Transfer on a Uniform Flux Surface. *Heat Transfer Phenomena in Radiation, Combustion, and Fires*, ASME HTD-106, 523–530.

Liu, X., J. H. Lienhard V, and J. S. Lombara (1991). Convective Heat Transfer by Impingement of Circular Liquid Jets. *J. Heat Transfer*, 113, 571–582.

Liu, X., L. A. Gabour, and J. H. Lienhard V (1993). Stagnation-Point Heat Transfer During Impingement of Laminar Liquid Jets: Analysis Including Surface Tension. *J. Heat Transfer*, 115, 99–105.

Liu, X., and J. H. Lienhard V (1993). The Hydraulic Jump in Circular Jet Impingement and in Other Thin Liquid Films. *Exp. Fluids*, 15, 108–116.

Ma, C. F., and A. E. Bergles (1988). Convective Heat Transfer on a Small Vertical Surface in an Impinging Circular Liquid Jet. *Heat Transfer Science and Technology*, B. X. Wang, Ed., Hemisphere, 193–200.

Ma, C. F., H. Sun, H. Auracher, and T. Gomi (1990). Local Convective Heat Transfer from Vertical Heated Surfaces to Impinging Liquid Jets of Large Prandtl Number Fluids. *Proc. Ninth Int. Heat Transfer Conf.*, Hemisphere, Vol. 2, 441–446.

MacGregor, R. K., and A. P. Emery (1969). Free Convection Through Vertical Plane Layers: Moderate and High Prandtl Number Fluids. *J. Heat Transfer*, 91, 391–396.

Maddox, D. E., and I. Mudawar (1988). Single and Two-Phase Convective Heat Transfer from Smooth and Enhanced Microelectronic Heat Sources in a Rectangular Channel. ASME HTD–Vol. 96, 533–541.

Mahalingam, M. (1985). Thermal Management in Semiconductor Device Packaging. *Proc. IEEE*, 73, 1396–1404.

Mahaney, H. V. (1989). Mixed Convection Heat Transfer from Discrete Heat Sources Mounted in a Rectangular Duct. Ph.D. Thesis, School of Mechanical Engineering, Purdue University, West Lafayette, IN.

Mahaney, H. V., S. Ramadhyani, and F. P. Incropera (1989). Numerical Simulation of Three-Dimensional Mixed Convection Heat Transfer from an Array of Discrete Heat Sources in a Horizontal Rectangular Duct. *Numerical Heat Transfer*, Part A, 16, 267–286.

Mahaney, H. V., F. P. Incropera, and S. Ramadhyani (1990a). Comparison of Predicted and Measured Mixed Convection Heat Transfer from an Array of Discrete Sources in a Horizontal Rectangular Channel. *Int. J. Heat Mass Transfer*, 33, 1233–1245.

Mahaney, H. V., F. P. Incropera, and S. Ramadhyani (1990b). Measurement of Mixed-Convection Heat Transfer from an Array of Discrete Sources in a Horizontal Rectangular Channel with and without Surface Augmentation. *Experimental Heat Transfer*, 3, 215–237.

Mahaney, H. V., S. Ramadhyani, and F. P. Incropera (1991). Numerical Simulation of Three-Dimensional Mixed Convection Heat Transfer from a Finned Array of Discrete Sources. *Numerical Heat Transfer*, Part A, 19, 125–149.

Mangler, W. (1948). Zusammenhang Zwischen Ebenen und Rotationssymmetrischen Grenzschichten in Kimpressiblen. *Z. Angew. Math. Mech.*, 28, 97–103.

Mansour, A., and N. Chigier (1994). Turbulence Characteristics in Cylindrical Liquid Jets. *Phys. Fluids*, 6, 3380–3391.

Martin, H. (1977). Heat and Mass Transfer between Impinging Gas Jets and Solid Surfaces. *Advances in Heat Transfer*, J. P. Hartnett and T. F. Irvine, Eds., Vol. 13, Academic Press, New York, 1–60.

McMurray, D. C., P. S. Meyers, and O. A. Uyehara (1966). Influence of Impinging Jet Variables on Local Heat Transfer Coefficients Along a Flat Surface with Constant Heat Flux. *Proc. Third Int. Heat Transfer Conf.*, Vol. 2, 292–299.

Metzger, D. E., K. N. Cummings, and W. A. Ruby (1974). Effects of Prandtl Number on Heat Transfer Characteristics of Impinging Liquid Jets. *Proc. Fifth Int. Heat Transfer*, Vol. 2, 20–24.

Milne-Thomson, L. M. (1955). *Theoretical Hydrodynamics*. 3rd ed., Macmillan, New York, 279–289.

Misumi, T., and K. Kitamura (1993). Heat Transfer Enhancement of Natural Convection and Development of a High Performance Heat Transfer Plate. *JSME Int. J. Ser. B*, 36, 143–149.

Miyasaka, Y., and S. Inada (1980). The Effect of Pure Forced Convection on the Boiling Heat Transfer between a Two-Dimensional Subcooled Water Jet and a Heated Surface. *J. Chem. Eng. Jpn.*, 13, 22–28.

Moffatt, D. F. (1985). Convection Heat Transfer from Discrete Heat Sources in the Wall of a Rectangular Channel. MSME Thesis, Purdue University, West Lafayette, IN.

Moffatt, D. F., S. Ramadhyani, and F. P. Incropera (1986). Conjugate Heat Transfer from Wall Embedded Sources in Turbulent Channel Flow. ASME HTD–Vol. 57, 177–182.

Morris, G.K., and S. V. Garimella (1996). Correlations for Single-Phase Convective Heat Transfer from an Array of Three-Dimensional Obstacles in a Channel. *Proc. IEEE Inter-Society Conf. on Thermal Phenomena in Electronic Systems*, 292–298.

Mouromtseff, I. E. (1942). Water and Forced Air Cooling of Vacuum Tubes. *Proc. IRE*, 30, 190–205.

Nakatogawa, T., N. Nishiwaki, M. Hirata, and K. Torii (1970). Heat Transfer of Round Turbulent Jet Impinging Normally on Flat Plate, *Proc. Fourth Int. Heat Transfer Conf.*, Paper FC 5.2.

Nayak, D., L. T. Hwang, I. Turlik, and A. Reisman (1987). A High-Performance Thermal Module for Computer Packaging. *J. Electronic Materials*, 16, 357–364.

Osborne, D. G., and F. P. Incropera (1985). Laminar, Mixed Convection Heat Transfer for Flow between Horizontal Parallel Plates with Asymmetric Heating. *Int. J. Heat Mass Transfer*, 28, 207–217.

Ostrach, S. (1953). An Analysis of Laminar Free Convection Flow and Heat Transfer about a Flat Plate Parallel to the Direction of the Generating Body Force. NACA Report 1111.

Ostrach, S. (1988). Natural Convection in Enclosures. *J. Heat Transfer*, 110, 1175–1190.

Pan, Y., J. Stevens, and B. W. Webb (1992). Effect of Nozzle Configuration on Transport in the Stagnation Zone of Axisymmetric Impinging Free-Surface Jets: Part 2. Local Heat Transfer. *J. Heat Transfer*, 114, 880–886.

Pan, Y., and B. W. Webb (1994a). Heat Transfer Characteristics of Arrays of Free-Surface Liquid Jets. ASME HTD-217, 23–28.

Pan, Y., and B. W. Webb (1994b). Visualization of Local Heat Transfer Under Arrays of Free-Surface Liquid Jets. *Proc. 10th Int. Heat Transfer Conf.*, Vol. 3, 77–82.

Pan, Y., and B. W. Webb (1995). Heat Transfer Characteristics of Arrays of Free-Surface Liquid Jets. *J. Heat Transfer*, 117, 878–883.

Peng, X. F., G. P. Peterson, and B. X. Wang (1994a). Frictional Flow Characteristics of Water Flowing Through Rectangular Microchannels. *Experimental Heat Transfer*, 7, 249–264.

Peng, X. F., G. P. Peterson, and B. X. Wang (1994b). Heat Transfer Characteristics of Water Flowing Through Microchannels. *Experimental Heat Transfer*, 7, 265–283.

Pfahler, J. N., J. Harley, H. H. Bau, and J. Zemel (1991). Gas and Liquid Flow in Small Channels. *Micromechanical Sensors, Actuators and Systems*, ASME DSC-32, 49–60.

Phillips, R. J. (1988). Forced-Convection, Liquid-Cooled, Microchannel Heat Sinks. Technical Report 787, Lincoln Laboratory, Massachusetts Institute of Technology.

Phillips, R. J. (1990). Microchannel Heat Sinks. *Advances in Thermal Modeling of Electronic Components and Systems*, A. Bar-Cohen and A. D. Kraus, Eds., Vol. 2, ASME Press, New York, 109–184.

Phillips, R. J., L. R. Glicksman, and R. Larson (1988). Forced-Convection, Liquid-Cooled, Microchannel Heat Sinks for High-Power-Density Microelectronics. *Cooling Technology for Electronic Equipment*, W. Aung, Ed., Hemisphere, New York, 295–316.

Polentini, M. S., S. Ramadhyani, and F. P. Incropera (1993). Single-Phase Thermosyphon Cooling of an Array of Discrete Heat Sources in a Rectangular Cavity. *Int. J. Heat Mass Transfer*, 36, 3983–3996.

Prasad, V., M. Keyhani, and R. Shen (1990). Free Convection in a Discretely Heated Vertical Enclosure: Effects of Prandtl Number and Cavity Size. *J. Electronic Packaging*, 112, 63–74.

Priedeman, D., V. Callahan, and B. W. Webb (1994). Enhancement of Liquid Jet Impingement Heat Transfer with Surface Modifications. *J. Heat Transfer*, 116, 486–489.

Ramadhyani, S., D. F. Moffatt, and F. P. Incropera (1985). Conjugate Heat Transfer from Small Isothermal Heat Sources Embedded in a Large Substrate. *Int. J. Heat Mass Transfer*, 28, 1945–1952.

Ramadhyani, S., and F. P. Incropera (1987). Forced Convection Cooling of Discrete Heat Sources with and without Surface Enhancement. *Proc. Int. Symp. on Cooling Technology for Electronic Equipment*, Honolulu, March 17–21, 249–264.

Saad, N. R., A. S. Mujumdar, and W. J. M. Douglas (1980). Heat Transfer Under Multiple Turbulent Slot Jets Impinging on a Flat Plate. *Drying 80*, A. S. Mujumdar, Ed., Hemisphere, New York, 422–430.

Samalam, V. K. (1989). Convective Heat Transfer in Microchannels. *J. Electronic Materials*, 18, 611–617.

Samant, K. R., and T. W. Simon (1986). Heat Transfer from a Small, High-Heat-Flux Patch to a Subcooled Turbulent Flow. ASME Paper 86–HT-22.

Sasaki, S., and T. Kishimoto (1986). Optimal Structure for Microgrooved Cooling Fin for High-Power LSI Devices. *Electron. Lett.*, 22, 1332–1333.

Sathe, S. B., and Y. Joshi (1991). Natural Convection Arising from a Heat Generating Substrate-Mounted Protrusion in a Liquid-Filled, Two-Dimensional Enclosure. *Int. J. Heat Mass Transfer*, 34, 2149–2163.

Sathe, S. B., and Y. Joshi (1992). Natural Convection Liquid Cooling of a Substrate-Mounted Protrusion in a Square Enclosure: A Parametric Study. *J. Heat Transfer*, 114, 401–409.

Saylor, J. R., A. Bar-Cohen, T. -Y. Lee, T. W. Simon, W. Tong,, and P. -S. Wu (1988). Fluid Selection and Property Effects in Single- and Two-Phase Immersion Cooling. *IEEE CHMT Trans.*, 11(4), 557–565.

Schafer, D. M. (1990). Planar Liquid Jet Impingement Cooling of Multiple Discrete Heat Sources. MSME Thesis, Purdue University, West Lafayette, IN.

Schafer, D. M., F. P. Incropera, and S. Ramadhyani (1991). Planar Liquid Jet Impingement Cooling of Multiple Discrete Heat Sources. *ASME J. Elec. Packaging*, 113, 359–366.

Schafer, D. M., S. Ramadhyani, and F. P. Incropera (1992). Numerical Simulation of Laminar Convection Heat Transfer from an In-Line Array of Discrete Sources to a Confined Rectangular Jet. *Numerical Heat Transfer*, Part A., 21, 121–142.

Schlichting, H. (1960). *Boundary Layer Theory*. McGraw–Hill, New York.

Scholtz, M. T., and O. Trass (1970). Mass Transfer in a Nonuniform Impinging Jet. *AIChE J.*, 16, 82–96.

Shah, R. K., and A. L. London (1978). *Laminar Flow Forced Convection in Ducts*. Academic Press, New York.

Shen, R, V. Prasad, and M. Keyhani (1989). Effect of Aspect Ratio and Size of Heat Source on Free Convection in a Discretely Heated Vertical Cavity. *Numerical Simulation of Convection in Electronic Equipment Cooling*, ASME HTD–Vol. 121, A. Ortega and D. Agonafer, Eds., 45–54.

Sheriff, H. S., and D. A. Zumbrunnen (1994). Effect of Flow Pulsations on the Cooling Effectiveness of an Impinging Jet. *J. Heat Transfer*, 116, 886–895.

Sitharamayya, S., and K. S. Raju (1969). Heat Transfer Between an Axisymmetric Jet and a Plate Held Normal to the Flow. *Can. J. Chem. Eng.*, 47, 365–368.

Slayzak, S. J., R. Viskanta, and F. P. Incropera (1994a). Effects of Interactions Between Adjoining Rows of Circular, Free-Surface Jets on Local Heat Transfer from the Impingement Surface. *J. Heat Transfer*, 116, 88–95.

Slayzak, S. J., R. Viskanta, and F. P. Incropera (1994b). Effects of Interaction between Adjacent Free-Surface Planar Jets on Local Heat Transfer from the Impingement Surface. *Int. J. Heat Mass Transfer*, 37, 269–282.

Sparrow, E. M., and L. Lee (1975). Analysis of Flow Field and Impingement Heat/Mass Transfer due to a Nonuniform Slot Jet. *J. Heat Transfer*, 97, 191–197.

Sparrow, E. M., R. J. Goldstein, and R. A. Rouf (1975). Effect of Nozzle-Surface Separation Distance on Impingement Heat Transfer for a Jet in a Crossflow. *J. Heat Transfer*, 97, 528–533.

Stevens, J., and B. W. Webb (1991). Local Heat Transfer Coefficients under an Axisymmetric Single-Phase Liquid Jet. *J. Heat Transfer*, 113, 71–78.

Stevens, J., and B. W. Webb (1992). Measurements of the Free Surface Flow Structure under an Impinging Free Liquid Jet. *J. Heat Transfer*, 114, 79–84.

Stevens, J., Y. Pan, and B. W. Webb (1992). Effect of Nozzle Configuration on Transport in the Stagnation Zone of Axisymmetric, Impinging Free-Surface Liquid Jets. Part 1. Turbulent Flow Structure. *J. Heat Transfer*, 114, 874–879.

Stevens, J., and B. W. Webb (1993). Measurements of Flow Structure in the Stagnation Zone of Impinging Free-Surface Liquid Jets. *Int. J. Heat Mass Transfer*, 36, 4283–4286.

Sullivan, P.F., S. Ramadhyani, and F. P. Incropera (1992a). Use of Smooth and Roughened Spreader Plates to Enhance Impingement Cooling of Small Heat Sources with Single Circular Liquid Jets. ASME HTD-206(2), 103–110.

Sullivan, P.F., S. Ramadhyani, and F. P. Incropera (1992b). Extended Surfaces to Enhance Impingement Cooling with Single Circular Jets. *Proc. Joint ASME/JSME Conf. on Electronic Packaging*, W. T. Chen and H. Abe, Eds., ASME EEP–Vol. 1–1, 207–215.

Sun, H., C. F. Ma, and W. Nakayama (1993). Local Characteristics of Convective Heat Transfer from Simulated Microelectronic Chips to Submerged Round Water Jets. *J. Heat Transfer*, 115, 71–77.

Sutera, S. P., P. F. Maeder, and J. Kestin (1963). On the Sensitivity of Heat Transfer in the Stagnation Point Boundary Layer to Free Stream Vorticity. *J. Fluid Mech.*, 16, 497–520.

Suzuki, M., Y. Udagawa, and H. Yamamoto (1989). Conductive Liquid Cooling for the FACOM VP2000 Supercomputer. *Proc. Ninth Int. Elec. Packaging Conference*.

Teuscher, K. L. (1992). Packaging Methods for Jet Impingement Cooling of an Array of Discrete Heat Sources. MSME Thesis, Purdue University, West Lafayette, IN.

Teuscher, K. L., S. Ramadhyani, and F. P. Incropera (1993). Jet Impingement Cooling of an Array of Discrete Heat Sources with Extended Surfaces, ASME HTD-263, 1–10.

Tou, K. W., G. P. Xu, and C.P. Tso (1998). Direct Liquid Cooling of Electronic Chips by Single-Phase Forced Convection of FC-72. *Experimental Heat Transfer*, 11, 121–134.

Tuckerman, D. B. (1984). Heat Transfer Microstructures for Integrated Circuits. Ph.D. Thesis, Stanford University, Stanford, CA.

Tuckerman, D. B., and R.F. Pease (1981). High Performance Heat Sinking for VLSI. *IEEE Electron. Device Lett.*, EDL-2, 126–129.

Tuckerman, D. B., and R. F. Pease (1982). Optimized Convective Cooling Using Micromachined Structure. *J. Electrochemical Soc.*, 129, C98.

Vader, D. T., F. P. Incropera, and R. Viskanta (1991). Local Convective Heat Transfer from a Heated Surface to an Impinging Planar Jet of Water. *Int. J. Heat Mass Transfer*, 34, 611–623.

VanFossen, G. J., Jr., and R. J. Simoneau (1987). A Study of the Relationship Between Free-Stream Turbulence and Stagnation Region Heat Transfer. *J. Heat Transfer*, 109, 10–15.

Wadsworth, D. C., and I. Mudawar (1990). Cooling of a Multichip Electronic Module by Means of Confined Two-Dimensional Jets of Dielectric Liquid. *J. Heat Transfer*, 112, 891–898.

Wadsworth, D. C., and I. Mudawar (1992). Enhancement of Single-Phase Heat Transfer and Critical Heat Flux from an Ultra-High Flux Simulated Microelectronic Heat Source to a Rectangular Impinging Jet of Dielectric Liquid. *J. Heat Transfer*, 114, 764–768.

Wang, B. X., and X. F. Peng (1994). Experimental Investigation on Liquid Forced Convection Heat Transfer Through Microchannels. *Int. J. Heat Mass Transfer*, 37, 73–82.

Webb, B. W., and C. F. Ma (1995). Single-Phase Liquid Jet Impingement Heat Transfer. *Advances in Heat Transfer*, J. P. Hartnett, T. F. Irving, Y. I. Cho, and G. A. Greene, Eds., Vol. 26, Academic Press, New York, 105–217.

Weisberg, A., H. H. Bau, and J. N. Zemel (1992). Analysis of Microchannels for Integrated Cooling. *Int. J. Heat Mass Transfer*, 35, 2465–2474.

Wirtz, R. A., and Y. Zhong (1991). Enhancement of Natural Convection Heat Transfer Using a Novel Rib-Turbulator. *Fundamentals of Natural Convection*, ASME HTD-178, 49–54.

Wolf, D. H., R. Viskanta, and F. P. Incropera (1990). Local Convective Heat Transfer from a Heated Surface to a Planar Jet of Water with a Nonuniform Velocity Profile. *J. Heat Transfer*, 112, 899–905.

Wolf, D. H., F. P. Incropera, and R. Viskanta (1995a). Measurement of the Turbulent Flow Field in a Free-Surface Jet of Water. *Exp. Fluids*, 18, 397–408.

Wolf, D. H., R. Viskanta, and F. P. Incropera (1995b). Turbulence Dissipation in a Free-Surface Jet of Water and Its Effect on Local Impingement Heat Transfer from a Heated Surface: Part 1—Flow Structure. *J. Heat Transfer*, 117, 85–94.

Wolf, D. H., R. Viskanta, and F. P. Incropera (1995c). Turbulence Dissipation in a Free-Surface Jet of Water and Its Effect on Local Impingement Heat Transfer from a Heated Surface: Part 2—Local Heat Transfer. *J. Heat Transfer*, 117, 95–103.

Womac, D. J., Aharoni, G., S. Ramadhyani, and F. P. Incropera (1990). Single Phase Liquid Jet Impingement Cooling of Small Heat Sources. *Proc. Ninth Int. Heat Transfer Conf.*, Vol. 4, 149–154.

Womac, D. J., S. Ramadhyani, and F. P. Incropera (1993). Correlating Equations for Impingement Cooling of Small Heat Sources with Single Circular Liquid Jets. *J. Heat Transfer*, 115, 106–115.

Womac, D. J., F. P. Incropera, and S. Ramadhyani (1994). Correlating Equations for Impingement Cooling of Small Heat Sources with Multiple Circular Liquid Jets. *J. Heat Transfer*, 116, 482–486.

Wroblewski, D. E., and Y. Joshi (1993). Computations of Liquid Immersion Cooling for a Protruding Heat Source in a Cubical Enclosure. *Int. J. Heat Mass Transfer*, 36, 1201–1218.

Wroblewski, D. E., and Y. Joshi (1994). Liquid Immersion Cooling of a Substrate-Mounted Protrusion in a Three-Dimensional Enclosure: The Effects of Geometry and Boundary Conditions. *J. Heat Transfer*, 116, 112–119.

Wu, P. Y., and W. A. Little (1983). Measurement of Friction Factor for the Flow of Gases in Very Fine Channels Used for Microminiature Joule-Thompson Refrigerators. *Cryogenics*, 24, 273–277.

Xu, G. P., K. W. Tou, and C. P. Tso (1998). Numerical Modelling of Turbulent Heat Transfer from Discrete Heat Sources in a Liquid-Cooled Channel. *Int. J. Heat Mass Transfer*, 41, 1157–1166.

Yamamoto, H., Y. Udagawa, and M. Suzuki (1988a). Cooling System for FACOM M-780 Large Scale Computer. *Cooling Technology for Electronic Equipment*, W. Aung, Ed., Hemisphere, New York, 701–730.

Yamamoto, H., M. Suzuki, Y. Udagawa, M. Nakata, K. Katsuyama, I. Uno, and S. Kikuchi (1988b). United States Patent No. 4,783,721.

Yang, K. T. (1987). Natural Convection in Enclosures. *Handbook of Single-Phase Heat Transfer*, S. Kakac, R. Shah, and W. Aung, Eds., Wiley–Interscience, New York, Chapter 13.

Yin, X., and H. H. Bau (1995). Micro Heat Exchangers Consisting of Pin Arrays. *Cooling and Design of Electronic Systems*, ASME HTD-319/EEP-15, 59–66.

Yonehara, N., and I. Ito (1982). Cooling Characteristics of Impinging Multiple Water Jets on a Horizontal Plane. Kansai University Technical Report 24, 267–281.

Yovanovich, M. M., and V. W. Antonetti (1988). Application of Thermal Contact Resistance Theory to Electronic Packages. *Advances in Thermal Modeling of Electronic Components and Systems*, Vol. 1, A. Bar-Cohen and A. D. Kraus, Eds., Hemisphere, New York, 79–128.

Zhukauskas, A. (1972). Heat Transfer from Tubes in Cross Flow. *Advances in Heat Transfer*, Vol. 8, J. P. Hartnett and T. F. Irvine, Jr., Eds., Academic Press, New York.

Zumbrunnen, D. A., F. P. Incropera, and R. Viskanta (1989). Convective Heat Transfer Distribution on a Plate Cooled by Planar Water Jets. *J. Heat Transfer*, 111, 889–896.

Zumbrunnen, D. A., and M. Aziz (1993). Convective Heat Transfer due to Intermittency in an Impinging Jet. *J. Heat Transfer*, 115, 91–98.

Zumbrunnen, D. A., and M. Balasubramanium (1995). Convective Heat Transfer Enhancement due to Gas Injection into an Impinging Liquid Jet. *J. Heat Transfer*, 117, 1011–1017.

Zumbrunnen, D. A. (1996). Convective Heat Transfer Modifications due to Flow Pulsations in Impinging Jets. *Proc. A. E. Bergles Festschrift on Process, Enhanced and Multiphase Heat Transfer*, R. M. Manglik and A. D. Kraus, Eds., Begell House, New York, 307–318.

AUTHOR INDEX

Aharoni, G., 175
Aihara, T., 225
Akfirat, J.C., 183, 190–192, 199
Al-Sanea, S., 190
Anand, N.K., 223
Antonetti, V.W., 19, 20
Armaly, B.F., 87
Aung, W., 87
Auracher, H., 183
Aziz, M., 215

Baker, E., 131
Baker, K., 154
Balasubramanium, M., 215
Bar-Cohen, A., 2, 9, 20, 26, 30, 44, 45, 89, 157, 218, 223, 252
Barron, R.R., 153
Bau, H.H., 153, 160, 236
Bejan, A., 223, 241
Benedict, T.J., 118
Bergles, A.E., 183, 252
Berkovsky, B.M., 46
Berry, R.D., 199
Besserman, D.L., 212–214, 244, 245
Bhavani, S.H., 252
Bhunia, S.K., 57
Bier, W., 152
Biskeborn, R.G., 217
Brinkman, R., 231
Brost, J., 5

Callahan, V., 241
Campo, A., 225
Chen, L., 117, 118
Chen, T.S., 87
Chigier, N., 57
Choi, S.B., 153
Chu, H.H.S., 42, 102
Chu, R.C., 2
Churchill, S.W., 42, 102
Cobonpue, J., 184, 199
Copeland, D., 154, 160
Cowen, G.H., 46
Cox, R., 102, 112
Cummings, K.N., 174

Dagan, Z., 174, 196, 197
Danielson, R.D., 5
DeWitt, D.P., 15, 16, 18, 19. 24, 26, 34, 41–43, 45, 48, 49, 51–53, 74, 75, 77, 81–83, 141, 154
DiMarco, P., 169
Douglas, W.J.M., 200
Dropkin, D., 47

Eibeck, P.A., 146
Elenbaas, W., 44
Emery, A.P., 46
Eninger, J.E., 160
Evans, H.L., 179

Faggiani, S., 169
Falkner, V.M., 166, 178
Fitzgerald, J.A., 211
Fletcher, L.S., 223
Fox, R.W., 76, 77, 83

Gabour, L.A., 59, 168, 169, 172, 239, 240
Gardon, R., 183, 184, 190–192, 199
Garimella, S.V., 146, 211
Gauntner, J.W., 65, 66
Gebhart, B., 40, 45
Geisler, K.J., 223
Gerner, F.M., 154, 160
Gersey, C.O., 134, 145–147
Glauert, M.G., 66
Globe, S., 47
Goldstein, R.J., 205
Gomi, T., 183
Goodson, K.E., 154, 160
Graham, K.M., 215
Grassi, W., 57, 169
Grimison, E.D., 52, 228
Gudapati, S., 146

Harley, J., 153
Harms, T.M., 154, 160
Harpole, G.M., 160
Heindel, T.J., 92, 93, 95, 97–113, 115–117,
    148–151, 218–223, 225, 244–246
Henderson, H.T., 154, 160
Hirata, M., 183, 185
Hölke, A., 154, 160
Hollworth, B.R., 199
Hoopman, T.L., 152
Horvath, J.L., 217
Hrycak, P., 65, 66
Hultmark, E.B., 217
Hwang, U.P. 2

Inada, S., 179
Incropera, F.P., 15, 16, 18, 19, 24, 26, 34,
    41–43, 45, 48, 49, 51–53, 57, 60, 63, 74,
    75, 77, 81–83, 89, 93, 95, 97–120,
    127–136, 138–142, 145, 148–151, 154,
    174–176, 179–182, 185–187, 196–199,
    201–205, 209, 210, 212–214, 218–223,
    225, 227–231, 235–246, 248–250
Ito, I., 196
Izumi, R., 179

Jaluria, Y., 40, 45
Jiji, L.M., 174, 196, 197
Junes, O.C., Jr., 77

Joshi, Y., 106, 107, 118
Joy, R.C., 157

Katsuyama, I., 3, 212, 246
Kazmierczak, M., 154, 160
Kelecy, F.J., 227–229
Kelleher, M.D., 107, 118
Keller, W., 152
Kendall, S.R., 153
Kerby, J., 132–134, 145, 209, 210
Kercher, D.M., 199
Kestin, J., 61, 168
Keyhani, M., 101, 102, 112–114, 116–118
Kikuchi, S., 3, 212, 246
Kim, S.H., 223
Kishimoto, T., 3, 153, 236
Kitamura, K., 252
Kitching, D., 223
Know, A.L., 135
Kobayakawa, S., 225
Korger, M., 199, 200, 205
Krajewski, N., 5
Kraus, A.D., 20, 26, 30, 157, 218, 252
Krizek, F., 199, 200, 205
Kuhn, D., 92
Kurabayaski, K., 154, 160

Ledezma, G., 223, 241
Lee, D.T., 65
Lee, J.J., 107
Lee, L., 60, 179, 190
Lee, T.-Y., 9
Levy, S., 179
Lienhard, J.H.V., 57, 59, 61–64, 167–169,
    171, 172, 178, 239, 240
Linder, G., 152
Little, W.A., 153, 154
Liu, K.V., 107
Liu, X., 59, 64, 168, 169, 171, 172
Livengood, J.N.B., 65, 66
Lombara, J.S., 62, 171
London, A.L., 72, 75–81, 130, 156, 159

Ma, C.F., 168, 183, 184, 215
MacGregor, R.K., 46
Maddox, D.E., 132, 147, 231, 232
Maeder, P.F., 61, 168
Magrini, A., 57, 169
Mahajan, R.L., 40, 45
Mahalingam, M., 153
Mahaney, H.V., 127, 128, 135, 136, 138–142,
    149, 235, 236
Mangler, W., 166

Mansour, A., 57
Martin, H., 66, 184, 186, 191, 199, 200, 208
Maruyama, S., 225
Maughan, J.R., 135
McDonald, A.T., 76, 77, 83
McMurray, D.C., 179, 182
Metzger, D.E., 174
Meyers, P.S., 179, 182
Milne-Thomson, L.M., 60
Misumi, T., 252
Miyasaka, Y., 179
Moffatt, D.F., 129–134, 145, 209, 210
Morega, A.M., 241
Morris, G.K., 146
Mouromtseff, I.F., 9
Mudawar, I., 132, 134, 145–147, 191–194, 208–210, 231, 232, 247, 249
Mujumdar, A.S., 200

Nakata, M., 3, 212, 246
Nakatogawa, T., 183, 185
Nakayama, W., 160, 183, 184
Nayak, D., 153
Nenaydykh, B., 211
Nishiwaki, N., 183, 185

Ohsaki, T., 3
Oosthuizen, P.H., 92
Osborne, D.G. 135
Ostrach, S., 40, 45

Pak, B.-C., 160
Pan, Y., 60, 169, 196
Patterson, C.V.S., 102
Pease, R.W.F., 5, 152–154, 160, 231
Peng, X.F., 153, 154
Peterson, G.P., 153, 154
Pfahler, J.N., 153
Phillips, R.J., 5, 78–82, 152, 156–160
Pilchowski, J., 154, 160
Pitts, D.R., 117, 118
Polentini, M.S., 114–120, 219
Polevikov, V.K., 46
Prasad, V., 101, 102, 112–114, 116, 117
Priedeman, D., 241

Raju, K.S., 185
Ramadhyani, S., 93, 95, 97–120, 127–136, 138–142, 145, 148–151, 174–176, 185–187, 196–199, 201–205, 209, 212–215, 218–223, 225, 227–231, 235, 236, 237–246, 248–250
Rao, H.V., 153

Reisman, A., 153
Rice, R.A., 211
Rohsenow, W.M., 44, 45, 223
Rouf, R.A., 205
Ruby, W.A., 174

Saad, N.R., 200
Samant, K.R., 131
Sammakia, B., 40, 45
Sasaki, S., 3, 153, 236
Sathe, S.B., 106
Saylor, J.R., 9
Schafer, D.M., 201–205
Schlichting, H., 238
Schlitz, D.J., 146, 157
Schubert, K., 152
Schultz, M.T., 59, 168
Shah, R.K., 72, 77–81, 130, 156
Shen, R., 101, 102, 113, 114, 116, 117
Sheriff, H.S., 215
Siedel, D., 152
Simor, T.W., 9, 131
Simoneau, R.J., 168
Simons, R.E., 2
Sitharamayya, S., 185
Skar, S.W., 166, 178
Slayzak, S.J., 196, 197
Sparrow, E.M., 60, 179, 190, 205
Stevens, J., 57, 60, 169, 172, 176, 177
Sullivan, P.F., 237–244, 245, 246, 250
Sun, H., 183, 184
Sutera, S.P., 61, 168
Suzuki, M., 3, 212, 217, 246

Tabakoff, W., 199
Takahira, H., 160
Teuscher, K.L., 205–211, 248–250
Tong, W., 9
Torii, K., 183, 185
Tou, K.W., 131, 134, 147
Trass, O., 59, 168
Tso, C.P., 131, 134, 147
Tuckerman, D.B., 5, 152, 153, 160, 231, 236
Turlik, 153

Udagawa, Y., 3, 212, 217, 246
Uno, I., 3, 212, 246
Uyehara, O.A., 179, 182

Vader, D.T., 63, 179, 181, 182
Van Fossen, G.J., Jr., 168
Viskanata, R., 57, 60, 63, 179–182, 196, 197

Wadsworth, D.C., 191–194, 208–210, 247, 249
Wang, B.X., 153, 154
Warrington, R.O., 153
Webb, B.W., 57, 60, 168, 169, 172, 176, 177, 196, 215, 241
Weisberg, A., 160
Wirtz, R.A., 252
Wolf, D.H., 57, 60, 179, 180, 182
Womac, D.J., 174–176, 185–187, 196–199, 239, 241
Wong, T.T., 101, 113, 116, 117
Wroblewski, D.E., 107
Wu, P.-S., 9
Wu, P.-Y., 153, 154

Xu, G.P., 131, 134, 147

Yamamoto, H., 3, 212, 217, 246
Yang, K.T., 40, 45, 107
Yin, X., 236
Yonehara, N., 196
Yovanovich, M.M., 19, 20

Zemel, J., 153, 160
Zhong, Y., 252
Zhukauskas, A., 51–53, 228
Zumbrunnen, D.A., 215

SUBJECT INDEX

Adiabatic heat transfer coefficient, 147
Advection, 35, 96

Boundary layers
  approximations, 40, 41, 48
  hydrodynamic, *see* velocity
  separation, 51
  thermal, 37–38, 46, 61, 167, 170, 217
  thickness, 37–38, 39, 58, 61, 63, 165
  velocity, 39, 40–41, 47, 61–63, 170
Boussinesq approximation, 35, 91
Buoyancy, 35, 40, 42, 43, 87, 91, 132, 133, 135, 137, 138, 150, 218, 222, 223

Channel flow
  forced convection, 125–135, 145–148, 226–235
  microchannels, 151–163, 235–237
  mixed convection, 135–143, 148–151, 235, 236
  protruding sources, 145–151
  theoretical considerations, 125–131, 135–139
Colburn j factor, 38
Cold plate, 2–5, 70, 83–86, 125, 218
Conduction heat transfer, 16–35, 47
  plane wall, 16–19
  transient, 34
Conjugate heat transfer, 89, 91, 127–131, 226
Contact resistance, *see* Thermal resistance

Convection heat transfer
  coefficient, 9, 13, 20, 37–38
  forced, 9, 11, 26, 35, 36, 40, 47
    circular cylinder, 51–53
    flat plate, 47–51
    internal flow, 72, 77, 79–83; *see also* Channel flow
    jets, 167; *see also* Jet impingement
  free, *see* natural
  mixed, 35, 37, 87, 235; *see also* Channel flow
  natural, 10–11, 26, 35, 36, 40–47, 89, 218–226
    horizontal surface, 42–43
    parallel-plate channel, 43–45
    rectangular cavity: horizontal, 46–47, 91, 118–120
    rectangular cavity: inclined, 118–120
    rectangular cavity: vertical, 45–46, 90–91
      flush-mounted heat sources, 92–102, 102–106, 112–117
      mathematical model, 91–92
      protruding heat sources, 106–112, 117–118
    vertical surface, 40–42

Darcy friction factor, *see* Friction factor
Diffusion, 35, 96
Direct liquid cooling, 2, 5–8

283

Elenbaas equation, 43–45, 219
Energy generation, 16
Energy storage, 16, 34
Extended surfaces, *see* Fins
Extensive medium, 40
External flow, 47

Fanning friction factor, *see* Friction coefficient
Film temperature, 51, 52
Fins, 23–30, 31, 217, 218, 237
  arrays, 26–30, 218–236
  effectiveness, 25, 26, 217
  efficiency, 25–27, 29, 30, 219–222, 225, 241
  heat rate, 24, 29, 219, 227
  longitudinal, *see* rectangular
  micro (milli) studs, 231, 232, 236, 238, 241, 247, 248
  pin, 23, 26–28, 223–232, 241–244, 248–250
  plate, *see* rectangular
  rectangular, 23, 26–27, 218–223, 235, 236, 248–250
  resistance, 26, 29–30
  strip, 231–232
Fluid properties, 12
Fluorinerts, 8
Fluorocarbons, 8
Forced convection, *see* Convection heat transfer
Fourier's law, 16, 18, 37
Free convection, *see* Convection heat transfer
Friction coefficient, 39, 76–79
Friction factor, 39, 153
Froude number, 56

Grashof number, 36

Heat equation, 16, 17
Heat flux, 9, 18, 20, 167
Heat generation, *see* Energy generation
Heat rate, 9, 18, 20, 47, 49, 52, 74, 75, 173
Heat sink, 31, 151, 152, 217
Hydraulic jump, *see* Jet impingement

Immersion cooling, *see* Direct liquid cooling
Impinging jets, *see* Jet impingement
Indirect liquid cooling, 2–5
Internal flow, 69; *see also* Convection heat transfer
  energy balance relations, 74–75
  entrance region, 71–74
  fully developed region, 71–74
  pressure losses, 75–79

Jet impingement,
  atomization, 57
  circular, 55–69, 165–177, 183–187, 195–197, 199, 211–214, 237–247
  cooling, 3, 4, 165
  free-surface, 55–64, 165, 239, 241
    cross-flow effects, 197
    multiple jets, 195–199
    unconfined, 55–64, 165–183
  hydraulic jump, 64
  inclination, 215
  planar, *see* rectangular
  pulsations, 214, 215
  rectangular, 56–61, 63, 66–69, 177–183, 190–193, 197–211, 247–251
  slot, *see* rectangular
  splattering, 57, 63–64
  stagnation zone, 58–66, 68–69, 165–170, 177–181, 183
  submerged
    cross-flow effects, 68–69, 205
    free jet, 64–65
    multiple jets, 199–200, 205–211
    semi-confined, 67–69, 200–214
    unconfined, 64–67, 183–187, 190–193
  wall jet, 61–64, 66, 67

Laminar flow, 10, 11
Log-mean temperature difference, 53, 75, 159, 229

Microgrooves, *see* Fins
Micro (milli) studs, *see* Fins
Microchannel cooling, 5, 125
  see Channel flow
Mixed convection, *see* Convection heat transfer and Channel flow
Moody friction factor, *see* Friction factor

Natural convection, *see* Convection heat transfer
Newtonian fluid, 39
Newton's law of cooling, 9, 20, 72, 173
Nusselt number, 9, 37, 167, 178

Passive Cooling, 217
Perfluorinated liquids, *see* Fluorinerts or Fluorocarbons
Prandtl number, 9, 36
Pressure drop, 153, 154, 158, 159
Properties, *see* Fluid properties
Pump power, 53

Quiescent medium, 40

SUBJECT INDEX

Rayleigh number, 10, 36, 42, 92
Resistance, *see* Thermal resistance
Reynolds number, 9, 36, 167

Similarity parameters and regions, 36, 41, 48, 61, 166, 177, 178
Similarity solution, 166, 177, 178
Spreader plate, 19, 23, 30, 237–244
Stanton number, 38
Strip heater, 91
Substrate, 19, 23, 30

Thermal conduction module, 2
Thermal constriction, 155

Thermal resistance, 2, 3, 13, 18, 34
   areal or unit, 2, 5, 13, 20, 83, 86, 87, 91, 152, 153, 157, 165, 221–223, 235, 236, 238–241, 243–246, 249, 250
   conduction, 18–19, 155
   contact, 20, 29–30, 155, 157, 218
   convection, 9, 19–20
   fin, 26, 29–30, 229–232
Turbulent flow, 10, 11, 38, 42, 46–49, 57, 58, 61, 62, 67, 70–71, 73–74, 77, 142, 149–151, 156, 160, 168–170, 172, 179–184, 190, 191, 200

Weber number, 63–64